Encyclopaedia of Mathematical Sciences

Volume 26

Editor-in-Chief: R.V. Gamkrelidze

S. M. Nikol'skiĭ (Ed.)

Analysis III

Spaces of Differentiable Functions

With 22 Figures

Springer-Verlag
Berlin Heidelberg NewYork
London Paris Tokyo
Hong Kong Barcelona

Scientific Editors of the Series:
A.A. Agrachev, E.F. Mishchenko, N.M. Ostianu, L.S. Pontryagin
Editors: V.P. Sakharova, Z.A. Izmailova

Title of the Russian edition:
Itogi nauki i tekhniki, Sovremennye problemy matematiki,
Fundamental'nye napravleniya, Vol. 26, Analiz 3
Publisher VINITI, Moscow 1988

Mathematics Subject Classification (1980): 46Exx

ISBN 3-540-51866-5 Springer-Verlag Berlin Heidelberg New York
ISBN 0-387-51866-5 Springer-Verlag New York Berlin Heidelberg

Library of Congress Cataloging-in-Publication Data
Analysis: integral representations and asymptotic methods.
(Encyclopaedia of mathematical sciences; v. 13)
Translation of: Analiz, issued as part of the serial: Itogi nauki i tekhniki.
Seriia sovremennye problemy matematiki. Fundamental'nye napravleniia.
Includes bibliographical references.
1. Operational calculus. 2. Integral representations.
3. Integral transforms. 4. Asymptotic expansions.
I. Gamkrelidze, R.V. II. Series. III. Series: Encyclopaedia of mathematical sciences; v. 13, etc.
QA432.A6213 1991 515'.72 89-6163
ISBN 0-387-17008-1 (U.S.: v. 1: alk. paper)

Printed in the United States of America
Media conversion: Universitätsdruckerei H. Stürtz AG, D-8700 Würzburg
2141/3140-543210 – Printed on acid-free paper

List of Editors, Contributors and Translators

Editor-in-Chief

R.V. Gamkrelidze, Academy of Sciences of the USSR, Steklov Mathematical Institute, ul. Vavilova 42, 117966 Moscow, Institute for Scientific Information (VINITI), Baltiiskaya ul. 14, 125219 Moscow, USSR

Consulting Editor

S.M. Nikol'skiĭ, Steklov Mathematical Institute, ul. Vavilova 42, 117333 Moscow, USSR

Contributors

L.D. Kudryavtsev, Steklov Mathematical Institute, ul. Vavilova 42, 117333 Moscow, USSR

V.G. Maz'ya, Institute for Research on Machines, Leningrad Department of the Academy of Sciences of the USSR, Bolshoĭ prospekt V.O. 61, 199179 Leningrad, USSR

S.M. Nikol'skiĭ, Steklov Mathematical Institute, ul. Vavilova 42, 117333 Moscow, USSR

Translator

J. Peetre, University of Stockholm, Department of Mathematics, Box 6701, S-113 85 Stockholm, Sweden

Contents

I. Spaces of Differentiable Functions of Several Variables and Imbedding Theorems
L.D. Kudryavtsev, S.M. Nikol'skiĭ
1

II. Classes of Domains, Measures and Capacities in the Theory of Differentiable Functions
V.G. Maz'ya
141

Author Index
213

Subject Index
217

I. Spaces of Differentiable Functions of Several Variables and Imbedding Theorems

L.D. Kudryavtsev, S.M. Nikol'skiĭ

Translated from the Russian
by J. Peetre

Contents

Introduction 4

Chapter 1. Function Spaces 4

§1. The Concept of Space 4
§2. Linear Spaces 6
§3. Topological Spaces 8
§4. Metric Spaces 9
§5. Normed and Seminormed Spaces. Banach Spaces 12
§6. Inner Product Spaces. Hilbert Spaces 17
§7. Complete Function Spaces. Bases 19
§8. Dual Spaces 23
§9. Linear Operators 27
§10. Lebesgue Spaces 28
§11. Morrey Spaces 30

Chapter 2. Sobolev Spaces 31

§1. Generalized Derivatives 31
§2. Boundary Values (Traces) of Functions 36
§3. Sobolev Spaces 37
§4. Sobolev Spaces of Infinite Order 42
§5. Derivatives and Integrals of Fractional Order 44

Chapter 3. The Imbedding Theorems of Nikol'skiĭ 46

§1. Inequalities for Entire Functions 46
§2. Imbedding Theorems for Isotropic *H*-Classes 47
§3. Imbedding Theorems for Anisotropic *H*-Classes 50

Chapter 4. Nikol'skiĭ-Besov Spaces 53
§1. Sobolev-Slobodetskiĭ Spaces 53
§2. The Besov Spaces $B_{p,q}^{(r)}$ 54
§3. Imbedding Theorems for Nikol'skiĭ-Besov Spaces. Solution of the Problem of Traces of Functions in Sobolev Spaces 55

Chapter 5. Sobolev-Liouville Spaces 58
§1. The Spaces L_p^r 58
§2. Study of Function Spaces with the Aid of Methods of Harmonic Analysis 62
§3. Lizorkin-Triebel Spaces 64
§4. Generalizations of Nikol'skiĭ-Besov and Lizorkin-Triebel Spaces . 65

Chapter 6. Spaces of Functions Defined in Domains 69
§1. Integral Representations of Functions 69
§2. Imbedding Theorems for Domains 73
§3. Traces of Functions on Manifolds 75
§4. Extension of Functions from Domains to the Whole Space with Preservation of Class 76
§5. Density of Infinitely Differentiable Functions 79
§6. Spaces with a Dominating Mixed Derivative and Other Spaces . 83
§7. Multiplicative Estimates for Derivatives and Differences 85
§8. Stabilizing Functions at Infinity to Polynomials 85
§9. Compact Imbeddings 86

Chapter 7. Weighted Function Spaces 87
§1. The Role of Weighted Spaces in the General Theory 87
§2. Weighted Spaces with a Power Weight 88
§3. Applications of the Theory of Weighted Spaces 91
§4. General Weighted Spaces 94
§5. The Weighted Spaces of Kipriyanov 101

Chapter 8. Interpolation Theory of Nikol'skiĭ-Besov and Lizorkin-Triebel Spaces 103
§1. Interpolation Spaces 103
§2. Lorentz and Marcinkiewicz Spaces 106
§3. Interpolation of the Nikol'skiĭ-Besov and Lizorkin-Triebel Spaces 108

Chapter 9. Orlicz and Orlicz-Sobolev Spaces 110
§1. Orlicz Spaces 110
§2. The Spaces $E_\Phi(X)$ 114
§3. Sobolev-Orlicz Spaces 115
§4. Imbedding Theorems for Sobolev-Orlicz Spaces 116

Chapter 10. Symmetric and Nonsymmetric Banach Function Spaces . 118
§1. Ideal Spaces and Symmetric Banach Spaces
of Measurable Functions 118
§2. Spaces with Mixed Norms 119
§3. Sobolev and Nikol'skiĭ-Besov Spaces Induced by Metrics
of General Type 120
§4. Some Problems 126

References 128

Introduction

In the Part at hand the authors undertake to give a presentation of the historical development of the theory of imbedding of function spaces, of the internal as well as the externals motives which have stimulated it, and of the current state of art in the field, in particular, what regards the methods employed today.

The impossibility to cover all the enormous material connected with these questions inevitably forced on us the necessity to restrict ourselves to a limited circle of ideas which are both fundamental and of principal interest. Of course, such a choice had to some extent have a subjective character, being in the first place dictated by the personal interests of the authors. Thus, the Part does not constitute a survey of all contemporary questions in the theory of imbedding of function spaces. Therefore also the bibliographical references given do not pretend to be exhaustive; we only list works mentioned in the text, and a more complete bibliography can be found in appropriate other monographs.

O.V. Besov, V.I. Burenkov, P.I. Lizorkin and V.G. Maz'ya have graciously read the Part in manuscript form. All their critical remarks, for which the authors hereby express their sincere thanks, were taken account of in the final editing of the manuscript.

Chapter 1
Function Spaces

§ 1. The Concept of Space

Around the turn of the century mathematicians formulated the notion of *function space*. By this is meant a set of functions of similar type in a suitable sense, equipped with additional structure: algebraic, topological or metric. Already in the 17th and 18th centuries, in the very period of birth of mathematical analysis and the creation of its methods, much use was made of the linearity of the operations of differentiation and integration. In the sequel it was realized that many important classes of functions, analytic, differentiable or integrable, in some sense or other, enjoy the property of being closed under linear operations: if we take any pair of functions of the class in question then any linear combination of them belongs to the same class. In the second half of the 19th century, after one had rigorously defined the concepts of continuity and continuous differentiability and begun to use these notions more widely, one also began to consider classes of continuous, respectively continuously differentiable functions. In this way ground was prepared for the appearance of the notion of linear (vector) space or, as one often says, linear manifold. However, in the beginning it was not necessary to consider such classes as a kind of spaces, that is, to consider them as in a certain sense homogeneous sets, that is, sets consisting of elements all regarded as equivalent, otherwise put, to substitute points for functions, because in this epoch the individual functions were the central objects of study. But already isolated results of this period, established essentially for function classes, (as, for instance, the Cauchy-Bunyakovskiĭ(-Schwarz) inequality (cf. § 6 of Chap. 1) pointed to an analogy between the class of square integrable functions and the geometry of an infinite dimensional vector space) inexorably foreshadowed a time when the notion of function space unfolded itself in a natural way.

An important step in the development of function theory, accelerating the clarification of this notion, was the proof of various limit theorems, asserting the validity of a property of the limit function if this property holds true for all the functions whose limit we consider. Thus one obtained classes which are closed not only under linear combinations but also under a passage to the limit. As an example of such a set we quote the family of continous functions under passage to the limit in the sense of uniform convergence. From this one might infer that only a small step remained for the introduction of the notion of function space. The difficulty was connected with the fact that in the solution of many problems with the aid of a suitable passage to the limit within a given class (for instance, the search of an exact solution of a differential equation as a limit of an approximate solution) the required smoothness properties are not always preserved. For instance, one cannot find a continuously differentiable curve of minimal length passing through three given points, not all three lying on a line; the minimal curve in the case at hand is a two linked broken line consisting of two links passing through the given points. In view of this, in the process of solving a problem one can sometimes recover the function from the set over which the limit is taken but in other cases one has to modify the sense in which the limit is taken (in such a case one uses essentially methods which regularize illposed problems). All these circumstances hindered the clarification of the notion of function space in mathematics. Before one could arrive at the idea of the soundness of such a concept, one had to realize the value and the sense of various modes of transition to the limit in the function classes under consideration. Thus, for instance, uniform convergence or convergence in an integral metric lead subsequently to the notion of metric and, in particular, normed space, while pointwise convergence required the more general notion of topological space.

As an example we may refer to Riemann's solution of the first boundary problem with the aid of the so-called *Dirichlet's principle*, which amounts to minimizing the energy integral, which at the case at hand is the *Dirichlet integral*, i.e. the integral of the squares of all first order partial derivatives extended over the domain. Riemann started out with a pointwise converging sequence of piecewise differentiable functions, the limit of which, generally speaking, is not a piecewise differentiable, which is required in the statement of the problem. In order to maintain the required properties of the limit function Riemann tried to use a kind of regularization process but he never succeeded to carry out this idea to the end. Subsequently, passing to an integral metric, Hilbert gave a foundation of Dirichlet's principle in the assumption of the existence of at least one admissible function, i.e. a function which takes given boundary values and has a finite Dirichlet integral in the domain.

Similar circumstances made mathematicians realize that the problems of mathematical physics may be illposed in some spaces (for instance, in function spaces with pointwise or uniform convergence) and wellposed in others (for instance, in spaces with an appropriate integral metric).

Essentially the first function space, to occupy a more permanent position in mathematics, was the space of square integrable functions. The notion of this space arose in the first place in the work of Hilbert and Schmidt, and later the study of its properties was pursued in investigations by Fischer, F. Riesz (let us for instance recall the famous Riesz-Fischer theorem on the completeness of the space of square integrable functions), J. v. Neumann and others. We underline that to great extent the completion of the general theory of the space of square integrable functions became possible only after the introduction of the notion of Lebesgue integral.

In the 30's of this century one began a systematic study of function spaces, both in their own right (more exactly in connection with the internal needs of function theory) and in connection with the demands of the theory of differential equations and of probability theory. Before we pass to an account of the current state of the theory of function spaces let us briefly recall the basic types of abstract spaces (i.e. spaces consisting of elements of an arbitrary nature) to which belong the function spaces that one encounters most frequently in modern mathematics. More specifically, we will recall the notion of linear space, of topological space, of metric space, of normed space, of inner product space. As for examples of such spaces we restrict our attention to function spaces only.

§ 2. Linear Spaces

A set $X = \{x, y, z, \dots\}$ is called a *real (complex) linear space* or a *vector space* over the field of real (complex) numbers if

1) it is an Abelian group (writing the group operation additively);

2) to each element $x \in X$ and each real (complex) number λ there corresponds a unique element of X, called the product of λ and x and denoted by λx, assuming that

a) $1x = x \quad \forall x \in X$;

b) $\lambda(\mu x) = (\lambda\mu)x \quad \forall x \in X, \forall \lambda \in \mathbb{R}, \forall \mu \in \mathbb{R}$ (respectively $\forall \lambda \in \mathbb{C}, \forall \mu \in \mathbb{C}$);

c) $\lambda(x + y) = \lambda x + \lambda y \quad \forall x \in X, \forall y \in X, \forall \lambda \in \mathbb{R}$ (respectively $\forall \lambda \in \mathbb{C}$).

This definition extends in a natural way to the definition of linear space over an arbitrary field.

A subset Y of a linear space X is a *subspace* of X if for any elements $x \in Y, y \in Y$ and any numbers λ and μ (respectively real or complex) the element $\lambda x + \mu y$ also belongs to Y. It is clear that a subspace of a linear space itself is a linear space.

If we have in a linear space X a finite number of vectors such that each vector in X is a linear combination of these vectors, we say that *the space X is finite dimensional*; the minimal number of vectors whose linear combinations give the entire space is called the *dimension* of the space. If X does not contain any finite system of vectors with this property then we say that X is *infinite dimensional.*

If E is any set, then the family $\mathbb{R}(E)$ (respectively $\mathbb{C}(E)$) of real- (complex-) valued functions on E forms a linear space. Then set $B(E)$ of all bounded (on E) functions $f \in \mathbb{R}(E)$ ($f \in \mathbb{C}(E)$) is a subspace of the linear space $\mathbb{R}(E)$ (respectively $\mathbb{C}(E)$). If E is infinite then the spaces $\mathbb{R}(E), \mathbb{C}(E), B(E)$ (of real- or complex-valued functions) are infinite dimensional.

Let $(E, \mathscr{F}, \mu)$ be a space E with a σ-algebra $\mathscr{F}$ of measurable subsets and a σ-finite measure μ. (Here and in the sequel we always assume that the measure is *complete*, i.e. if $E_1 \in \mathscr{F}, \mu(E) = 0$ and $E_2 \subset E_1$ then $\mu(E_2) = 0$.) Then the set $M(E, \mu)$ of all measurable functions, the set $BM(E, \mu)$ of all bounded measurable functions, the set $L_\infty(E, \mu)$ of all *essentially bounded* functions on E, i.e. functions f such that

$$\operatorname*{vrai\,sup}_{x \in E} |f(x)| = \min\{y : \mu\{x : |f(x)| > y\} = 0\} =$$

$$= \sup\{y : \mu\{x : |f(x)| > y\} > 0\} = \inf_{\{X : \mu X = 0\}} \sup_{x \in E \setminus X} |f(x)| < +\infty,$$

the set $L_p(E, \mu)$ of all measurable functions whose p-th powers, $1 \le p < +\infty$, are integrable over E are all linear subspaces of $\mathbb{R}(E)$ (respectively $\mathbb{C}(E)$). The spaces $L_p(E, \mu)$, $1 \le p \le +\infty$, are called *Lebesgue spaces.* If there is no ambiguity about which measure we have in mind, for instance, if E is a measurable subset with respect to n-dimensional Lebesgue measure of the n-dimensional space and μ denotes this measure then we usually write $M(E), BM(E), L_p(E)$ instead of $M(E, \mu), BM(E, \mu), L_p(E, \mu)$. In place of $L_1(E)$ we often write simply $L(E)$.

A function $f : X \to \mathbb{R}$ (or $f : X \to \mathbb{C}$), given on a Lebesgue measurable subset X of $\mathbb{R}^n$, is said to belong locally to the space L_p and we write then $f \in L_p(X, \text{loc})$ if for each compact subset $E \subset X$ the restriction of f to E belongs to the Lebesgue space $L_p(E)$, $1 \le p \le +\infty$. A function f belonging to $L_1(X, \text{loc})$ is also termed *locally integrable* on X.

The set of all functions in $L_p(X, \text{loc})$ is a linear space.

If E is a topological space (cf. § 3) then the set $C(E)$ of all continuous functions on E and likewise the set $BC(E)$ of all bounded and continuous functions on E are linear subspaces of $\mathbb{R}(E)$ ($\mathbb{C}(E)$).

Finally, if E is a domain in n-dimensional Euclidean space, then the set $\mathscr{D}^{(l)}(E)$ of all l times differentiable functions on E, the set $C\mathscr{D}^{(l)}(E)$ of all l times continuously differentiable functions on E, the set $W_p^{(l)}(E)$ of l times differentiable functions in a generalized sense whose partial derivatives (generalized) of orders up to l ($l \geq 0$) belong to $L_p(E)$ all are subspaces of $\mathbb{R}(E)$ ($\mathbb{C}(E)$); it is clear that $W_p^{(0)}(E) = L_p(E)$. In the following sections we will encounter also other examples of linear function spaces.

In the sequel we will only deal with numerical functions, generally speaking, complex valued. All sets of functions to be considered below will always be assumed to consist either of functions, defined by suitable properties, which take only real values (i.e. they are subsets of $\mathbb{R}(E)$), or else they may take arbitrary complex values (i.e. we have a subset of $\mathbb{C}(E)$). When the notions considered make sense in the first as well as in the second case (for example, boundedness, measurability, integrability, continuity, differentiability etc.), we will speak only of functions, with the tacit understanding that we in reality can only deal with each of the two cases at a time.

If the set E, on which the functions are given, is a segment of the real axis, for example the open interval (a, b) or the closed interval $[a, b]$, then we write $\mathbb{R}[a, b], \mathbb{C}(a, b)$ etc. in place of $\mathbb{R}([a, b]), \mathbb{C}((a, b))$ etc.

§ 3. Topological Spaces

A set X consisting of any kind of elements is called a *topological space* if on it is given a system $\Omega = \{G\}$ of subsets satisfying the following conditions:

1°. The intersection of finitely many sets in Ω belongs to the same system;

2°. The union of arbitrary many sets in Ω is in the system;

3°. $X \in \Omega, \emptyset \in \Omega$.

The sets in Ω will be called *open* sets of the topological space and the system Ω itself the *topology of the space X*.

For each point $x \in X$, any set $G \in \Omega$ such that $x \in G$, i.e. any open set containing x, will be termed a *neighborhood* of this *point*.

If any two points of a topological space have nonintersecting neighborhoods, then the space is said to be *Hausdorff*.

Every metric space is an example of a Hausdorff topological space (cf. § 4 of Chap. 1).

A subsystem B of Ω of open subsets of a topological space X is called a *base for the topology* of the space, if each nonempty open set is a union of a suitable family in B.

A system $B(x)$ of neighborhoods of a point x of a topological space X is called a *local base for the topology* at the point, if for each neighborhood U of x in X there exists a neighborhood $V \in B(x)$ such that $V \subset U$.

The union of local bases of the topology at all points forms a base of the topology of the entire space, as every nonempty open set can be written as a union of neighborhoods of its points contained in it, where the neighborhoods in question can be taken from the given local basis of the topology. By the same token, one can define the topology of a set by first writing down the bases of the topology at all of its points.

In terms of topological spaces one can describe the *pointwise convergence of sequences of functions*. Let X be a family of functions defined on some set E and let $f_0 \in X$. A local basis $B(f_0)$ of the topology at the point f_0 is defined by the system of all possible sets of the form

$$U(f_0) = \{f \in X : |f(a_i) - f_0(a_i)| < \varepsilon, i = 1, 2, \ldots, k\},$$

where $a_i \in E$, $i = 1, 2, \ldots, k$, is any finite family of points in E and $\varepsilon > 0$ an arbitrary fixed number. If now $f_n \in X$, $n = 1, 2, \ldots$, and for each point $a \in E$ holds the identity $\lim_{n\to\infty} f_n(a) = f_0(a)$ then this is equivalent to the statement that for each neighborhood $U(f_0)$ of the function f_0 there exists a number n_0 such that for each index $n > n_0$ there holds the inclusion $f_n \in U(f_0)$.

§4. Metric Spaces

A set X is said to be a *metric space* if on the set of order pairs (x, y) in X there is given a nonnegative function, called the *distance* or the *metric* and written $\varrho(x, y)$, satisfying the following conditions:

1) $\varrho(x, y) = 0$ iff $x = y$, $\forall x \in X, y \in X$;
2) $\varrho(x, y) = \varrho(y, x) \quad \forall x \in X, y \in X$;
3) $\varrho(x, y) \leq \varrho(x, z) + \varrho(z, y) \quad \forall x \in X, y \in X, z \in X$.

These conditions are called the *distance axioms*.

Each subset of a metric space is a metric space in its own right.

If $B(E)$ is the family of bounded function on a set E (cf. §2 of Chap. 1) then the function

$$\varrho(f, g) = \sup_{x \in E} |f(x) - g(x)|, \quad \forall f \in B(E), g \in B(E), \tag{1.1}$$

is a metric and, consequently, $B(E)$ can be viewed as a metric space.

An important example of a metric in function spaces are *integral metrics*. Let E be a space equipped with a complete measure μ. Then

$$\varrho(f, g) = \left(\int_E |f(x) - g(x)|^p dx\right)^{\frac{1}{p}}, \quad \forall f \in L_p(E), g \in L_p(E), \tag{1.2}$$

is a metric in the space $L_p(E), 1 \le p < +\infty$. Note that the metric in $L_\infty(E)$ (cf. § 2 of Chap. 1) is given by

$$\varrho(f, g) \stackrel{\text{def}}{=} \text{vrai} \sup_{x \in E} |f(x) - g(x)|. \tag{1.3}$$

Two metric spaces are said to be *isometric* if there exists a bijective correspondence which preserves the distance.

A *sequence* of points $x_n \in X$, $n = 1, 2 \ldots$, of a metric space X is said to *converge* to the point $x \in X$ if

$$\lim_{n \to \infty} \varrho(x_n, x) = 0.$$

In this case x is said to be the *limit of the sequence* $\{x_n\}$ and one writes $\lim_{n \to \infty} x_n = x$ or $x_n \to x$. Convergence in the metric of the space $B(E)$ is just uniform convergence.

A basic notion is the notion of fundamental sequence. A sequence $\{x_n\}$ of points of a metric space is termed *fundamental* if

$$\lim_{n,m \to \infty} \varrho(x_n, x_m) = 0.$$

It is clear that if a sequence converges then it is fundamental; the converse is not true. If every fundamental sequence in a metric space converges to an element of the space, then the space is said to be *complete*.

As examples of complete metric spaces we mention the spaces $L_p(E)$, $1 \le p \le +\infty$. If $1 \le p < +\infty$ the completeness follows from the properties of the Lebesgue integral (this circumstance, together with the fact that with the aid of the Lebesgue integral one can obtain the primitives for functions in a sufficiently wide class, also explains the great usefulness of the Lebesgue integral in various branches of mathematics).

Another important example of a complete metric space is the space $BC(E)$ of bounded continuous functions on a topological space E, with the metric (1.1) (that is, $BC(E)$ is considered as a subspace of $B(E)$). In this case the completeness follows from the observation that convergence in the metric (1.1) is the same as uniform convergence, that the limit of a uniformly converging sequence of continuous functions is again a continuous function and that for a sequence in $C(E)$ to be fundamental means that Cauchy's condition for uniform convergence holds for the sequence in question.

For a point x in a metric space one defines in a natural way the notion of an *ε-neighborhood* $U(x, \varepsilon)$ (here $\varepsilon > 0$ is an arbitrary positive number) as the set of all points y in this space such that $\rho(y, x) < \varepsilon$.

If X is a metric space and E a subset of X, we say that a point $x \in E$ is an *interior* point of E if there exists an $\varepsilon > 0$ such that the ε-neighborhood $U(x, \varepsilon)$ of x is contained in E.

A set $E \subset X$ is said to be open if all points of E are interior. The system of all open sets is a topology in the sense of the definition in § 3 of Chap. 1.

Thus, each metric space is topological. On the other hand, the space of all functions with pointwise convergence is a topological space where it is not possible to introduce a metric which generates there the topology of pointwise convergence, this provided the underlying set is uncountable.

A point x of a metric space X is called a *limit* point for the set $E \subset X$ if each neighborhood of x contains points in E different from x itself.

A set which contains all its limit points is called *closed.* In a metric space X the complement $X \backslash E$ of any open set E is closed and, conversely, the complement of a closed set is open.

The minimal closed set containing a given set $E \subset X$ is called the *closure* of E and is written $\bar{E}$ (the closure $\bar{E}$ of E is also the intersection of all closed sets containing E).

A set $E \subset X$ is called *dense* in X if the closure of E coincides with X, i.e. $\bar{E} = X$. Every metric space is contained as a dense subset in a suitable complete metric space, called the *completion* of the given one.

The construction of the completion of a given metric space X can be effectuated in the following manner. Two fundamental sequences $\{x_n\}$ and $\{y_n\}$ are termed equivalent if $\lim_{n\to\infty} \rho(x_n, y_n) = 0$. Let X^* be the set of all equivalence classes of fundamental sequences in X. If $x^* \in X^*$, $y^* \in X^*$ and $\{x_n\} \in x^*$, $\{y_n\} \in y^*$, then the limit

$$\rho^*(x^*, y^*) \stackrel{\text{def}}{=} \lim_{n\to\infty} \rho(x_n, y_n)$$

exists, does not depend on the choice of the sequences $\{x_n\}$, $\{y_n\}$ and defines a metric in X^*. The given space X is isometric to the subset of all classes which contain a stationary sequence $\{x, x, \dots, x, \dots\}$, $x \in X$, and so can be identified with this subset. Thus X can be viewed as a subset of X^*. It turns out that X is dense in X^* and that X^* is complete, that is, X^* is the completion of X.

If a metric space contains a countable dense set then we say that it is *separable.*

An important notion in the theory of metric spaces is compactness.

A metric space is called *compact* if each sequences of points contains a subsequence which converges to a point of the space.

In order to formulate *compactness criterion for a space* we first formulate the notions of bounded and totally bounded metric spaces.

The number

$$\operatorname{diam} X = \sup_{x \in X, y \in X} \rho(x, y)$$

is called the *diameter* of the space X. If the diameter of X is finite then X is termed *bounded.*

Let A and B be given subsets of a metric space X and let ε be a number > 0. We say that B is an *ε-net* for A if for each $x \in A$ there exists a point $y \in B$ such that $\rho(x, y) \leq \varepsilon$.

A set $A \subset X$ is called *totally bounded* if for each $\varepsilon > 0$ there exists a finite ε-net for it.

In n-dimensional Euclidean space $\mathbb{R}^n$ the notions of bounded and totally bounded sets are *equivalent*. In an infinite dimensional space each totally bounded set is clearly bounded; the converse does not hold true. Let us further mention that if a metric space is totally bounded then it is separable.

A necessary and sufficient condition for a metric space to be compact is that it is bounded and complete.

From this it follows that a subset of a complete metric space is compact iff it is totally bounded and complete. In particular, a set in finite dimensional Euclidean space is compact iff it is bounded and closed.

Let us remark that by what was just said it follows from our compactness criterion that each compact metric space is separable.

Another compactness criterion is connected with the notion of covering of a set. Let $E \subset X$. A family $\Omega = \{E_\alpha\}$, $\alpha \in \mathscr{A}$ of sets $E_\alpha \subset X$ ($\mathscr{A} = \{\alpha\}$ is a suitable set of indices α) is called a *covering* of E if

$$E \subset \bigcup_{\alpha \in \mathscr{A}} E_\alpha.$$

A covering consisting of finitely many sets E_α is called *finite*.

It turns out that a metric space is compact iff each open covering of it contains a finite subcovering.

In the case of finite dimensional Euclidean space this entails the well-known *Borel-Lebesgue lemma*: every covering of a bounded closed subset of $\mathbb{R}$ contains a finite covering.

Concluding our discussion of compactness, let us mention that that a subset A of a metric space X is called *pre-compact* if its closure $\bar{A}$ in X is compact.

The notions of closed set, closure, density of a set and compactness make sense also for topological spaces but in the sequel we will only encounter them in the context of metric spaces. Therefore we content ourselves with what has been said.

§ 5. Normed and Seminormed Spaces. Banach Spaces

A linear space (real or complex) is said to be a *normed space* if on the set of its points x there is defined a nonnegative function called, the *norm*, written $\|x\|_X$ or simply $\|x\|$, enjoying the following properties:

1) $\|x + y\| \leq \|x\| + \|y\| \ \forall x \in X, \forall y \in X$;
2) $\|\lambda x\| = |\lambda| \|x\| \ \forall x \in X$;
3) if $\|x\| = 0$, $x \in X$, then $x = 0$.

Here λ is a real or complex number according to whether we consider a real or a complex linear space.

If we on the set of points of a linear space X give a nonnegative function $\|x\|$ satisfying only 1) and 2), we say that X is a *seminormed* space and that $\|x\|$ is a *seminorm*.

Two linear normed spaces are said to be *isomorphic* if there exists a bijective correspondence between its points which preserves the linear operation and the value of the norm.

Two norms (seminorms) $\|x\|$ and $\|x\|^*$ in a normed space X are said to be *equivalent* if there exist constants $c_1 > 0$ and $c_2 > 0$ such that for all $x \in X$ holds

$$c_1\|x\| \leq \|x\|^* \leq c_2\|x\|.$$

In a finite dimensional space *all* semi-norms are equivalent but in infinite dimensional spaces there exist also nonequivalent norms. It is an important problem for infinite dimensional spaces, in particular for function spaces, to investigate when various norms are equivalent.

Every normed linear space is a metric space with the metric

$$\rho(x, y) = \|x - y\|, \quad x \in X, y \in X.$$

(Instead of convergence in metric one likewise says *convergence in norm.*) Therefore the notions of open and closed sets, closure of a set, compact set, density of a set in a space, separability and completeness are defined for normed spaces. A complete normed linear space is termed a *Banach space.* Every normed linear space is contained as a dense subspace in a Banach space.

The construction of the completion of a given normed linear space X can be obtained by the same scheme as for the completion of a metric space. Let us denote by X^* the set of all classes x^* of equivalent fundamental sequences $\{x_n\}$ of points of a given normed linear space X. If $x^* \in X^*$, $y^* \in X^*$ and $\{x_n\} \in x^*$, $\{y_n\} \in X^*$ then for any numbers λ and μ we set $\lambda x^* + \mu y^* = \lim_{n\to\infty}(\lambda x_n + \mu y_n)$ and $\|x^*\| = \lim_{n\to\infty}\|x_n\|$. These limits exist, the linear operation $\lambda x^* + \mu y^*$ does not depend on the choice of the sequences $\{x_n\}$, $\{y_n\}$ in respectively the classes x^*, y^*, the value $\|x^*\|$ does not depend on the choice of $\{x_n\} \in x^*$ and is a norm in X^*, which space turns out to be complete and contains a dense subset isomorphic to X, with which this subset therefore can be identified. Thus X^* is the completion of X.

There exist also metric linear spaces where the metric is not generated by a norm. Such is the case for the space of measurable functions on a set E with the metric

$$\rho(f, g) = \int_E \frac{|f(x) - g(x)|}{1 + |f(x) - g(x)|}\,dx, \quad \mu E < +\infty.$$

Convergence in this metric is equivalent to *convergence in measure,* that is $f_n \to f$ in this metric means that for any $\varepsilon > 0$ holds

$$\lim_{n\to\infty} \mu\{x : |f_n(x) - f(x)| \geq \varepsilon\} = 0.$$

Examples of normed function spaces.

1) The space $B(E)$ of bounded functions on a set E (cf. § 2) gets normed if we define the norm by the formula

$$\|f\|_{B(E)} \stackrel{\text{def}}{=} \sup_{x\in E} |f(x)|. \tag{1.4}$$

This norm is called the *uniform norm.*

2) Let E be a space with a complete measure μ. On the subset $L_\infty(E)$ of the set $M(E)$ consisting of all measurable functions $f : E \to \mathbb{R}$ (or $f : E \to \mathbb{C}$) such that the functional

$$\|f\|_\infty = \operatorname{vrai}\sup_{x\in E}|f(x)| \tag{1.5}$$

is finite, the latter constitutes a seminorm. On the subset $L_p(E) \subset M(E)$, $1 \le p < +\infty$ of measurable functions such that the functional

$$\|f\|_p = \left(\int_E |f(x)|^p dx\right)^{1/p}, \tag{1.6}$$

is finite, this functional too is a seminorm (here the integral is taken in Lebesgue's sense with respect to μ).

The norms (1.5) and (1.6) are called *p-norms,* $1 \le p \le +\infty$.

Two functions $f \in M(E)$ and $g \in M(E)$ are said to be *equivalent* (with respect to μ) and we write $f \sim g$ if $\mu\{x : f(x) \neq g(x)\} = 0$. If we identify equivalent functions in the seminormed space $L_p(E)$, $1 \le p \le +\infty$, then they become normed spaces, which we likewise denote by the same symbols $L_p(E)$, $1 \le p \le +\infty$.

3) If E is a topological space we denote as before by $BC(E)$ the space of continuous bounded numerical functions; clearly this is a normed linear space and in this case the norm (1.4) usually will be written $\|\cdot\|_C$. If E is compact then every continuous numerical function is bounded and its absolute value is attained on E. Therefore, in this case (1.4) coincides with the norm

$$\|f\|_C = \max_{x\in E} |f(x)|. \tag{1.7}$$

If X and Y are any sets, $E = E(X)$ is a space of certain functions $X \to \mathbb{R}$ and $f(x, y) : X \times Y \to \mathbb{R}$ a function such that for every fixed $y \in Y$ one has $f(x, y) \in E(X)$, then the norm of f considered as a function of a single variable x will be denoted by $\|f\|_{E,X}$. This is clearly a function of $y \in Y$.

In the study of the spaces L_p and related function spaces a major role is played by Hölder's inequality and the generalized integral Minkowski's inequality. By Minkowski's inequality or the *usual integral Minkowski's inequality* we intend the first property of the norm, written for integral p-norms:

$$\left(\int_E |f(x) + g(x)|^p dx\right)^{1/p} \le \left(\int_E |f(x)|^p dx\right)^{1/p} + \left(\int_E |g(x)|^p dx\right)^{1/p}. \tag{1.8}$$

Hölder's inequality has the form

$$\int_E |f(x)g(x)|dx \leq \|f\|_p \|g\|_{p'}, \tag{1.9}$$

where $1/p + 1/p' = 1$, $f \in L_p(E)$, $g \in L_{p'}(E)$, $1 \leq p \leq +\infty$.

If E_1 and E_2 are spaces with the complete measures μ_1 and μ_2 and f is a numerical function defined on the product $E_1 \times E_2$ of these spaces ($f = f(x, y)$, $x \in E_1$, $y \in E_2$), then the following inequality is in force

$$\left(\int_{E_1} \left|\int_{E_2} f(x,y)dy\right|^p dx\right)^{1/p} \leq \int_{E_2} \left(\int_{E_1} |f(x,y)|^p dx\right)^{1/p} dy,$$

or in norm notation

$$\left\|\int_{E_2} f(x,y)dy\right\|_{L_p(E_1),x} \leq \int_{E_2} \|f(x,y)\|_{L_p(E_1),x} dy \tag{1.10}$$

(the norm of the integral does not exceed the integral of the norm), where we assume that for almost every $y \in E_2$ holds $f(x, y) \in L_p(E_1)$, $x \in E_1$ and that $\|f(x,y)\|_{L_p(E_1),x} \in L_p(E_2)$, $y \in E_2$. Inequality (1.10) is called the *generalized Minkowski's inequality*.

At the hand of p-norms one can illustrate the connection between different norms. Namely, in certain cases one can obtain estimates of some p-norms in terms of others (such statements are usually called *imbedding theorems*). If $f \in L_p(E)$, $1 < p < +\infty$, then it follows from Hölder's inequality that

$$\|f\|_1 \leq (\mu E)^{1/p'} \|f\|_p, \quad 1/p + 1/p' = 1, \tag{1.11}$$

and if $f \in L_\infty(E)$ then

$$\|f\|_p \leq (\mu E)^{1/p} \|f\|_\infty, \quad 1 \leq p < +\infty. \tag{1.12}$$

Of course, these inequalities are of interest only if $\mu E < +\infty$.

In terms of normed function spaces one can describe *uniform convergence, convergence in mean* and *mean square convergence*.

A sequence of function $f_n \in B(E)$, $n = 1, 2, \ldots$, converges *uniformly* to a function $f \in B(E)$ iff $f_n \to f$ in $B(E)$.

A sequence of functions $f_n \in L(E)$, $n = 1, 2, \ldots, \infty$, converges to a function $f \in L(E)$ *in mean* iff $f_n \to f$ in $L(E)$.

A sequence of functions $f_n \in L_2(E)$, $n = 1, 2, \ldots$, converges in *square mean* iff $f_n \to f$ in $L_2(E)$.

Examples of Banach spaces of functions are the spaces of bounded continuous functions $BC(E)$ and the Lebesgue spaces $L_p(E)$, $1 \leq p \leq +\infty$.

For some function spaces one has special compactness criteria, which usually are simpler to verify than the general compactness criteria discussed above (cf. § 4 of Chap. 1).

A family of function $\mathcal{F} = \{f\}$, $f : E \to \mathbb{R}$, $E \subset \mathbb{R}^n$, is said to be *uniformly bounded* if there exists a constant $c > 0$ such that $|f(x)| < c$ for all $x \in E$ and all $f \in \mathcal{F}$. In other words, a family $\mathcal{F}$ of functions f is uniformly bounded iff it constitutes a bounded subset of the Banach space $B(E)$.

The family $\mathcal{F} = \{f\}$ is said to be *equicontinuous* if for each $\varepsilon > 0$ there exists a $\delta > 0$ such that

$$|f(x_1) - f(x_2)| < \varepsilon$$

for all $x_1 \in E$, $x_2 \in E_2$ with $\rho(x_1, x_2) < \delta$ and all $f \in \mathcal{F}$.

For instance, a family of differentiable functions on a convex domain is equicontinuous if all the partial derivatives form a uniformly bounded family.

The following compactness criterion, usually called the *Arcela-Ascoli theorem*, holds true for the space $C(X)$.

A necessary and sufficient condition for a family $\mathcal{F} = \{f\}$ of continuous functions on a compact set $X \subset \mathbb{R}^n$ to be pre-compact is that it is uniformly bounded and equicontinuous.

This criterion can be generalized to the case of maps from an arbitrary compact metric space into another.

An analogous criterion for the spaces L_p, $1 \le p < +\infty$, was obtained by F. Riesz in 1910.

Let E be a measurable subset of $\mathbb{R}^n$, $\mu < +\infty$. A set of functions $\mathcal{F} \subset L_p(E)$ is precompact in $L_p(E)$ iff it is bounded in $L_p(E)$ and *equicontinuous in the L_p-metric*, i.e. for each $\varepsilon > 0$ there exists an $\delta > 0$ such that for all $f \in \mathcal{F}$ subject to $|h| < \delta$ holds

$$\|f(x+h) - f(x)\|_{L_p(\mathbb{R}^n)} < \varepsilon$$

(all functions $f \in \mathcal{F}$ are assumed to be continued by zero outside E).

Another compactness criterion for the space L_p was obtained by A.N. Kolmogorov in 1931 for $1 < p < \infty$ and two years afterwards for $p = 1$ by A.N. Tulaĭkov. It is based on the approximation of functions $f \in L_p$ by so-called Steklov averaging functions.

Let f be defined in a measurable set $E \subset \mathbb{R}^n$. We continue it by zero outside E. Denote by $Q_h(x)$ the n-dimensional ball of radius $h > 0$ and center at the point $x \in \mathbb{R}^n$ and put

$$f_h(x) \stackrel{\text{def}}{=} \frac{1}{\mu Q_h(x)} \int_{Q_h(x)} f(x)dx. \tag{1.13}$$

The function $f_h(x)$ is called the *Steklov averaging function* or, simply, the *Steklov average* (of the function f).

If f is locally in L_p (cf. § 2 of Chap. 1) then for every $h > 0$ and arbitrary $x \in \mathbb{R}^n$ holds

$$f \in L_p(Q_h(x)),$$

and its Steklov averaging function f_h is continuous in the whole of $\mathbb{R}^n$, vanishes off the h-neighborhood of E and satisfies

$$\|f_h\|_p \le \|f\|_p, 1 \le p \le +\infty, \quad \lim_{h \to 0} \|f_h - f\|_p = 0, 1 \le p < +\infty.$$

The *compactness criterion of Kolmogorov* amounts to the following. A necessary and sufficient condition for a set of functions $\mathscr{F} \subset L_p(E)$, where $E \subset \mathbb{R}^n$ is compact, $1 \leq p < +\infty$, to be precompact in $L_p(E)$ is that $\mathscr{F}$ be bounded in $L_p(E)$ and that the difference $\|f_h - f\|_p$ for $h \to 0$ tends to zero uniformly for $f \in \mathscr{F}$.

In conclusion let us remark that often one also encounters *quasi-normed spaces*, which differ from normed ones in the respect that the triangle inequality holds with a suitable constant $c > 1$:

$$\|x + y\| \leq c(\|x\| + \|y\|), \quad x \in X, y \in X.$$

For instance, $L_p(E)$ is such a space if $0 < p < 1$ (here $c = 2^{1/p'}$, $1/p + 1/p' = 1$).

§ 6. Inner Product Spaces. Hilbert Spaces

A real function defined on the set of ordered pairs of a real linear space X, written (x, y), $x \in X, y \in X$, is called an *inner product* (or a *scalar product*) if it enjoys the following properties:

1) $(x, y) = (y, x) \quad \forall x \in X, y \in X$;
2) $(\lambda x + \mu y, z) = \lambda(x, z) + \mu(y, z) \quad \forall x \in X, y \in X, z \in X, \lambda \in \mathbb{R}, \mu \in \mathbb{R}$;
3) $(x, x) \geq 0 \quad \forall x \in X$;
4) if $(x, x) = 0$ then $x = 0$.

A real function, defined on the set of ordered pairs of elements of a real vector space X and satisfying only conditions 1), 2), 3) is called a *semi-* (or *quasi-*)*inner product*.

In an analogous way we introduce the notion of inner and semi-inner product in a complex linear space X. In this case a complex function (x, y) is said to be an inner product if it satisfies 2) for arbitrary complex numbers λ and μ, 3) and 4) and, instead of 1), the condition

1′) $(x, y) = \overline{(y, x)}$,

where, as usual, a bar over a complex number stands for the conjugate. For a semi-inner product in a complex linear space, as in the real case, one does not require 4).

For every pair of points $x \in X, y \in X$ in a linear space X with an inner product (an inner product space) holds the inequality

$$|(x, y)|^2 \leq (x, x)(y, y), \tag{1.14}$$

called *Cauchy's inequality*.

If we in a semi-inner space X set

$$\|x\| \stackrel{\text{def}}{=} \sqrt{(x, x)}, \quad x \in X, \tag{1.15}$$

then the function $\|x\|$ enjoys the properties 1), 2), 3) of a seminorm and is called the *seminorm induced by the given inner product*. If the semi-inner product is an inner product then the seminorm (1.15) is a norm. Thus, each inner product space is a normed space. The converse is *not* true: the norm in a normed linear space does not always come from an inner product. Namely, a necessary and sufficient condition for the existence of an inner product in a normed linear space X induced by the given norm is that the *"parallelogram rule"* holds for any two elements $x \in X, y \in X$:

$$\|x+y\|^2 + \|x-y\|^2 = 2(\|x\|^2 + \|y\|^2). \tag{1.16}$$

If this equality holds true for all $x \in X, y \in X$ then the function

$$(x, y) = \frac{1}{4}(\|x+y\|^2 + \|x-y\|^2)$$

is an inner product induced by a given norm.

The parallelogram rule is not true in the space of continuous functions $C([a, b])$. Therefore its norm is not induced by an inner product.

As every inner product space is a normed space and each normed space is a metric space, it follows that each inner product is a metric space.

An inner product space which is complete in the metric induced by the given inner product is called *Hilbert space.* A space with an inner product is often also termed a *pre-Hilbert space.* This term is justified by the following fact: every inner product space is contained as a dense subset in some Hilbert space, called its *completion.*

The construction of this completion is achieved by the same method as for the completion of a normed linear space. Let X be an inner product space. Let X^* be its completion (§5 of Chap. 1) with respect to the norm (1.15). The inner product of two elements of X^* is defined by the formula $(x^*, y^*) = \lim_{n\to\infty}(x_n, y_n)$ where $\{x_n\} \in x^* \in X^*$, $\{y_n\} \in x^* \in X^*$. This limit exists, does not depend on the choice of the fundamental sequences x_n, y_n in respectively the classes x^*, y^* and enjoys the properties 1)-4) of an inner product. The space X^* is a Hilbert space and, thus, constitutes the completion of X.

As an example of a Hilbert space let us mention the space $L_2(E)$ where E is a space with a complete measure. The inner product is the bilinear function

$$(f, g) = \int_E f(x)\overline{g(x)}dx, \tag{1.17}$$

and the norm induced by it has the form

$$\|f\|_2 = \left(\int_E |f(x)|^2 dx\right)^{1/2}. \tag{1.18}$$

In this case Cauchy's inequality (1.14) takes the form

$$\int_E |f(x)\overline{g(x)}|dx \leq \|f\|_2 \|g\|_2$$

and is often called the *Cauchy-Bunyakovskiĭ*[1] *inequality*. It is a special case of Hölder's inequality (cf. (1.19) of Chap. 1) for $p = p' = 2$.

Let us remark that also in the case $E = [a, b]$ the subset $CL_2[a, b]$ of $L_2[a, b]$ consisting of all continuous functions on $[a, b]$ with the inner product (1.17) is a pre-Hilbert space, but not a Hilbert space; it is dense in the Hilbert space $L_2[a, b]$.

§ 7. Complete Function Spaces. Bases

The proof of many theorems in mathematical analysis is first carried out for functions which are simple in some sense (often smooth) and is then carried over with the help of a passage to the limit to functions in a wider class. In order to apply this method it is necessary to study the question of approximation of "simple" functions by more "complicated" ones, i.e. in terms of function spaces one has to study the density of suitable systems of functions in corresponding function spaces. In the numerical solution of problems the possibility to substitute, up to any given degree of accuracy in the metric of the space under consideration, a "complicated" function by a "simpler" makes it possible to apply simpler numerical algorithms for the execution of the calculations in question.

In what follows we will in general deal only with function spaces consisting of functions (mostly only taking real values) defined on subsets of Euclidean spaces. By the measure μ in a n-dimensional Euclidean space $\mathbb{R}^n$, unless we say explicitly something else, we always intend the n-dimensional Lebesgue measure.

Let E be a Lebesgue measurable set in $\mathbb{R}^n$. An example of a dense set of functions in $L_p(E)$, $1 \leq p < +\infty$, is the set of all continuous functions on E. A sequence of continuous functions tending to a given function $f \in L_p(E)$ can be gotten with the aid of averages (of functions of several variables); cf. (1.13).

If $p = +\infty$ the continuous functions do not form a dense subset in $L_{+\infty}(E)$, as starting with continuous functions one can reach with the aid of uniform convergence only continuous functions.

In view of the Weierstrass theorem on the approximation of continuous functions by polynomials, every continuous function on a compact set E can be approximated in the uniform metric (i.e. in $C(E)$) with arbitrary accuracy by algebraic polynomials. This means that the set of algebraic polynomials forms a dense subset in $C(E)$ and, consequently, also in the spaces $CL_p(E) \subset L_p(E)$, consisting of all continuous functions viewed as elements of $L_p(E)$ ($1 \leq p \leq +\infty$, E a compact set), for inequality (1.12) holds for continuous functions.

[1] *Translator's Note.* In Western literature: Cauchy-Schwarz.

A notion closely connected with the notion of density of a family of functions in a space is the concept of complete system of functions. First let us mention that the notion of density of a set in a metric space also can be generalized to a seminormed space: a set $A \subset X$ is said to be *dense in the seminormed space* X if for any element $x \in X$ and any $\varepsilon > 0$ one can find an element $a \in A$ such that $\|x - a\| < \varepsilon$.

It is clear that if the space X is normed then the above definitions means the density of A in X in the sense of the metric induced by the norm of X.

Now we can define the notion of *complete* system in a seminormed space.

A system $\Omega = \{x_\alpha\}$, $\alpha \in \mathscr{A}$ ($\mathscr{A}$ being a suitable index set) of elements in a seminormed space is termed complete if the set of all finite linear combinations of its elements is dense in X in the sense of the given seminorm in X. This means that for each $x \in X$ and each $\varepsilon > 0$ there exists elements $x_{\alpha_k} \in \Omega$ and numbers λ_k, $k = 1, \ldots, n$, such that

$$\left\| x - \sum_{k=1}^{n} \lambda_k x_{\alpha_k} \right\|_x < \varepsilon.$$

As an example of a complete system we quote the nonnegative integer powers of the independent variable

$$1, x, x^2, \ldots, x^n, \ldots, \tag{1.19}$$

which is a complete system, for instance, in $C[a, b]$ on an arbitrary closed interval $[a, b]$.

Indeed, as finite linear combinations of the functions in the system (1.19) are polynomials, the completeness of this system clearly follows at once from Weierstrass's theorem on uniform approximation of continuous function by polynomials.

Another classical example of a complete system is the *trigonometric system*

$$1, \cos x, \sin x, \ldots, \cos nx, \sin nx, \ldots, \tag{1.20}$$

which forms a complete system in the space $\tilde{C}(\mathbb{R})$ of all continuous functions periodic of period 2π.

In order to desribe the connection between complete systems in various function spaces let us introduce the important notion of imbedding of one function space into another.

A seminormed space X is *imbedded* in a seminormed space Y and we write $X \hookrightarrow Y$ if

1) $X \subset Y$;

2) there exists a constant $c > 0$ such that the inequality

$$\|x\|_Y \leq c\|x\|_X \tag{1.21}$$

is fulfilled. In other words, X is imbedded in Y if X forms a subset of Y and the identity operator is bounded on X. The constant $c > 0$ in (1.21) is called the *imbedding constant.*

It turns out that if a system $\Omega = \{x_\alpha\}$, $\alpha \in \mathscr{A}$, is complete in a seminormed space X and X is dense in Y in the seminorm of the latter space, then Ω is complete in Y.

From this fact and inequalities (1.11) and (1.12) (cf. § 5 of Chap. 1) follows at once that the system of nonnegative integer powers of the independent variable (1.19) is dense not only in $C[a,b]$ but also in the space $L_p[a,b]$, while the trigonometric system (1.20) is dense in $L_p[-\pi,\pi]$, $1 \le p < \infty$.

Finally, another important notion in normed linear spaces is the concept of basis of a space.

A system of elements $\{e_\alpha\}$, $\alpha \in \mathscr{A}$, of a normed linear space X is called a *basis* (or, more precisely, a *Schauder basis*) in X if every element $x \in X$ can be expanded in a unique way with respect to the elements of the system:

$$x = \sum_{\alpha \in \mathscr{A}} \lambda_\alpha e_\alpha. \tag{1.22}$$

It is assumed that for each $x \in X$ there exist in the expansion (1.22) at most a countable number of λ_α with $\lambda_\alpha \neq 0$, so that the right hand side is an ordinary sum.

A *system* of elements is said to be *linearly independent* if all its finite subsystems are linearly independent. If the system of elements $\{e_\alpha\}$, $\alpha \in \mathscr{A}$, is a Schauder basis in the normed linear space X, then it is linearly independent.

A Schauder basis $\{e_n\}$ in a Banach space is called *unconditional* if from the convergence in X of any series $\sum_{n=1}^{\infty} \lambda_n e_n$ it follows that the series $\sum_{n=1}^{\infty} \lambda_{\pi(n)} e_{\pi(n)}$, where π is any surjective map from the set of natural integers $\mathbb{N}$ into itself, is convergent.

Along with the notion of Schauder basis we have also the notion of *Hamel basis*: any maximal linearly independent subset of a linear space is said to be a Hamel basis. In other words: a linearly independent set of elements such that each element of the space is a finite linear combination of elements of the system. In contrast to the notion of Schauder basis, the notion of Hamel basis makes sense for an arbitrary linear space without the assumption of a norm being defined there.

An infinite dimensional normed linear space may have a countable Schauder basis but then it automatically does not admit a countable Hamel basis (otherwise it would consist of a countable number of points). Similarly not every separable normed linear space has a Schauder, but there always exists a Schauder basis in each Hilbert space and even an *orthogonal* one, i.e. a basis such that any two elements are orthogonal (their inner product is zero).

All bases of a Hilbert space have the same cardinality and this cardinal number is called the *dimension* of the given Hilbert space. Two Hilbert spaces are isomorphic iff they have the same dimension. In particular, all separable (infinitely dimensional) Hilbert spaces are isomorphic.

As an example of an orthogonal basis we have the trigonometric system (1.20) in $L_2(-\pi, \pi)$.

Orthogonal bases can be obtained from complete systems with the help of an orthogonalization process. For example, if $\{x_n\}$ is a linearly independent system then the elements

$$y_1 = x_1, \; y_{k+1} = \alpha_{k+1,1}y_1 + \cdots + \alpha_{k+1,k}y_k - x_{k+1},$$

$$\alpha_{k+1,i} = \frac{(y_i, x_{k+1})}{(y_i, y_i)},$$

form an orthogonal system: $(y_i, y_j) = 0, i \neq j$.

We remark that each orthogonal system, whose elements do not vanish, is a linearly independent system.

If one applies the orthogonalization process in $L_2[-1, 1]$ to the system of nonnegative integer powers of the independent variable x (1.19) then, up to a constant factor, one gets the so-called *Legendre polynomials*

$$P_0(x) = 1, \quad P_n(x) = \frac{1}{2^n n!}\frac{d^n(x^2-1)^n}{dx^n}.$$

The system (1.19) being complete, the Legendre polynomials therefore form an orthogonal basis in $L_2[-1, 1]$.

If we give on $[-1, 1]$ not the Lebesgue measure but another measure μ such that (1.19) is linearly independent in the corresponding space with the scalar product

$$(f, g) = \int_{-1}^{1} f(x)g(x)d_\mu x, \tag{1.23}$$

then the orthogonalization process applied to (1.19) in this space yields a certain system of orthogonal polynomials $\{Q_n(x)\}$, which of course depends on the choice of μ. If μ is defined for Lebesgue measurable subsets of $[-1, 1]$ by the formula

$$\mu E = \int_E \varphi(x)dx, \tag{1.24}$$

where φ is a fixed nonnegative summable function, then the orthogonality condition for polynomials $Q_n(x)$ takes the form

$$(Q_m, Q_n) = \int_{-1}^{1} Q_m(x)Q_n(x)\varphi(x)dx = 0, \quad m \neq n. \tag{1.25}$$

The function φ defining the measure (1.24) is called a *weight function* or, in brief, a *weight* and the corresponding space with the inner product (1.24) will be denoted by $L_{2,\varphi}[-1, 1]$. This is a Hilbert space.

If condition (1.25) is fulfilled for some polynomials, we say that these polynomials are orthogonal with respect to the weight $\varphi(x)$ or that they are orthogonal in $L_{2,\varphi}[-1, 1]$.

For instance, if $\varphi(x) = 1/\sqrt{1-x^2}$ then the polynomials corresponding to these weights differ only by a constant factor from the *Chebyshev polynomials*

$$T_n(x) = \cos n \arccos x, \quad n = 0, 1, 2, \dots,$$

which give an orthogonal basis in $L_{2,1/\sqrt{1-x^2}}[-1, 1]$.

In the case of an infinite interval $(-\infty, +\infty)$ a complete system in $L_2(-\infty, +\infty)$ is given by the system of functions $\{x^n e^{-x^2/2}\}$ and the functions obtained from the ones in this system by the orthogonalization process in the inner product (1.23) have the form

$$f_n(x) = Q_n(x) e^{-x^2/2},$$

where the polynomials $Q_n(x)$ differ at most by a constant factor from the so-called *Hermite polynomials*

$$H_n(x) = e^{x^2} \frac{d^n e^{-x^2}}{dx^n}.$$

Thus, the Hermite polynomials form an orthogonal basis in $L_{2,e^{-x^2}}(-\infty, +\infty)$.

In the case of the halfaxis $(0, +\infty)$ one has the complete system $\{x^n e^{-x}\}$ in $L_2(0, \infty)$. Applying the orthogonalization process to them yields functions of the form

$$Q_n(x) e^{-x},$$

where the $Q_n(x)$ are polynomials which differ by a constant factor from the polynomials

$$L_n(x) = e^x \frac{d^n x^n e^{-x}}{dx^n},$$

called *Laguerre polynomials*. Consequently, Laguerre polynomials determine a basis in $L_{2,e^{-x}}(0, +\infty)$.

§ 8. Dual Spaces

A map from a linear space X into the set of real numbers $\mathbb{R}$ (or the set of complex numbers $\mathbb{C}$) is called a *functional* defined on this space. The value of a functional f at the point x of a linear space X is written (f, x), that is, as the inner product of two elements f and x of a pre-Hilbert space. This notation is justified, in particular, by the fact that the inner product (a, x) of a fixed element a is a functional defined on the pre-Hilbert space X, $a \in X$, $x \in X$.

Functionals, exactly as all numerical functions, can be added and multiplied with each other, in particular, with numbers. For example, if f and g are functionals on a space, then the value of the functional $\alpha f + \beta g$ (α and β being numbers) is defined at the point x by the formula

$$(\alpha f + \beta g, x) = \alpha(f, x) + \beta(g, x), \quad x \in X. \tag{1.26}$$

Let X be a linear space. A functional f defined on this space is called *linear* (or, more precisely, *linear homogeneous*) if for any two elements $x \in X, y \in X$ and arbitrary numbers α, β holds

$$(f, \alpha x + \beta y) = \alpha(f, x) + \beta(f, y). \tag{1.27}$$

It follows from (1.26) and (1.27) that a linear combination of linear functionals likewise is a linear functional and, that, consequently, the set of all linear functionals on a linear space X is a linear space. This space is called the *dual space* of X and is denoted by X^*.

Linear functionals on a seminormed linear space fall into two categories: bounded and unbounded ones.

A linear functional on a seminormed linear space is said to be *bounded* if there exists a constant $c > 0$ such that for all elements $x \in X$ holds the inequality

$$|(f, x)| \leq c\|x\|. \tag{1.28}$$

If there is no such constant, f is said to be *unbounded.*

In the space of bounded linear functionals on a normed linear space one can introduce a norm putting

$$\|f\| \stackrel{\text{def}}{=} \inf_{c \in \mathbf{R}^+} \{c : |(f, x)| \leq c\|x\|_X, x \in X\}. \tag{1.29}$$

The following obvious identities hold true:

$$\|f\| = \sup_{x \neq 0} \frac{|(f, x)|}{\|x\|_X} = \sup_{\|x\| \leq 1} |(f, x)| = \sup_{\|x\| = 1} |(f, x)|. \tag{1.30}$$

We observe that from the definition (1.29) of the norm of a functional follows an estimate for the value of the functional,

$$|(f, x)| \leq \|f\| \|x\|, \tag{1.31}$$

generalizing Cauchy's inequality.

The set of all bounded linear functionals on a normed space X likewise forms a normed linear space with the norm (1.29). It is called the *dual* space of the given space X and is written X^*. One can show that the dual X^* always is a *complete* space, regardless of whether X is complete or not.

Every bounded linear functional is continuous in the sense that if

$$\lim_{n \to \infty} x_n = x \text{ in } X, \tag{1.32}$$

then

$$\lim_{n \to \infty} (f, x_n) = (f, x). \tag{1.33}$$

Also the converse is true: if a linear functional satisfies (1.32)-(1.33) then it is bounded. Therefore one often says continuous linear functional in place of bounded linear functional – in the case at hand the two notions are equivalent.

Each element $x \in X$ defines a bounded linear functional on X^*, the dual of X, via the formula

$$(x, f) \stackrel{\text{def}}{=} (f, x), \quad f \in X^*, \tag{1.34}$$

and, consequently, each element $x \in X$ is an element of the space $(X^*)^* = X^{**}$ dual to X^*.

Thus

$$X \subset X^{**}. \tag{1.35}$$

If $X = X^{**}$ we say that X is *reflexive*. As the dual of any normed linear space is complete, a reflexive space is always complete, that is, it is a Banach space.

As an example of reflexive spaces we quote Hilbert spaces, in particular, the spaces $L_2(E)$. This follows from the fact that each bounded linear functional f on a Hilbert space X has the form

$$(f, x) = (a, x), \quad x \in X, \tag{1.36}$$

where to the right stands the inner product, a being a fixed element of X. It follows also that each Hilbert space is not only reflexive but indeed coincides with its dual.

In the Banach space of continuous functions $C[a, b]$ each bounded linear functional is defined with the aid of a suitable function $y(t)$ defined on $[a, b]$, of total variation $\overset{b}{\underset{a}{\vee}} y(t)$, via the formula

$$(f, x) = \int_a^b x(t) dy(t), \quad x = x(t) \in C[a, b]. \tag{1.37}$$

Moreover, two functions y_1 and y_2 of bounded variation on $[a, b]$ give the same functional iff they differ by a constant at all continuity points and, therefore, their difference can differ from this constant only at an at most countable set of points. It follows that the elements of the dual $C^*[a, b]$ of $C[a, b]$ can be identified with classes of functions of bounded variation, each class consisting of functions which, up to a constant, differ from each other only on the set where both are not simultaneously continuous.

For the functional f defined by (1.37) holds

$$\|f\|_{C^*} \leq \overset{b}{\underset{a}{\vee}} y(t), \tag{1.38}$$

and there exists in the class of functions y defining f a function such that equality holds in (1.38).

Each bounded linear functional on $L_1(E)$ is of the form

$$(f, x) = \int_E y(t)x(t)dt, \tag{1.39}$$

where $y = y(t)$ is a suitable function (depending on f) in $L_\infty(E)$. The converse is trivial: if $y(t) \in L_\infty(E)$ then the functional f defined by (1.39) is a bounded linear functional on $L_1(E)$. Thus now the bounded linear functionals f can be identified with the corresponding functions in $L_\infty(E)$ (more exactly, with equivalence classes of functions with respect to the measure μ).

On the space $L_p(E)$, $1 < p < \infty$, each bounded linear functional f admits the representation

$$(f, x) = \int_E y(t)x(t)dt, \quad x = x(t) \in L_p(E), \tag{1.40}$$

where

$$y = y(t) \in L_{p'}(E), \tag{1.41}$$

$$1/p + 1/p' = 1 \tag{1.42}$$

(that the functional given by (1.41) is bounded follows from Hölder's inequality (1.9)). If we again identify the linear functionals f with the corresponding functions $y(t)$, it follows from (1.40)-(1.42) that the dual of $L_p(E)$ is $L_{p'}(E)$ and thus, in view of (1.42), the spaces $L_p(E)$, $1 < p < +\infty$, must be reflexive.

Notice that if $p \neq 2$ then $L_p(E)$ is not isomorphic to $L_{p'}(E)$ and, consequently, does not coincide with its dual.

In a normed linear space it is often expedient to consider, besides convergence in norm, also another kind of convergence, generated by bounded linear functionals and called weak convergence.

A *sequence of points* $\{x_n\}$ of a normed linear space X is said to be *weakly convergent* to the point $x \in X$ if for each bounded linear functional f holds

$$\lim_{n \to +\infty} (f, x_n) = (f, x). \tag{1.43}$$

If the *sequence* $\{x_n\}$ converges in the sense of the norm of the space X, then we also say that it is *strongly convergent*. If a sequence converges strongly then it converges also weakly (this follows at once from (1.28)); the converse is in general not true. However, it is true that weak and strong convergence are *equivalent* in the finite dimensional case. It is of interest to remark that each weakly convergent sequence is bounded in norm and $\|x\| \leq \underline{\lim}_{n \to \infty} \|x_n\|$.

Pointwise convergence in function spaces can be described in terms of weak convergence. Consider, for instance, the space $C(E)$ of continuous functions on a compact set E (cf. § 5 of Chap. 1). Then a sequence $x_n = x_n(t) \in C(E)$, $n = 1, 2, \ldots$, is weakly convergent in $C(E)$ to a function $x(t)$ iff it is bounded in norm in $C(E)$ and for each point $t \in E$ holds

$$\lim_{n \to \infty} x_n(t) = x(t). \tag{1.44}$$

Of course, weak convergence in function spaces can be described also starting with the general form of the bounded linear functionals in such spaces. Thus, for example, weak convergence of a sequence $\{x_n(t)\}$ in $L_p(E)$, $1 \leq p < +\infty$, to the functions $x(t)$ means that for each function $y(t) \in L_{p'}(E)$, $1/p + 1/p' = 1$, holds the formula

$$\lim_{n\to\infty} \int_E x_n(t)y(t)dt = \int_E x(t)y(t)dt. \tag{1.45}$$

If X^* is the dual of a normed linear space X, then, along with convergence in norm of functionals, also called strong convergence of sequences of functionals, one can define weak convergence: a sequence $x_n^* \in X$, $n = 1, 2, \ldots$, is called weakly convergent to the functional $x^* \in X^*$ if for each element $x^{**} \in X^{**}$ holds

$$\lim_{n\to\infty} (x^{**}, x_n^*) = (x^{**}, x^*). \tag{1.46}$$

The topology generated by this convergence is called the *weak topology* (induced by X^{**}). Besides this topology one considers in X^* also the *weak-* topology* (a topology induced by X), corresponding to convergence of points in X^* defined only with the aid of those functionals on X^* which belong to X: a sequence $x_n^* \in X^*$, $n = 1, 2, \ldots$, is called *weak-* convergent* to the point $x^* \in X^*$ if for each $x \in X$ holds $\lim_{n\to\infty}(x_n^*, x) = (x^*, x)$. In other words, weak-* convergence is pointwise convergence of functionals on X.

The weak-* topology of X^* is weaker than the weak topology of the space, i.e. if a sequence of points is weakly convergent then it is weak-* convergent. The weak-* convergence in a dual space X^* is interesting in particular because if X a separable normed space then each closed ball of the space is compact in this topology.

§9. Linear Operators

A more general notion than the notion of functional is the notion of operator. By an *operator* we intend any transformation from a linear space X into another Y, i.e. if $A : X \to Y$ then A is an operator. An operator A is called *linear* if it sends each linear combination of elements of X into a linear combination of elements of Y. An operator A is called *continuous* if from $\lim_{n\to\infty} x_n = x$, $x \in X$, $x_n \in X$, $n = 1, 2, \ldots$, it follows that $\lim_{n\to\infty} Ax_n = Ax$. We say that A is *bounded* if there exists a constant $c > 0$ such that for all $x \in X$ holds

$$\|Ax\|_Y \leq c\|x\|_X. \tag{1.47}$$

As in the case of functionals, the notions of continuity and boundedness for operators are equivalent.

In analogy to the case of functionals we define the *norm of an operator*: the norm of a bounded operator A is the infimum of all constants $c > 0$ which enter into the inequality (1.47).

We can also carry over inequality (1.30); we have only to replace the modulus of the value of the functional by the norm of the operator at the point $x \in X$.

The set of all continuous linear operators from a normed linear space X into another linear space Y likewise forms a linear space, which will be denoted by $\mathscr{L}(X, Y)$.

§ 10. Lebesgue Spaces

In our discussion of various types of abstract spaces we have used function spaces as examples of many properties. Here we turn to a few more properties of the Lebesgue spaces $L_p(E)$, $1 \le p \le +\infty$, where E is a Lebesgue measurable set of the n-dimensional Euclidean space $\mathbb{R}^n$, with $\mu E > 0$. If $p \neq 2$, $L_p(E)$ is not a Hilbert space, which follows at once, as the parallelogram rule (1.16) fails. The spaces $L_p(E)$, $1 \le p < +\infty$, are separable Banach spaces.

The spaces $L_p(E)$ enjoy the so-called *(global) average continuity* property. By this we intend the following. Let $f \in L_p(E)$. We extend f outside E by zero to the entire space $\mathbb{R}^n$ and use for simplicity the same symbol f for the extension. Then

$$\lim_{h \to 0} \|f(x+h) - f(x)\|_{L_p(E)}, \quad 1 \le p < +\infty \tag{1.48}$$

or, in integral notation,

$$\lim_{h \to 0} \int_E |f(x+h) - f(x)|^p dx = 0. \tag{1.49}$$

In order to show that "good" functions are dense in Lebesgue spaces one often employs the method of averaging functions. An example of an average (or, more precisely, an averaging function) is the Steklov average used above (cf. § 5 of Chap. 1) for the approximation of functions $f \in L_p(E)$ by continuous functions. For approximation of functions in an integral metric (i.e. in the sense of the norm $\|\cdot\|_p$, $1 \le p < +\infty$) by more smooth functions one uses so-called *means with a kernel.*

Let $\psi(t)$ be an even infinitely differentiable function in one variable t, $-\infty < t < +\infty$, equal to zero for $|t| \ge 1$ and such that

$$\kappa_n \int_{\mathbb{R}^n} \psi(|x|) dx = 1, \tag{1.50}$$

where κ_n is the area of the $(n-1)$-dimensional unit sphere.

As the function ψ, called a *kernel*, one can take

$$\psi(t) = \begin{cases} \dfrac{1}{c_n} e^{-\dfrac{t^2}{t^2-1}}, & 0 \le |t| < 1, \\ 0, & |t| \ge 1, \end{cases}$$

where the constant c_n is chosen so as to meet condition (1.50).

The function

$$\psi_h(x) = \frac{1}{h^n}\psi\left(\frac{|x|}{h}\right), \quad h > 0, \tag{1.51}$$

is infinitely differentiable in $\mathbb{R}^n$, has its support in the ball $Q_h(0)$ of radius h and center at the origin 0, and satisfies

$$\int_{\mathbb{R}^n} \psi_h(x)dx = \frac{1}{h^n}\int_{\mathbb{R}^n} \psi\left(\frac{|x|}{h}\right) dx = 1.$$

Let $f \in L_p(E)$, again putting f equal to 0 off the set E. Then the function

$$f_h(x) = f_{h,\psi}(x) \stackrel{\text{def}}{=} \frac{1}{h^n}\int_{\mathbb{R}^n} \psi\left(\frac{|x-y|}{h}\right) f(y)dy = \frac{1}{h^n}\int_{\mathbb{R}^n} \psi\left(\frac{|y|}{h}\right) f(x-y)dy, \tag{1.52}$$

or, in brief,

$$f_h = f_{h,\psi} = f * \psi_h,$$

is called the *Sobolev mean* of f with step h. Clearly it is an infinitely differentiable function on the whole space $\mathbb{R}^n$. It enjoys properties analogous to the Steklov average:

$$\|f_{h,\psi}\|_p \leq \|f\|_p, \quad 1 \leq p \leq +\infty, \tag{1.53}$$

$$\lim_{h\to 0} \|f_{h,\psi} - f\|_p = 0, \quad 1 \leq p < +\infty. \tag{1.54}$$

It follows from (1.54) that for each measurable set E the set of restrictions of infinitely differentiable functions to this set (we denote it by $C^\infty(E)$) is dense in $L_p(E)$, $1 \leq p < +\infty$.

If G is an open set in $\mathbb{R}^n$ and if we denote by $\overset{\circ}{C}{}^\infty(G)$ the set of all infinitely differentiable functions with compact supports contained in G (such functions are also said to be *finite*[2] on G) then $\overset{\circ}{C}{}^\infty(G)$ too is dense in $L_p(G)$, $1 \leq p < +\infty$. In other words, the set of finite infinitely differentiable functions is dense in $L_p(G)$, $1 \leq p < +\infty$.

If G is a nonempty open set in $\mathbb{R}^n$ then the space $L_p(G)$ admits a Schauder basis (an unconditional one if $p > 1$). A Schauder basis in $L_p(0,1)$ is, for instance, the so-called *orthogonal system of Haar*.

In conclusion, let us remark that in the formulation of the preceding series of properties of the spaces $L_p(E)$ we assumed throughout that $p < +\infty$. This is not without cause: these properties do not hold true for the space $L_{+\infty}(E)$ of essentially bounded measurable functions, which is the dual of $L(E) = L_1(E)$. For example, if $\mu E > 0$ then $L_{+\infty}(E)$ is not separable and the functions in this space do not, in general, possess the analogue of the property of average continuity, i.e. the property

$$\lim_{h\to 0} \|f(x+h) - f(x)\|_{+\infty} = 0. \tag{1.55}$$

[2] *Translator's Note.* Rarely used in the Western literature.

This property distinguishes in $L_{+\infty}(E)$ the subset of uniformly continuous functions on E. The set of infinitely differentiable functions on G likewise is not dense in $L_{+\infty}(G)$.

Along with the space L_p, where p is a scalar, one also considers L_p, where p is a vector, $p = (p_1, \ldots, p_n)$. In this case for $p_j \geq 1$, $j = 1, 2, \ldots, n$, the norm $\|f\|_p$ of f, called a *mixed norm*, is given by

$$\|f\|_p = \|\ldots\|\|f\|_{p_1,x_1}\|_{p_2,x_2}\ldots\|_{p_n,x_n}.$$

§ 11. Morrey Spaces

Along with Lebesgue spaces one encounters in various applications in mathematics also several other spaces, which will be considered in the following divisions of this Part. Many of them consist of functions with given smoothness properties: they possess derivatives satisfying Hölder type conditions etc. Now we turn to yet another type of spaces, the definition of which does not involve any smoothness requirements. Of course, this does not mean that functions in these spaces are not smooth in any sense. We have seen, for example, that all functions in L_p, $1 \leq p \leq +\infty$, have smoothness in the sense of average continuity.

Let G be a bounded domain in $\mathbb{R}^n$, $\delta = \operatorname{diam} G$, $Q_r(x) = \{y : \varrho(y, x) < r\}$, $1 \leq p < +\infty$, $\lambda \geq 0$. Denote by $L^M_{p,\lambda}(G)$ the set of all functions $f \in L_p(G)$ such that

$$^M\|u\|_{p,\lambda} \overset{\text{def}}{=} \{\sup_{x\in G, r\in(0,\delta)} \frac{1}{r^\lambda} \int_{Q_r(x)\cap G} |f(y)|^p dy\}^{1/p} < +\infty. \tag{1.56}$$

The functional $^M\|u\|_{p,\lambda}$ is a norm in $L^M_{p,\lambda}(G)$.

The space $L^M_{p,\lambda}(G)$ with the norm (1.56) is complete, that is, a Banach space. It is called a *Morrey space*.

In the hypothesis that G is such that there exists a constant $c > 0$ such that for $r < \delta = \operatorname{diam} G$ holds

$$\mu(Q_r(x) \cap G) \geq cr^n, \quad x \in G, \tag{1.57}$$

the norm (1.56) is equivalent to the norm

$$\left\{\sup_{x\in G, 0<r<\delta} [\mu(Q_r(x) \cap G)]^{-\frac{\lambda}{n}} \int_{Q_r(x)\cap G} |f(y)|^p dy\right\}^{1/p}.$$

Let us list a few properties of Morrey spaces for domains subject to (1.57).

The space $L^M_{p,0}(G)$ coincides with $L_p(G)$, and $L^M_{p,n}(G)$ coincides with $L_{+\infty}(G)$, $1 \leq p < +\infty$.

If $1 \le p \le q < +\infty$ and λ and ν are numbers such that

$$\frac{\lambda - n}{p} \le \frac{\nu - n}{q},$$

then

$$L^M_{q,\nu}(G) \hookrightarrow L^M_{p,\lambda}(G).$$

Thus, for fixed p, the Morrey spaces form a continuous family (in λ) of function spaces, successively contained one in the other, between the Lebesgue spaces $L_p(G)$ and $L_{+\infty}(G)$, $0 \le \lambda \le n$. Let us further mention the Campanato spaces and their special case, the John-Nirenberg space (see Kufner-John-Fucik [1977]) with a definition in a sense close to the definition of Morrey spaces. All these spaces have important applications in the theory of partial differential equations, in linear as well as in non-linear theory.

Chapter 2

Sobolev Spaces

§ 1. Generalized Derivatives

Spaces of differentiable functions with uniform convergence enjoy the property of completeness. This is connected with the fact that it follows from the uniform convergence of the sequence of functions and the sequence of their derivatives that the limit function is differentiable and that its derivative is the limit of the sequence of derivatives. In the case of integral metrics the situation is different: the use derivatives in the ordinary sense does not lead to complete function spaces, which would have been desirable from many points of view and, in particular, from the point of view of applications of function spaces to the solution of boundary problems for differential equations.[1] Therefore one is lead to consider various generalizations of the derivative.

Let us set forth a definition of the derivative which goes back to B. Levi. Let f be defined in the interval $[a, b]$. The function f' is called the *generalized derivative* of f on $[a, b]$ if there exists an equivalent (in the sense of Lebesgue measure) to f function f_0, which is absolutely continuous on $[a, b]$ and,

[1] Cf. e.g. V.M. Fillipov, *Variational principles for non-potential operators* (Izd. Univ. Druzhby Narodov 1985).

consequently, has almost everywhere on $[a, b]$ an ordinary derivative f_0', and if f' is equivalent to f_0' on $[a, b]$.

One can show that, in view of this definition, the derivative f' of a given function f is unique up to an equivalent function: if f' is a generalized derivative of f then each function equivalent to f' is also a generalized derivative of f.

The basic properties of ordinary derivatives remain in force for generalized derivatives: linearity, the rule for the derivation of a product and a quotient of functions. It is of great importance that the *Newton-Leibniz formula* carries over to the generalized derivative, i.e. a function f can be reconstructed almost everywhere with the aid of an integral with a variable limit of integration with respect to its derivative f'; more exactly we have:

$$f_0(x) = f_0(a) + \int_a^x f'(t)dt, \quad x \in [a, b]$$

(here f_0 is a function equivalent to the function f on $[a, b]$ and absolutely convergent there). Thanks to this, one can in the study of generalized derivatives apply all the usual methods of differential calculus. In particular, the rule for partial integration is available (and of great importance).

Moreover, with this definition of the derivative the corresponding function spaces with an integral metric are complete spaces (more about this below). This follows from the fact that if we understand all derivations as generalized derivations then the *analogue of the theorem on the limit of a uniformly convergent sequence of functions with uniformly convergent sequence of derivatives* holds true in Lebesgue function spaces: if the functions in a sequence $f_n : [a, b] \to \mathbb{R}$ take on the interval $[a, b]$ the generalized derivatives f_n', $n = 1, 2, \ldots$, and if for some p, $1 \leq p < +\infty$, in $L_p[a, b]$ holds

$$\lim_{n\to\infty} f_n = f, \quad \lim_{n\to\infty} f_n' = \varphi, \tag{2.1}$$

in other words, if

$$\|f_n - f\|_p \to 0, \quad \|f_n' - \varphi\|_p \to 0,$$

then the function f has a generalized derivative on $[a, b]$ and we have almost everywhere

$$f' = \varphi. \tag{2.2}$$

Recall that for $p = +\infty$ it is here question of uniform convergence in the sense of the essential supremum.

The generalized derivatives of higher orders are defined as follows: the function $f^{(r)}$, $r \in \mathbb{N}$, (and any equivalent function) is called the *generalized derivative of order* r of the function $f : [a, b] \to \mathbb{R}$ if there exists a function f_0 equivalent to f admitting on $[a, b]$ an ordinary absolutely continuous derivative $f_0^{(r-1)}$ of order $r-1$ and such that $f^{(r)}$ at almost all points coincides with the generalized derivative of $f_0^{(r-1)}$.

It is essential to remark that for generalized derivatives, as for ordinary ones, intermediate derivatives can be estimated in terms of the function itself and the top derivative.

If $f \in L_p[a,b]$ has a generalized derivative of order r, $r \geq 1$ on $[a,b]$ and if $f^{(r)} \in L_p[a,b]$, then there exists a constant $c > 0$ such that for each derivative $f^{(k)}$ of order $k = 1, 2, \ldots, r-1$ (their existence follows from the definition of derivative of order r) holds the inequality:

$$\|f^{(k)}\|_p \leq c(\|f\|_p + \|f^{(r)}\|_p). \tag{2.3}$$

From this inequality and the properties (2.1)-(2.2) of generalized derivatives follows the following fact, which is often used in the theory of function spaces itself as well as in various questions of mathematical analysis.

Assume that the functions $f_n \in L_p[a,b]$ admit on $[a,b]$ generalized derivatives of order r, $r \geq 1$, and that $f_n^{(r)} \in L_p[a,b]$, $1 \leq p \leq +\infty$, $n = 1, 2, \ldots$, and assume that there exist functions f and φ such that in $L_p[a,b]$ holds

$$\lim_{n\to\infty} f_n = f, \quad \lim_{n\to\infty} f_n^{(r)} = \varphi,$$

then f too admits on $[a,b]$ a generalized derivative of order r and this derivative coincides almost everywhere on this interval with φ.

The notion of generalized derivative can also be carried over to the case of partial derivatives of functions of several variables. Let us formulate this for functions defined on open sets.

Let $f(x_1, \ldots, x_n)$ be a function defined on an open set G in n-dimensional space $\mathbb{R}^n$ and let G_i be the projection of G onto the hyperplane $x_i = 0$ (i takes its values among the integers $1, \ldots, n$). If there exists a function f_0 equivalent in the sense of n-dimensional Lebesgue measure to f such that for almost all (now with respect to $(n-1)$-dimensional Lebesgue measure) points $x^{(n-1)} = (x_1, \ldots, x_{n-1}) \in G_i$ there exists on each segment parallel to the x_i axis and projecting onto the point $x^{(n-1)}$ a generalized derivative in the previous sense, this derivative is called the *generalized derivative* $\partial f/\partial x_i$ of the given function in the variable x_i. The *generalized derivatives* of f *of higher order* (mixed or pure)

$$f^{(s)} = \frac{\partial^{|s|} f}{\partial x_1^{s_1} \ldots \partial x_n^{s_n}}, \qquad s = (s_1, \ldots, s_n), |s| = s_1 + \cdots + s_n,$$

are defined in an analogous way as in the case of ordinary partial derivatives of functions of several variables: the pure derivatives $\partial^{s_i} f/\partial x_i^{s_i}$, $i = 1, 2, \ldots, n$, are defined as was done for higher order derivatives of functions of one variable, and mixed derivatives in the standard way by induction.

There is also another approach to the definition of the notion of generalized derivative, due to S. L. Sobolev and based on the notion of *generalized function*, likewise introduced by him, as a functional on a suitable space of ordinary functions.

Let f and g be locally summable functions on an open set $G \subset \mathbb{R}^n$. Then if for each infinitely differentiable finite function φ on G holds the identity

$$\int_G g\varphi dx = (-1)^{|s|} \int_G f\varphi^{(s)} dx, \quad s = (s_1, \dots, s_n), \tag{2.4}$$

we say that g is the *generalized derivative* $f^{(s)}$ of f.

For pure locally summable partial derivative of a locally summable function both definitions are *equivalent*, but for mixed derivatives this is not so: if a function admits a mixed derivative in the sense of the first definition it has also a derivative in the sense of the second definition, but the converse is *not* true. A function may have a mixed derivative in the second definition and fail to have such a derivative in the sense of the first definition.

This follows, for instance, from the fact that there exist functions having generalized mixed derivatives in the second definition and do not possess generalized first order derivatives. As an example of such a function we mention the function $f(x, y) = \varphi(x) + \psi(y)$, $0 \le x \le 1$, where φ is continuous but not absolutely continuouson $[0, 1]$. The function f does not admit first order derivative, as φ is not absolutely continuous. On the other hand the mixed derivative $\partial^2 f/\partial x\partial y$ in the sense of the second definition vanishes identically.

The generalized derivatives in the sense of both definitions coincide, at any rate, for the derivatives $f^{(s)}$, $s = (s_1, \dots, s_n)$, in the hypothesis of local summability of the generalized derivatives

$$f^{(s_1,0,\dots,0)}, f^{(s_1,s_2,0,\dots,0)}, \dots, f^{(s_1,\dots,s_{n-2},0,0)}, f^{(s_1,s_2,\dots,s_{n-1},0)}.$$

It is of importance that the mixed generalized derivative, as well as ordinary derivatives, do not depend on the order of differentiation in the various variables.

In the case of an open set G with sufficiently smooth bounadry one has for generalized partial derivatives, similarly as for generalized derivatives in one variable, estimates in the space $L_p(G)$ of intermediate derivatives in terms of the top derivative and the function itself. It is also true that the set of all functions admitting generalized derivatives of given order belonging to $L_p(G)$ (cf. § 3 of Chap. 2) is complete in that metric.

Another important fact is further the possibility of estimating the L_p-norms of functions and moduli of continuity in the L_p-metric in terms of the norms of the corresponding generalized partial derivatives. Let $f : G \to \mathbb{R}$, where G is an open set. For $\eta > 0$ set

$$G_\eta \stackrel{\text{def}}{=} \{x \in G : \varrho(x, \mathbb{R}^n \backslash G) > \eta\}.$$

Then the difference

$$\Delta_h f = \Delta_h f(x) = f(x + h) - f(x), \quad x \in G_\eta, |h| < \eta,$$

where h is a vector, makes sense. By induction one defines the kth difference $\Delta_h^{(k)} f$ of f, $k \in \mathbb{N}$:

$$\Delta_h^{(k)} f = \Delta_h^{(k)} f(x) \overset{\text{def}}{=} \Delta_h \Delta_h^{(k-1)} f(x), \quad x \in G_\eta, k|h| < \eta, k = 1, 2, \ldots. \tag{2.5}$$

If $e_1, \ldots, e_n$ denote the basis vectors, $k = (k_1, \ldots, k_n) \in \mathbb{N}^n$, $h = (h_1, \ldots, h_n) \in \mathbb{R}^n$, then

$$\Delta_h^{(k)} f(x) \overset{\text{def}}{=} \Delta_{h_1 e_1}^{(k_1)} \Delta_{h_2 e_2}^{(k_2)} \ldots \Delta_{h_n e_n}^{(k_n)} f(x). \tag{2.6}$$

For two multi-indices $s = (s_1, \ldots, s_n)$ and $k = (k_1, \ldots, k_n)$ such that

$$s_i \leq k_i, \quad i = 1, 2, \ldots, n,$$

we write, as usual, $s \leq k$.

If the function f has on the open set G locally summable partial derivatives $f^{(s)}$, $s \leq k$, $s \neq k$, and $f^{(k)} \in L_p(G)$ then (cf. Nikol'skiĭ [1977])

$$\left\| \frac{\Delta_h^k f}{h^k} \right\|_{L_p(G_{k|h|})} \leq \|f^{(k)}\|_{L_p(G)}, \tag{2.7}$$

$$h^k = h_1^{k_1} \ldots h_n^{k_n}, k|h| = k_1|h_1| + \ldots + k_n|h_n|.$$

It is of importance that in a certain sense also the converse statement (loc. cit.; see further Lizorkin-Nikol'skiĭ [1969]) is true: if the function $f : G \to \mathbb{R}^n$ belongs locally to L_p: $f \in L_p(G, \text{loc})$, $1 < p < +\infty$, and

$$\left\| \frac{\Delta_h^k f}{h^k} \right\|_{L_p(G_{k|h|})} \leq c, \quad \forall h \in \mathbb{R}^n, k = (k_1, \ldots, k_n), \tag{2.8}$$

then the generalized derivative $f^{(k)}$ exists on G and

$$\|f^{(k)}\|_{L_p(G)} \leq c. \tag{2.9}$$

Note that this statement is a generalization of the following fact (corresponding to the case $n = 1, p = +\infty$): if the function f satisfies on the interval (a, b) a Lipschitz condition with constant $c > 0$ then it has almost everywhere on (a, b) a derivative satisfying the inequality $|f'(x)| \leq c$.

Let us define the moduli of continuity of a function. By the *modulus of continuity of order* $k = 0, 1, 2, \ldots$ *of a function* f *in the metric of the space* $L_p(G)$ *in the direction of the vector* $h \in \mathbb{R}^n$ we intend the quantity

$$\omega_{h,p}^{(k)}(f; \delta) = \sup_{|t| \leq \delta} \|\Delta_{th}^{(k)} f(x)\|_{L_p(G_{k|t|})}, \quad |h| = 1, \tag{2.10}$$

and by the *modulus of continuity*, taking account of all directions, the number

$$\omega_p^{(k)}(f; \delta) = \sup_{|h|=1} \omega_{h,p}^{(k)}(f; \delta), \quad h \in \mathbb{R}^n. \tag{2.11}$$

§ 2. Boundary Values (Traces) of Functions

If the function $f \in L_p(G)$, $1 \leq p \leq +\infty$, $G \subset \mathbb{R}^n$, G a domain, Γ^m a sufficiently smooth compact manifold of dimension $m < n$ (possibly with boundary), $\Gamma^m \subset \bar{G}$, then it does not make sense to speak of the values of f on Γ^m in the usual sense. In fact, the measure of Γ^m equals zero and therefore for a given function $f \in L_p(G)$ one can change it arbitrarily on Γ^m – every function obtained in this way is equivalent to it, i.e. also an element in $L_p(G)$. Consequently, a reasonable introduction of the concept of boundary values of f requires a special definition. We consider three approaches to this notion. For simplicity we formulate the definitions in the case when $G = \mathbb{R}^n$, $\Gamma^m = \mathbb{R}^m$ (an m-dimensional plane in $\mathbb{R}^m$), $x = (x_1, \ldots, x_n) = (x^{(m)}, \tilde{x}^{(m)})$, $x^{(m)} = (x_1, \ldots, x_m)$, $\tilde{x}^{(m)} = (x_{m+1}, \ldots, x_n)$, $|\tilde{x}^{(m)}| = \sqrt{x_{m+1}^2 + \ldots + x_n^2}$, $\mathbb{R}^m = \{x : \tilde{x}^{(m)} = 0\}$.

A function φ defined on $\mathbb{R}^m$ is called *the trace on* $\mathbb{R}^m$ of the function f *in the sense of almost everywhere convergence* if there exists a function f_0 equivalent to f such that for almost all (in the sense of m-dimensional Lebesgue measure) points $x^{(m)}$ holds

$$\lim_{|\tilde{x}^{(m)}| \to 0} f_0(x^{(m)}, \tilde{x}^{(m)}) = \varphi(x^{(m)}).$$

Another approach to the notion of trace can be formulated in terms of the L_p-metric. A function φ defined on $\mathbb{R}^m$ is called the *trace of a function* f belonging to L_p (cf. § 10 of Chap. 1) *in the sense of* L_p*-convergence* if there exists a function f_0 equivalent to f such that

$$\lim_{\tilde{x}^{(m)} \to 0} \int_{\mathbb{R}^m} |f_0(x^{(m)}, \tilde{x}^{(m)}) - \varphi(x^{(m)})|^p dx^{(m)} = 0.$$

Finally, a third approach to the notion of trace, though applicable only to "sufficiently good" functions, is based on approximation of functions by continuous functions. A function φ defined on $\mathbb{R}^m$ is called the *trace of the function* $f \in L_p(\mathbb{R}^n)$ *in the sense of approximation by continuous functions* if there exists a sequence of continuous functions f_k, $k = 1, 2, \ldots$, on $\mathbb{R}^n$ such that

$$\lim_{k \to \infty} \|f_k - f\|_{L_p(\mathbb{R}^n)} = 0, \quad \lim_{k \to \infty} \|f_k|_{\mathbb{R}^m} - \varphi\|_{L_p(\mathbb{R}^m)} = 0$$

(in the second identity $f_k|_{\mathbb{R}^m}$ denotes the restriction of f_k to the m-dimensional plane $\mathbb{R}^m$).

In all three cases the trace of a function, if it exists, is defined uniquely up to a set of measure zero. This means that if two functions are traces on an m-dimensional plane of one and the same function then they are equivalent in the sense of m-dimensional measure.

In the sequel, in all cases when it is question of traces of functions, all three definitions of the trace of a function will be equivalent.

The trace of a function $f : G \to \mathbb{R}$, $G \subset \mathbb{R}^n$, on a sufficiently smooth manifold $\Gamma^m \subset G$ is defined with the aid of a "local straightening" of the manifold, i.e. by choosing sufficiently smooth locally diffeomorphic maps of n-dimensional neighborhoods of the points of Γ^m in $\mathbb{R}^n$ such that the portions of Γ^m lying in these neighborhoods are mapped onto portions of an m-dimensional plane. The trace of a function defined in this way exists in the hypothesis of suitable differential and global properties and is then unique up to a set of zero m-dimensional measure independent of the choice of the straightening maps for Γ^m.

Of course, far from every function in $L_p(\mathbb{R}^n)$ admits a trace on $\mathbb{R}^m$. Therefore there arises the question which supplementary conditions have to be imposed on a function in order to secure the existence of a trace on $\mathbb{R}^m$. It turns out that for example for $m = n - 1$ it is sufficient that the function has also derivatives in $L_p(\mathbb{R}^n)$. More about this in the next section. Here we just remark that the connection between the properties of functions and their traces in the case of function spaces with an integral metric is more complicated than in the case of spaces with the uniform metric. If a function is continuous on a closed domain then its trace on the boundary of the domain is continuous. If a function is k times continuously differentiable on a closed domain then the trace of the function on the boundary, assumed to be sufficiently smooth, is the same number of times differentiable. In the case of function spaces with integral metric the traces of functions have, in general, lesser smoothness than the function itself. This question will likewise be treated in greater detail below.

§ 3. Sobolev Spaces

Let $f : G \to \mathbb{R}$, $G \subset \mathbb{R}^n$, G an open set, $l \in \mathbb{N}_0$,[2] $1 \le p \le +\infty$. Recall that

$$\|f\|_{L_p(G)} = \begin{cases} \left(\int_G |f(x)|^p dx\right)^{1/p} & \text{for } 1 \le p < +\infty \\ \operatorname*{vrai\,sup}_{x \in G} |f(x)| & \text{for } p = +\infty \end{cases}$$

and set

$$\|f\|_{w_p^{(l)}(G)} \stackrel{\text{def}}{=} \sum_{|\alpha|=l} \|f^{(\alpha)}\|_{L_p(G)}. \tag{2.12}$$

The *Sobolev space* $W_p^{(l)} = W_p^{(l)}(G)$ consists of all functions f for which the norm

$$\|f\|_{W_p^{(l)}(G)} \stackrel{\text{def}}{=} \|f\|_{L_p(G)} + \|f\|_{w_p^{(l)}(G)} \tag{2.13}$$

makes sense and is finite. Here we understand the derivatives in the sense of Sobolev, thanks to which the spaces $W_p^{(l)}(G)$ come as complete normed

[2] $\mathbb{N}_0$ is the union of zero and the set of all natural numbers $\mathbb{N}$.

spaces. They form a discrete family in the parameter $l \in \mathbb{N}_0$ (by definition we set $W_p^{(0)} = L_p(G)$) and a continuous one in p. The Sobolev spaces $W_p^{(l)}(G)$ are separable if $1 \le p < +\infty$ and reflexive if $1 < p < +\infty$, while $W_\infty^{(l)}$ is not separable and $W_1^{(l)}$ and $W_\infty^{(l)}$ are nonreflexive. In the case of an arbitrary domain G the infinitely differentiable functions on its closure are, generally speaking, not dense in $W_p^{(l)}(G)$, this in contrast to the space $L_p(G)$, while the infinitely differentiable functions on G itself are as before dense in $W_p^{(l)}(G)$.

S.L. Sobolev developed the theory of the spaces $W_p^{(l)}(G)$ for domains with a cone condition and for domains starshaped with respect to a ball. By definition G satisfies the *cone condition* if there exists a cone

$$V = \{x : 0 \le \sqrt{\sum_{i=1}^{n-1} x_i^2} \le \varepsilon x_n \le h\}, \quad h > 0, \varepsilon > 0,$$

such that for each point $x \in G$ there exists a congruent cone with vertex at x and entirely contained in G.

If the boundary of a bounded domain is *Lipschitz*, i.e. it is locally given by the graph of functions of the form

$$x_k = f(x_1, \ldots, \hat{x}_k, \ldots, x_n),$$

defined on appropriate $(n-1)$-dimensional domains and satisfying a Lipschitz condition (the symbol $\hat{}$ signifies that the variable below it is omitted), then the domain has the cone property.

In his investigations S.L. Sobolev was the first to obtain and make an essential use of integral representations of functions in terms of their partial derivatives. More precise representations became in the sequel one of the basic instruments in the study of imbedding theorems in function spaces. S.L. Sobolev's *method of integral representations*, developed along these lines, includes estimating integrals of potential type. These important estimates (generalizing one dimensional estimates by Hardy-Littlewood) were found by S.L. Sobolev in 1938 and were subsequently completed by V.P. Il'in in 1954. As a result the following theorem (see Besov-Il'in-Nikol'skiĭ [1975]) was obtained.

Let $f \in L_p(\mathbb{R}^n)$, $1 < p < +\infty$, and set

$$(K_r f)(x) \stackrel{\text{def}}{=} \int_{\mathbb{R}^n} \frac{f(y)}{|x-y|^{n-r}} dy.$$

Then for $1 \le m \le n$, $n - m < pr < n$, the restriction of $K_r f$ to any m-dimensional plane is summable of power $q = mp/(n - pr)$ and one has the estimate

$$\|K_r f\|_{L_q(\mathbb{R}^m)} \le c \|f\|_{L_p(\mathbb{R}^n)},$$

with a constant $c > 0$ independent of f.

The method of integral representations found also in the sequel wide applications in the study of function spaces in the work of V.I. Kondrashov, O.V. Besov, A.D. Dzhabrailov, V.I. Il'in, Calderón, Yu.G. Reshetnyak, V.A. Solonnikov, Smith and others (cf. Besov-Il'in-Nikol'skiĭ [1975], Gol'dshteĭn-Reshetnyak [1983]).

Along with the method of integral representations S.L. Sobolev also applied the *method of smoothing of functions* (cf. (1.52)). The application of this method depends on two facts: first, the averaged functions of the generalized derivatives of f coincide with the derivatives of the averaged function at all points whose distance to the boundary is larger than the radius of averaging and, second, if f admits the derivative $f^{(l)} \in L_p(G)$, $l = (l_1, \ldots, l_n)$, then for each $\eta > 0$ holds

$$\lim_{h \to 0} \|f_h^{(l)} - f^{(l)}\|_{L_p(G_\eta)} = 0, \quad 0 < h < \eta, \quad 1 \le p < +\infty \tag{2.14}$$

(for the notation f_h and G_η, cf. § 10 of Chap. 1 and § 2 of Chap. 2; if $G = \mathbb{R}^n$ then $G_\eta = \mathbb{R}^n$ for all $\eta > 0$). As the averaged functions f_h are infinitely differentiable it follows from (2.14) that functions in Sobolev spaces $W_p^{(l)}$ in any domain, interior to the domain G, can be arbitrarily well approximated by differentiable functions in the L_p-metric.

To this type of results pertains also a result by V.I. Kondrashov [1938] showing that the *set* $C^{(l)}(G)$ of functions which are uniformly continuous together with their derivatives up to order l in a domain G, which is starshaped with respect to a ball, is dense in $W_p^{(l)}(G)$, $1 \le p < +\infty$.

Recall that a *domain* G in n-dimensional space is called *starshaped with respect to a ball* if there exists an n-dimensional ball $Q \subset G$ such that G is starshaped with respect to each point of this ball, that is, whenever $x \in Q$ then any ray issuing from this point intersects the boundary of G in not more than one point. Subsequently an analogous statement on the density of $C^{(l)}(G)$ in $W_p^{(l)}(G)$ was proved also for a significantly larger class of domains (Besov-Il'in-Nikol'skiĭ [1975]).

In a natural way one extends the definition of Sobolev spaces to the case of functions given on sufficiently smooth manifolds of various dimensions $m < n$ lying in $\mathbb{R}^n$.

Let us state *Sobolev's imbedding theorems* for domains satisfying a cone condition (the notation $\hookrightarrow$ is explained in § 7 of Chap. 1).

$$W_p^{(l)}(G) \hookrightarrow W_q^{([r])}(G), \quad 1 < p < q < +\infty, \tag{2.15}$$

$0 \le r \overset{\text{def}}{=} l - n\left(\frac{1}{p} - \frac{1}{q}\right)$, $[r]$ being the integer part of r.

$$W_p^{(l)}(G) \hookrightarrow C^{(\bar{r})}(G), \quad 0 < r \overset{\text{def}}{=} l - \frac{n}{p}, \tag{2.16}$$

$r = \bar{r} + \alpha$, $\bar{r} \ge 0$ integer, $0 < \alpha \le 1$.

Here $\bar{r}$ is the largest integer less than r. Thus $\bar{r} = [r]$ for noninteger r and $\bar{r} = r - 1$ for r integer.

$$W_p^{(l)}(G) \hookrightarrow W_q^{([r])}(\Gamma^m), \quad 1 < p < q < +\infty, \tag{2.17}$$

$$0 \leq r = l - \frac{n}{p} + \frac{m}{q}, \quad 0 \leq m < n.$$

$$W_p^{(l)}(G) \hookrightarrow W_p^{\bar{r}}(\Gamma^m), \quad 1 \leq p \leq +\infty, \quad 0 < r \stackrel{\text{def}}{=} l - \frac{n-m}{p}, \quad 0 \leq m < n. \tag{2.18}$$

The normed linear space $C^{(s)}(G)$ consists of functions having on G piecewise uniformly continuous derivatives of orders up to $s \in \mathbb{N}_0$, with the uniform norm.

For example, the imbedding (2.15) means that each function $f \in W_p^{(l)}(G)$ admits partial derivatives of order $[r]$, integrable on G with power q and that

$$\|f\|_{W_q^{([r])}(G)} \leq c\|f\|_{W_p^{(l)}(G)}.$$

By Γ^m in (2.17) we intend an m-dimensional sufficiently smooth manifold belonging to the closure of G ($\Gamma^m \subset \bar{G}$). The statement (2.17) means that $f \in W_p^{(l)}(G)$ has a trace $f|_{\Gamma^m} = \varphi$ on Γ^m and that

$$\varphi \in W_q^{([r])}(\Gamma^m), \quad 0 \leq m < n$$

and that

$$\|\varphi\|_{W_q^{([r])}(\Gamma^m)} \leq c\|f\|_{W_p^{(l)}(G)}.$$

The Sobolev imbedding theorems are *optimal* in the terms which were used there.

The inequality (generalization of inequality (2.3) to the case of several variables)

$$\|f^{(\alpha)}\|_{L_p(G)} \leq c\|f\|_{W_p^{(l)}(G)}, \quad |\alpha| \leq l, \quad 1 \leq p \leq +\infty, \tag{2.19}$$

where the constant $c > 0$ does not depend on f, adjoins to the above imbedding theorems. This inequality holds for domains G with locally continuous boundary (Nečas [1967]). For arbitrary domains it is, generally speaking, not true. It follows from it that for domain with locally continuous boundary the norm (2.13) is equivalent to the following norm

$$\|f\|^*_{W_p^{(l)}(G)} \stackrel{\text{def}}{=} \sum_{|\alpha| \leq l} \|f^{(\alpha)}\|_{L_p(G)}. \tag{2.20}$$

In S.L. Sobolev's investigations the *seminormed space* $w_p^{(l)}(G)$ with the seminorm $\|f\|_{w_p^{(l)}(G)}$ appears as the basic object of study. It becomes a normed space upon factoring with the set of polynomials of degree not higher than

$l-1$. In the space $w_p^{(l)}(G)$ one can introduce various norms and S. L. Sobolev studied the question of their equivalence (Sobolev [1962]).

We remark that the requirement that the domain satisfies a cone condition, which was sufficient for the validity of the above statements, is in fact close to being sufficient.

The space $W_p^{(l)}(G)$ suits for many applications to mathematical physics. For example, from the inclusion (2.18) applied to the partial derivatives of f it follows that a function $f \in W_p^{(l)}(G)$, $l-(n-m)/p > 0$, has on the manifold Γ^m the following boundary values:

$$\left.\frac{\partial^s f}{\partial n^s}\right|_{\Gamma^m} \in L_p(\Gamma^m), \quad s = 0, 1, \ldots, \bar{r},$$

where the derivation is in the normal direction to Γ^m. For $s > \bar{r}$ this statement is not true: there exists a function in $W_p^{(l)}(G)$ such that the derivatives of order $s > \bar{r}$ do not possess traces on Γ^m. Thus, one finds the number of boundary functions which one has to impose in the formulation of boundary problems with a boundary consisting of parts of various dimensions $m \le n-1$ with $l-(n-m)/p > 0$ in the class $W_p^{(l)}(G)$ (on portions of the boundary where $l-(n-m)/p \le 0$ no boundary conditions are given).

Let us remark that Theorem (2.17) for integer l was proved by V.I. Kondrashov [1950] and V.I. Il'in (see Besov-Il'in-Nikol'skiĭ [1975]). V.I. Kondrashov also obtained a series of sharpenings of these theorems with Hölder classes with integral metric.

The Sobolev spaces $W_p^{(l)}(G)$ have been generalized in many directions. First of all one has considered Sobolev spaces with mixed norm, i.e. the case when the parameter p is a vector (cf. § 10 of Chap. 1). Moreover, one has also considered domains with non-Lipschitz boundary (cf. Besov-Il'in-Nikol'skiĭ [1975], I. G. Globenko [1962], V. G. Maz'ya [1979], [1980], [1981], [1983], [1984] etc.). The theory of Sobolev spaces, in particular imbedding theorems for them, has further been generalized to the case of so-called *abstract functions*, i.e. functions taking their values in a suitable Banach space (cf. S.L. Sobolev [1959], V.B. Korotkov [1962], [1965], A.V. Bukhvalov [1979], [1981] etc.).

If G is any nonempty open set in $\mathbb{R}^n$ then $W_p^{(l)}(G)$ admits a basis. One can also prove that the spaces $W_p^{(l)}(G)$, $1 < p < +\infty$, are isomorphic to $L_p(0,1)$. It follows that Sobolev spaces with $1 < p < +\infty$ have an unconditional basis (Nikolsky-Lions-Lisorkin [1965], Kufner-John-Fucik [1977]). All these results have the character of pure existence theorems. Concrete bases for Sobolev spaces over the n-dimensional cube were found by Cisielski.

§ 4. Sobolev Spaces of Infinite Order

In the study of nonlinear differential equations of infinite order

$$\sum_{|\nu|=0}^{\infty} (-1)^{|\nu|} (A_\nu(x, D^{|\nu|}u))^{(\nu)} = h(x)$$

with boundary conditions, for example, of the form

$$u^{(\omega)}|\partial G = \varphi_\omega, \quad \omega = 0, 1, \ldots,$$

(ν and ω are multi-indices, $D^{|\nu|}u$ denoting the family of derivatives $u^{(\mu)}$ with $|\mu| \leq |\nu|$) there arises a natural need for introducing spaces of a new type. This was done by Yu.A. Dubinskiĭ ([1975], [1976], [1978], [1979]): he defined spaces $W^\infty\{a_\alpha, p_\alpha\}(G)$ and $\overset{\circ}{W}{}^\infty\{a_\alpha, p_\alpha\}(G)$, called *Sobolev spaces of infinite order* and developed their theory in a direction suitable for problems in the theory of differential equations of infinite order.

Let there be given multiple sequences

$$a_\alpha \geq 0, p_\alpha > 1, 1 \leq r_\alpha \leq +\infty, \alpha \in \mathbb{N}_0^r,$$

and let G be an open set in $\mathbb{R}^n$. The space $W^\infty = W^\infty\{a_\alpha, p_\alpha\}(G)$ consists of all infinitely differentiable functions f on G such that

$$\sum_{|\alpha|=0}^{\infty} a_\alpha \|f^{(\alpha)}\|_{L_{r_\alpha}}^{p_\alpha}(G) < +\infty$$

(the numbers r_α are not essential so we omit them in the notation).

The space $\overset{\circ}{W}{}^\infty\{a_\alpha, p_\alpha\}(G)$ consits of all infinitely differentiable finite functions g on G which belong to W^∞. It is clear that the zero function belongs to W^∞ and $\overset{\circ}{W}{}^\infty$. If there exist other such functions we say that these spaces are *nontrivial*. Yu. A. Dubinskiĭ established a criterion for the nontriviality of W^∞ and $\overset{\circ}{W}{}^\infty$.

One can define in W^∞ a norm using the notion of limit of a monotone sequence of Banach spaces. Let X_m be Banach spaces with norm $\|x\|_m$, $m = \pm 1, \pm 2, \ldots$. If

$$X_1 \supset X_2 \supset \ldots \supset X_m \supset \ldots$$

and

$$\|x\|_1 \leq \|x\|_2 \leq \ldots \leq \|x\|_m \leq \ldots, \quad x \in \bigcap_{m=1}^{\infty} X_m,$$

then the space

$$X_\infty = \lim_{m\to\infty} X_m \overset{\text{def}}{=} \{x \in \bigcap_{m=1}^{\infty} X_m : \|x\| \overset{\text{def}}{=} \lim_{m\to\infty} \|x\|_m < +\infty\}$$

is called the *projective limit of the Banach spaces* X_m, $m = 1, 2, \ldots$.

If

$$X_{-1} \subset X_{-2} \subset \dots \subset X_{-m} \subset \dots$$

and

$$\|x\|_{-1} \geq \|x\|_{-2} \geq \dots \geq \|x\|_{-m} \geq \dots,$$

we set

$$Y_{-\infty} \overset{\text{def}}{=} \{x \in \bigcap_{m=1}^{\infty} X_{-m} : \|x\|_{-\infty} \overset{\text{def}}{=} \lim_{m\to\infty} \|x\|_{-m}\}.$$

The completion $X_{-\infty}$ of $Y_{-\infty}$ in the norm $\|x\|_{-\infty}$ is called the *inductive limit of the sequence* $\{X_{-m}\}$ and is written

$$X_{-\infty} = \lim_{m\to\infty} X_{-m}.$$

If $X_{-m} = X_m^*$ and X_∞ is reflexive and locally convex and its dual $X_{-\infty}^*$ also is reflexive, then $X_{-\infty} = X_\infty^*$.

With the aid of these notions one can now investigate the properties of separability, reflexivity and uniform convexity of the spaces $\overset{\circ}{W}{}^{\infty}\{a_\alpha, p\} = \lim_{m\to\infty} \overset{\circ}{W}{}_p^{(m)}\{a_\alpha, p\}$ and $W^{-\infty}\{a_\alpha, p'\} = \lim_{m\to\infty} W^{(-m)}\{a_\alpha, p'\}$, where

$$\overset{\circ}{W}{}^{(m)}\{a_\alpha, p\} = \{u : \|u\|_{m,p}^p \overset{\text{def}}{=} \sum_{|\alpha|=0}^{m} a_\alpha \|u^{(\alpha)}\|_{L_p}^p, \quad u^{(\omega)}|_{\partial G} = 0, \quad |\omega| < m\},$$

and $W^{(-m)}\{a_\alpha, p'\}$ is the dual of $\overset{\circ}{W}{}^{(m)}\{a_\alpha, p\}$. In particular one sees that

$$W^{-\infty}\{a_\alpha, p'\} = (\overset{\circ}{W}{}^{\infty}\{a_\alpha, p\})^*.$$

In the solution of inhomogeneous boundary problems for differential equations of infinite order there arises the need for introducing the notion of trace of functions in W^∞.

A family of functions $\varphi_\infty : \partial G \to \mathbb{R}$, $|\omega| = 0, 1, \dots$, is called the *trace of the function* $f \in W^\infty\{a_\alpha, p_\alpha\}(G)$ if

$$f^{(\omega)}|_{\partial G} = \varphi_\omega, \quad |\omega| = 1, 2, \dots.$$

Let the boundary ∂G of G be infinitely differentiable and $(n-1)$-dimensional. Let us formulate, in the case $p_\alpha = r_\alpha$, a condition for the family $\{\varphi_\omega\}$ to be the trace of a function $f \in W^\infty\{a_\alpha, p_\alpha\}(G)$.

For each $N = 0, 1, \dots$ set

$$W_N = \bigcap_{|\nu|\leq N} W_{p_\alpha}^{(\nu)}(G), \quad \overset{\circ}{W}_N = \{f \in W_N : f^{(\nu)}|_{\partial G} = 0, \quad |\nu| \leq N-1\},$$

$$\varrho_N(f) = \sum_{|\nu|=0} a_\alpha \|f^{(\alpha)}\|_{p_\alpha}^{p_\alpha},$$

and for each fixed family of boundary functions $\{\varphi_\omega\}$

$$E_N \stackrel{\text{def}}{=} \{f(x) = g(x) + h(x) : g(x) \in W_N,\ h(x) \in \overset{\circ}{W}_N,\ g|_{\partial G} = \varphi_\omega,\ |\omega| \le N-1\}.$$

Then there exits a unique function (denote it by f_N) such that

$$\varrho_N(f_N) = \inf_{f \in E_N} \varrho_N(f).$$

A family of boundary functions $\{\varphi_\omega\}$ is the trace of a function $f \in W^\infty\{a_\alpha, p_\alpha\}(G)$ iff the following conditions hold true:

a) for each $N = 1, 2, \ldots$ the functions φ_ω, $|\omega| \le N-1$, admit extensions to W_N;

b) there exist a constant $k > 0$ such that for all $N = 0, 1, \ldots$ holds

$$\varrho_N(f_N) \le k.$$

Sobolev spaces of infinite order have applications not only in the study of boundary problems for elliptic equations of infinite order, but also in the study of hyperbolic ones.

Ya.V. Radyno [1985] gave a constructive definition of the inductive limit $X_{-\infty} = \lim_{m\to\infty} X_{-m}$. He showed that

$$X_{-\infty} = \{x \in X : x = \sum_{\mu=1}^{\infty} x_\nu,\ x_\nu \in X_\nu,\ \sum_{\nu=1}^{\infty} \|x_\nu\|_\nu < +\infty\}$$

and that

$$\|x\|_{-\infty} = \inf_{\substack{x=\sum_{\nu=1}^{\infty} x_\nu \\ x_\nu \in X_\nu, \nu=1,2,\ldots}} \sum_{\nu=1}^{\infty} \|x_\nu\|_\nu.$$

§ 5. Derivatives and Integrals of Fractional Order

It is clear that if f is a periodic function of period 2π, $f \in L(-\pi, \pi)$, $f' \in L(-\pi, \pi)$, with the Fourier series

$$\sum_{n=-\infty}^{\infty} c_n e^{inx},$$

then

$$\sum_{n=-\infty}^{\infty} inc_n e^{inx}$$

is the Fourier series of f'. If there exists a function $f^{(\alpha)} \in L(-\pi, \pi)$, $\alpha > 0$, with the Fourier series

$$\sum_{n=-\infty}^{\infty} (in)^{\alpha} c_n e^{inx},$$

then $f^{(\alpha)}$ is called the *generalized derivative in the sense of Weyl of order* α of f. Here

$$(in)^{\alpha} = |n|^{\alpha} \exp(\frac{\pi}{2} i\alpha \operatorname{sign} n)$$

and the number α may be an integer as well as a fraction.

Analogously, one defines, following Weyl, *fractional integration: the integral in the sense of Weyl of order* β of a function f, for which the Fourier coefficient c_0 equals 0, is the function $f_\beta \in L(-\pi, \pi)$ with the Fourier series

$$\sum_{|n|>0} \frac{c_n e^{inx}}{(in)^{\beta}}.$$

The operations of fractional differentiation and integration are each others inverses:

$$(f^{(\alpha)})_{\alpha} = f, \quad c_0 = 0, \alpha > 0.$$

Another approach to the notion of fractional derivative, originating from Liouville, is based on a generalization of the Newton-Leibniz formula, which recovers a function from its derivative. Define first *fractional integration*: if $f \in L(a, b)$ then the *integral* $I_{\alpha}^{a} f(x)$ *of order* $\alpha > 0$ with origin at the point a is the function

$$I_{\alpha}^{a} f(x) \stackrel{\text{def}}{=} \frac{1}{\Gamma(\alpha)} \int_a^x (x-t)^{\alpha-1} f(t)dt, \quad a \le x \le b.$$

The inverse operation to the notion of fractional integraltion I_{α}^{a} is called the operation of fractional integration. If $I_{\alpha}^{a} f(x) = g(x)$ then f is called the *fractional derivative* $g^{(\alpha)}$ *of order* α *of* g *in the sense of Liouville.* Thus $I_{\alpha}^{a} g^{(\alpha)} = g$, $\alpha > 0$.

The connection between the two types of fractional integrals is given by the following formula: if

$$\int_{-\pi}^{\pi} f(x)dx = 0$$

and f has period 2π then

$$I_{\alpha}^{-\infty} f(x) = f_{\alpha}(x).$$

Let us also indicate an approach to the concept of fractional derivative based on the use of the Fourier transform Ff of f. If $f \in L(\mathbb{R})$ and $f^{(\alpha)} \in L(\mathbb{R}^n)$, $\alpha \in \mathbb{N}_0^n$, then it is well-known that

$$(Ff^{(\alpha)})(y) = (iy)^{(\alpha)}(Ff)(y), \quad y = (y_1, \dots, y_n),$$

$$(iy)^\alpha = (iy_1)^{\alpha_1} \dots (iy_n)^{\alpha_n}, \alpha = (\alpha_1, \dots, \alpha_n). \tag{2.21}$$

In suitable assumption, starting with an appropriate class of generalized functions (cf. P. I. Lizorkin [1967]), the identity (2.21) can also be taken as a definition of derivatives also for fractional values of α.

It turns out that the scales (families) of spaces constructed by filling out the gaps between the Sobolev spaces $W_p^{(l)}$ with integer l with spaces whose derivatives of integral order satisfy Hölder conditions or spaces of functions having fractional derivatives do not suffice to solve the problem of the complete description of the traces of Sobolev spaces. The solution of this problem requires a different approach, which will be described below (Chap. 4). However, the spaces of functions with fractional derivatives provide a possibility of having a continuous (in the parameter l) family of spaces $W_p^{(l)}$, which for integer l reduce to the usual Sobolev spaces, and this with preservation of the previous imbedding theorems for the latter (cf. § 2 of Chap. 3).

Chapter 3

The Imbedding Theorems of Nikol'skiĭ

§ 1. Inequalities for Entire Functions

S.M. Nikol'skiĭ has in the study of imbedding theorems for differentiable functions of several variables made use of the method of approximating by entire functions of exponential type (or trigonometrical polynomial in the periodic case).

Let $\nu = (\nu_1, \dots, \nu_n) \in [0, +\infty)^n$, $z = (z_1, \dots, z_n)$, $z_j = x_j + iy_j$, $x_j \in \mathbb{R}$, $y_j \in \mathbb{R}$, $j = 1, 2, \dots, n$. A function $g_\nu = g_\nu(z) = g_\nu(z_1, \dots, z_n)$ is said to be an *entire function of exponential type* ν if

1) g_ν is an entire function in the variables $z_1, \dots, z_n$:

$$g_\nu(z) = \sum_{k \in \mathbb{N}_0^n} a_k z^k = \sum_{k_j \geq 0} a_{k_1 \dots k_n} z_1^{k_1} \dots z_n^{k_n},$$

where the series is convergent in the whole space $\mathbb{C}^n$;

2) for each $\varepsilon > 0$ there exists a constant $A_\varepsilon > 0$ such that

$$|g(z)| \leq A_\varepsilon \exp\left(\sum_{j=1}^n (\nu_j + \varepsilon)|z_j|\right), \quad z \in \mathbb{C}^n.$$

In 1951 S.M. Nikol'skiĭ (see [1951], [1953]) established the following *inequalities for entire functions* g_ν:

$$\|g_\nu\|_{L_q(\mathbb{R}^n)} \leq 2^n \left(\prod_{j=1}^{n} \nu_j\right)^{1/p-1/q} \|g_\nu\|_{L_p(\mathbb{R}^n)}, \quad 1 \leq p \leq q \leq +\infty; \tag{3.1}$$

$$\|g_\nu(x^{(m)}, \tilde{x}^{(m)})\|_{L_p(\mathbb{R}^m)} \leq 2^m \left(\prod_{j=1}^{m} \nu_j\right)^{1/p} \|g_\nu\|_{L_p(\mathbb{R}^n)}, \quad 1 \leq m < n, \tag{3.2}$$

$$x = (x^{(m)}, \tilde{x}^{(m)}), \quad x^{(m)} = (x_1, \dots, x_m) \in \mathbb{R}^m, \quad \tilde{x}^{(m)} = (x_{m+1}, \dots, x_n) \in \mathbb{R}^{n-m}.$$

In the same year S.M. Nikol'skiĭ also was the first to obtain imbedding theorems of a new kind for functions of several variables: they have a closed, inversible character and are thus exact. From them one obtains, in particular, a complete charaterization of the traces of functions belonging to these spaces. S.M. Nikol'skiĭ considered the classes composed by functions of several variables having derivatives of given order which satisfy a Hölder-Zygmund condition in L_p and studied these classes with the aid of approximation theory methods specially designed by him for this purpose. In these investigations, along with the well-known *Bernshteĭn's inequality*

$$\|g_\nu^{(\alpha)}\|_{L_p(\mathbb{R}^n)} \leq \nu^\alpha \|g_\nu\|_{L_p(\mathbb{R}^n)}, \quad 1 \leq p \leq +\infty, \quad \nu^\alpha = \nu_1^{\alpha_1} \dots \nu_n^{\alpha_n}, \tag{3.3}$$

estimating the norm of the derivatives of a function of exponential type g_ν by the norm of the function itself, a decisive role is played by the inequalities (3.1) and (3.2), obtained by S. M. Nikol'skiĭ, which estimate the norm of the function g_ν in different L_p metrics or on subspaces of lower dimension.

§ 2. Imbedding Theorems for Isotropic *H*-Classes

Let $f(x)$ be a function defined on an open set $G \subset \mathbb{R}^n$ and let $h = (h_1, \dots, h_n) \in \mathbb{R}^n$, $x = (x_1, \dots, x_n) \in \mathbb{R}^n$, $r > 0$, $k \in \mathbb{N}$, $s \in \mathbb{N}_0$,

$$k > r - s > 0, \tag{3.4}$$

$1 \leq p \leq +\infty$. By definition f belongs to the *class* $H_p^{(r)}(G)$ if

$$\|f\|_{L_p(G)} < +\infty,$$

and if there exists a constant $M > 0$ such that

$$\sum_{|\nu|=s} \|\Delta_h^{(k)} f^{(\nu)}\|_{L_p(G_{k|h|})} \leq M|h|^{r-s}, \quad \nu = (\nu_1, \dots, \nu_n) \tag{3.5}$$

(here $\Delta_h^{(k)}$ is the difference of order k with step h of the function under consideration; cf. (2.5) in § 1 of Chap. 2).

Condition (3.5) is equivalent to an analogous condition with the modulus on continuity $\omega_p^{(k)}(f^{(\nu)})$ (cf. (2.11) of Sect. 2.1 of Chap. 2) of the derivative $f^{(\nu)}$, i.e. the requirement that there exists a constant $M > 0$ such that

$$\omega_p^{(k)}(f^{(\nu)};\delta) \leq M\delta^{r-s}, \quad \delta > 0, |\nu| = s.$$

The norm of a function f in the class $H_p^{(r)}(G)$ is defined by the formula

$$\|f\|_{H_p^{(r)}(G)} = \|f\|_{L_p(G)} + \sup_h |h|^{s-r} \sum_{|\alpha|=s} \|\Delta_h^{(k)} f^{\alpha}\|_{L_p(G_{k|h|})}. \tag{3.6}$$

The definition of the class $H_p^{(r)}(G)$ given depends on the choice of the numbers k and s, but for the domains G to be considered (and especially for domains with sufficiently good boundaries) classes corresponding to different choices of pairs k, r subject to condition (3.4) coincide and the norms (3.6) are equivalent. Therefore we need not complicate the notation $H_p^{(r)}$ by appending the letters k and s. It is essential to bear in mind that the classes $H_p^{(r)}$ are defined for arbitrary positive values of r.

For $r > 0$ given pick the largest of all possible integers s subject to condition (3.4), i.e. if r is noninteger then $s = [m]$ and if r is an integer then $s = r - 1$. For the integer k in (3.4) we take its least possible value, i.e. $k = 1$ if r noninteger and $k = 2$ if r integer. In both cases we have $r = \bar{r} + \alpha$, $r > 0$, $\bar{r}$ an integer, $0 < \alpha \leq 1$. Then the definition of $H_p^{(r)}(G)$ can be formulated as follows: the function $f \in L_p(G)$ belongs to $H_p^{(r)}(G)$ iff it has partial derivatives of order $\bar{r}$ and there exists a constant $M > 0$ such that for noninteger r holds the *integral Hölder condition*

$$\sum_{|\nu|=\bar{r}} \|\Delta_h f^{(\nu)}\|_{L_p(G_{|h|})} \leq M|h|^{\alpha}, \quad 0 < \alpha < 1, \tag{3.7}$$

and for r integer the *integral Zygmund condition*

$$\sum_{|\nu|=\bar{r}} \|\Delta_h^{(2)} f^{(\nu)}\|_{L_p(G_{2|h|})} \leq M|h|. \tag{3.8}$$

From what has been said it is, in particular, clear that in the definition of H-classes one can always restrict attention just to differences of order not higher than two.

Let us remark that in the definition of the spaces $H_p^{(r)}(\mathbb{R}^n)$ we may restrict ourselves to consider only the pure generalized derivatives

$$\frac{\partial^{\bar{r}} f}{\partial x_1^{\bar{r}}}, \ldots, \frac{\partial^{\bar{r}} f}{\partial x_n^{\bar{r}}},$$

as in Nikol'skiĭ [1951] it is shown that a function f is in $H_p^{(r)}(\mathbb{R}^n)$ iff $f \in L_p(\mathbb{R}^n)$ and the partial derivatives

$$\frac{\partial^{\bar{r}} f}{\partial x_j^{\bar{r}}}, \quad j = 1, 2, \ldots, n,$$

satisfy an integral Hölder condition of exponent α or an integral Zygmund condition according to whether $\alpha < 1$ or $\alpha = 1$. Thus, in the case at hand it follows from the existence of pure derivatives of a given order also the existence of all mixed derivatives of this order. The study of the question when an analogous phenomenon takes place in other function spaces has been the object of further investigations by many authors.

The definition of $H_p^{(r)}(G)$ extends in a natural way to the case of functions defined on sufficiently smooth manifolds Γ^m of dimension m (with or without edges) contained in $\mathbb{R}^n$, $1 \leq m < n$.

For the spaces $H_p^{(r)}(G)$ the imbedding theorem of type (2.16) (cf. Sect. 2.3) takes a definitive form: it is invertible, i.e. we have an invertible imbedding (Nikol'skiĭ [1951])

$$H_p^{(r)}(G) \overset{\hookrightarrow}{\hookleftarrow} H_p^{(\rho)}(\Gamma^m), \quad 0 < \rho = r - (n-m)/p. \tag{3.9}$$

Here the symbol $\hookrightarrow$ has the same meaning as in (2.16). Now the symbol $\hookleftarrow$ means that any function $\varphi \in H_p^{(\rho)}(\Gamma^m)$ can be extended to the domain G in such a way that the extension is a function in $H_p^{(r)}(G)$, its trace to Γ^m equals φ and one has the inequality

$$\|f\|_{H_p^{(r)}(G)} \leq c \|\varphi\|_{H_p^{(\varrho)}(\Gamma^m)}, \tag{3.10}$$

with a constant $c > 0$ independent of φ. Statements of this type are called *invertible imbedding theorems*.

For the spaces $H_p^{(r)}(G)$ one also has imbeddings analogous to the imbeddings (2.14):

$$H_p^{(r)}(G) \hookrightarrow H_q^{(\varrho)}(G), \quad 1 \leq p < q \leq +\infty, \quad \varrho = 1 - r(1/p - 1/q), \tag{3.11}$$

where now r and ϱ can be arbitrary nonnegative numbers, not necessary integers as in the case of (2.14).

The classes $W_p^{(l)}$ and $H_p^{(r)}$ considered here (with scalar parmeters $l \in \mathbb{N}_0$, $r \in \mathbb{R}$, $r > 0$) are called *isotropic*, as the differentiability properties of their functions are the same in all directions of space. Next we pass to the definition and the description of anisotropic function classes with vectorial smoothness powers.

§ 3. Imbedding Theorems for Anisotropic H-Classes

In the case when the smoothness parameter is a vector we will write it in the notation for the class without parantheses, and not with parantheses as was done in the case of a scalar smoothness parameter.

The *anisotropic class* $W_p^l(G)$, $l = (l_1, \dots, l_n) \in \mathbb{N}_0^n$, consists of all functions defined on the open set G which make the norm

$$\|f\|_{W_p^l(G)} = \|f\|_{L_p(G)} + \|f\|_{w_p^l(G)}, \tag{3.12}$$

finite, where

$$\|f\|_{w_p^l(G)} = \sum_{j=1}^{n} \left\| \frac{\partial^{l_j} f}{\partial x_j^{l_j}} \right\|_{L_p(G)} . \tag{3.13}$$

The *anisotropic class* $H_p^r(G)$, $r = (r_1, \dots, r_n)$, $r_j > 0$, $j = 1, 2, \dots, n$, is given by the finiteness of the norm

$$\|f\|_{H_p^r(G)} = \|f\|_{L_p(G)} + \sum_{j=1}^{n} \sup_{h_j} |h_j|^{s_j - r_j} \|\Delta_{h_j}^{(k_j)}(G) f_{x_j^{s_j}}^{(s_j)}\|_{L_p(G)}, \tag{3.14}$$

where $k_j \in \mathbb{N}$, $s_j \in \mathbb{N}_0$, $k_j > r_j - s_j > 0$, h_j a vector directed along the x_j axis,

$$\Delta_{h_j}^{(k_j)}(G) f(x) = \Delta_{h_j}^{(k_j)} f(x), \tag{3.15}$$

provided the segment $[x, x + k_j h_j]$ is contained in G, and

$$\Delta_{h_j}^{(k_j)}(G) f(x) = 0 \tag{3.16}$$

in the opposite case, $j = 1, \dots, n$.

To have some variation we have defined the norm (3.14) in a somewhat different way compared to (3.6): instead of considering sets of the type G_η, $\eta > 0$, we have introduced the convention (3.15)-(3.16).

In the formulation of the definition there is some ambiguity in the choice of the pairs of numbers k_j, s_j corresponding to the numbers r_j. It turns out that for $G = \mathbb{R}^n$ or for domains with geometric properties adjusted to the vectorial smoothness indices this ambiguity is not essential and the norms (3.14) of functions $f \in H_p^r(G)$, defined with an arbitrary choice of pairs k_j, s_j satisfying the conditions $k_j > r_j - s_j > 0$, $j = 1, 2, \dots, n$, are all equivalent. We remark that the norm (3.14) may be viewed as a special case of the norm (4.4) (cf. Chap. 4).

If r is a number then $H_p^r(G) = H_p^{(r,\dots,r)}(G)$.

Let us now formulate the *imbedding theorems* (Nikol'skiĭ) *for anisotropic H-classes of functions* defined in the whole of $\mathbb{R}^n$.

1)
$$H_p^r(\mathbb{R}^n) \hookrightarrow H_q^\varrho(\mathbb{R}^n), \quad 1 \le p < q \le +\infty \tag{3.17}$$
$$\varrho = \kappa r, \kappa = 1 - (1/p - 1/q) \sum_{j=1}^n 1/r_j > 0; \tag{3.18}$$

2)
$$H_p^r(\mathbb{R}^n) \overset{\hookrightarrow}{\leftarrow} H_p^\varrho(\mathbb{R}^m), \quad 0 \le m < n, \quad \rho = \kappa r, \tag{3.19}$$
$$1 \le p \le +\infty, \quad \kappa = 1 - 1/p \sum_{j=m+1}^n 1/r_j > 0. \tag{3.20}$$

3) If $f \in H_p^r(\mathbb{R}^n)$ and
$$\kappa_1 = 1 - \sum_{j=1}^n \alpha_j/r_j > 0, \tag{3.21}$$
$$\rho = \kappa_1 r, \quad \alpha = (\alpha_1, \ldots, \alpha_n) \in \mathbb{N}_0^n,$$
then
$$f^{(\alpha)} \in H_p^r(\mathbb{R}^n) \tag{3.22}$$
and
$$\|f^{(\alpha)}\|_{H_p^\varrho(\mathbb{R}^n)} \le c\|f\|_{H_p^r(\mathbb{R}^n)}, \tag{3.23}$$
where the constant $c > 0$ does not depend on f.

In the one dimensional case Theorem (3.16) is a special case of an old result by Hardy and Littlewood.

The Theorems (3.9) and (3.19)-(3.20) solve completely the problem of the description of the properties of the traces of functions in H-classes: it is seen that their exact description can be done in terms of the H-classes themselves. Also the extension operator from functions on the subspace $\mathbb{R}^m$ to the whole of $\mathbb{R}^m$ is linear and bounded. Let us further remark that (3.22)-(3.23) contain, in particular, the statement concerning the existence of mixed partial derivatives of order $\alpha = (\alpha_1, \ldots, \alpha_n)$ for functions $f \in H_p^r(\mathbb{R}^n)$ in the hypothesis of (3.21).

The method by which these theorems are proved depends on expanding the given function $f \in H_p^r(\mathbb{R}^n)$ in a series
$$f(x) = \sum_{s=0}^{\infty} q_s(x) \tag{3.24}$$
in entire functions of exponential type
$$q_s(x) = q_{a^{s/r_1}, \ldots, a^{s/r_n}}(x), \quad a > 1,$$
of degree a^{s/r_j} in x_j. The functions $q_s(x)$ are chosen such that
$$\sup_{s=0,1,\ldots} a^s \|q_s\|_{L_p(\mathbb{R}^n)}$$
is equivalent to the norm $\|f\|_{H_p^r(\mathbb{R}^n)}$.

In order to arrive at the desired conclusions one requires various estimates for the terms of the series (3.24) based on the inequalities (3.1), (3.2), (3.3) for entire functions of exponential type. Note that the extension of functions in Theorems (3.9) and (3.19)-(3.20) can be realized by functions which are infinitely differentiable off $\mathbb{R}^m$ (Kudryavtsev [1959]) and even by solutions of appropriate differential equations of a special kind (Ya.S. Bugrov [1963], S.V. Uspenskiĭ [1968]).

An important fact in the whole theory is that one can describe the class $H_p^r(\mathbb{R}^n)$ in terms of best approximations by functions of exponential type (trigonometric polynomials in the periodic case). Let $E_{\nu_1,\ldots,\nu_n}(f)_p$ be the best approximation in the metric of the space $L_p(\mathbb{R}^n)$ of the function $f \in L_p(\mathbb{R}^n)$ with the aid of entire function $g_{\nu_1\ldots\nu_n}(x_1,\ldots,x_n) \in L_p(\mathbb{R}^n)$ of exponential type of degrees $\nu_1,\ldots,\nu_n$ in the variables $x_1,\ldots,x_n$. For the spaces $H_p^r(\mathbb{R}^n)$, $r = (r_1,\ldots,r_n)$, hold the following *direct and inverse theorems of the type of the theorems of Bernshteĭn, Jackson and Zygmund.*

If $f \in H_p^r(\mathbb{R}^n)$, $r = (r_1,\ldots,r_n)$, then for each $\nu_j > 0$ holds

$$E_{\nu_1\ldots\nu_n}(f)_p \le c\sum_{j=1}^{n} 1/\nu_j^{r_j}. \tag{3.25}$$

Conversely, if the function $f \in L_p(\mathbb{R}^n)$ satisfies (3.25) with $\nu_j = a_j^k$, $k = 0,1,2,\ldots$, $a_j > 1$, $j = 1,2,\ldots,n$, then $f \in H_p^r(\mathbb{R}^n)$.

Similar methods were applied in work by S.M. Nikol'skiĭ and M.K. Potapov [1964] in the study of properties of functions analytic in a strip on the boundary of the strip.

The imbedding theorems in the case $G = \mathbb{R}^n$ (Nikol'skiĭ [1951]) can be carried over in the same form to domains G for which functions $f \in H_p^r(G)$ can be extended with preservation of class to the whole of $\mathbb{R}^n$, i.e. there exists a function $f^* : \mathbb{R}^n \to \mathbb{R}$ such that $f^* = f$ on G and $f^* \in H_p^r(\mathbb{R}^n)$. This fact has lead mathematicians to investigate the extension of functions to the whole space with presevation of class or, if this is not possible, to the extension to the whole space in such a way that the extended function belongs to a in some sense optimal class among all possible classes (for this cf. § 3 of Chap. 6).

The theory of imbedding of H-classes was applied to the study of various concrete types of functions (harmonic, polyharmonic) in work by S.M. Nikol'skiĭ [1953], T.I.Amanov, O.V. Besov, Ya.S. Bugrov. For example, Ya.S. Bugrov [1957a] obtained theorems giving conditions for a harmonic function in a disk to belong to an H-class in terms of approximation by harmonic polynomials of special type.

Chapter 4

Nikol'skiĭ-Besov Spaces

§ 1. Sobolev-Slobodetskiĭ Spaces

As we remarked above, the problem of describing the properties of the traces of functions in the Sobolev spaces $W_p^{(l)}(G)$ on manifolds of dimension m, $1 \le m \le n$, can not be solved in terms of these classes alone. The solution of this problem in the case $m = n - 1$ first for $l = 1$, $p = 2$ and then for all real $p \ge 1$ and all integers $l > 0$ was given in the work of Aronszajn (1955), V.M. Babich-L.N. Slobodetskiĭ (1956-1958) and Gagliardo (1957). The description of the traces of functions in the Sobolev spaces $W_p^{(l)}(G)$, $l \in \mathbb{N}$, on $(n-1)$-dimensional manifolds was given in terms of spaces denoted by $W_p^{(r)}(G)$, with a fractional power of smoothness r. These spaces were introduced by L.N. Slobodetskiĭ (1958) and are therefore named after him. Let us define them.

Let G be an open subset of $\mathbb{R}^n$ and let $1 \le p < +\infty$, $r > 0$, $r = [r] + \alpha$, $0 < \alpha < 1$. The *Slobodetskiĭ space* $W_p^{(r)}(G)$ consists of all functions $f \in L_p(G)$ with

$$\|f\|_{W_p^{(r)}(G)} \stackrel{\text{def}}{=} \|f\|_{L_p(G)} + \|f\|_{w_p^{(r)}(G)} < +\infty, \tag{4.1}$$

where

$$\|f\|_{w_p^{(r)}(G)} = \sum_{|\nu|=[r]} \left(\int_G \int_G \frac{|f^{(\nu)}(x) - f^{(\nu)}(y)|^p}{|x-y|^{n+\alpha p}} dxdy \right)^{1/p}. \tag{4.2}$$

The functional $\|f\|_{W_p^{(r)}(G)}$ is a norm in $W_p^{(r)}(G)$. L.N. Slobodetskiĭ proved that if $\mathbb{R}^{n-1}$ is any $(n-1)$-dimensional hyperplane in $\mathbb{R}^n$ then

$$W_p^{(l)}(G) \overset{\hookrightarrow}{\leftarrow} W_p^{(l-1/p)}(\mathbb{R}^{n-1} \cap G), \quad l - 1/p > 0, 1 \le p < +\infty \tag{4.3}$$

To the spaces $W_p^{(l)}(G)$, $l > 0$, which for l integer coincide with the Sobolev spaces and for l noninteger are the Slobodetskiĭ spaces, we will apply the common appelation *Sobolev-Slobodetskiĭ spaces*.

The Slobodetskiĭ space $W_p^{(r)}$ ($r > 0$, r noninteger) coincides for sufficiently smooth domains, as will be explained in the sequel, with the Besov spaces $B_{p,p}^{(r)}$ (cf. § 2 of Chap. 4). Thus, although the term Slobodetskiĭ space is used in the contemporary literature, it has a historical character. Besides, recently one has begun to study various generalizations of these spaces, based on any one of the (equivalent) norms of these spaces, which in some cases leads to spaces which differ from each other.

§ 2. The Besov Spaces $B_{p,q}^{(r)}$

O.V. Besov introduced in 1959 spaces $B_{p,q}^{(r)}(\mathbb{R}^n)$, $1 \leq p \leq +\infty$, $1 \leq q \leq +\infty$, $r > 0$, which are more general than the spaces $H_p^{(r)}(\mathbb{R}^n)$, and developed their theory using the methods deviced by S.M. Nikol'skiĭ. These spaces generalize the H-classes in the sense that $H_p^{(r)} = B_{p,+\infty}^{(r)}$ and, similarly to the spaces $H_p^{(r)}$, they form a closed system of spaces with respect to the direct and converse imbedding theorems, formulated (for $q \in [1, +\infty]$ fixed) exactly in the same terms as for the spaces $H_p^{(r)}$. O.V. Besov introduced also *anisotropic spaces* $B_{p,q}^r$, where the smoothness index r is a vector: $r = (r_1, \ldots, r_n)$. The spaces $B_{p,q}^r$ are called *Nikol'skiĭ-Besov* or simply *Besov spaces.*

In order to build the theory of the spaces $B_{p,q}^{(r)}(\mathbb{R}^n)$ O.V. Besov investigated the equivalence of the convergence of series of best approximations and corresponding integrals of the L_p-moduli of smoothness of functions. These investiagtions were followed up by work in the one dimensional case by L.P. Ul'yanov [1967], [1968] in the case $p = 2$ and by A.A. Konyushkov [1958] for $p \geq 1$. It turned out that for integer l and $p = q = 2$ the space $B_{p,q}^{(l)}$ coincides with corresponding Sobolev spaces, namely one has $B_{2,2}^{(l)}(\mathbb{R}^n) = W_2^{(l)}(\mathbb{R}^n)$, $l \in \mathbb{N}$. If $p \neq 2$ then the analogous formula fails. For fractional $r > 0$, as already has been observed, one has $B_{p,p}^{(r)}(\mathbb{R}^n) = W_p^{(r)}(\mathbb{R}^n)$, i.e. the Nikol'skiĭ-Besov spaces coincide up to equivalence of norm with the Slobodetskiĭ spaces and, similarly, one obtains in terms of $B_{p,p}^{(r)}$ an exact description of the space of traces of functions in Sobolev spaces on $(n-1)$-dimensional manifolds. In 1959 O.V. Besov (see Nikol'skiĭ [1977]) gave the complete solution of the problem of the exact description of the traces of the same functions on planes $\mathbb{R}^m$ of arbitrary dimension m, $1 \leq m \leq n$. It was now realized that the space of traces of functions in the space $W_p^{(l)}(\mathbb{R}^n)$, $l \in \mathbb{N}$, on m-dimensional planes $\mathbb{R}^m$, $1 \leq m < n$, is described exactly in terms of the spaces $B_{p,p}^{(r)}(\mathbb{R}^n)$.

In a somewhat different but equivalent form this result was independently established by V.A. Solonnikov [1960] and it was also proved in a different way by P.I. Lizorkin [1960] and S.V. Uspenskiĭ (see Uspenskiĭ-Demidenko-Perepelkin [1984]).

The problem of the exact description of the traces in anisotropic Sobolev spaces can likewise be solved in terms of anisotropic Nikol'skiĭ-Besov spaces (cf. Besov-Il'in-Nikol'skiĭ [1975]).

Let us give the exact definitions and statements of the results.

Let G be an open set in $\mathbb{R}^n$ and let $1 \leq p \leq +\infty$, $1 \leq q \leq +\infty$, $r = (r_1, \ldots, r_n) \in \mathbb{R}^n$, $r_j > 0$, $k_j \in \mathbb{N}$, $s_j \in \mathbb{N}_0$, $k_j > r_j - s_j > 0$, $j = 1, \ldots, n$, $h = (h_1, \ldots, h_n) \in \mathbb{R}^n$.

The space $B_{p,q}^{(r)}(G)$ consists of all functions f such that the norm

$$\|f\|_{B_{p,q}^r(G)} = \|f\|_{L_p(G)} + \|f\|_{b_{p,q}^r(G)} \tag{4.4}$$

where

$$\|f\|_{b^r_{p,q}(G)} = \sum_{j=1}^{n} \left(\int_0^1 \left[\frac{\|\Delta_{h_j}^{(k_j)}(G) f_{x_j^{s_j}}^{(s_j)}\|_{L_p(G)}}{|h_j|^{r_j - s_j}} \right]^q \frac{dh_j}{h_j} \right)^{1/q}, \quad 1 \le q < +\infty, \tag{4.5}$$

$$\|f\|_{b^r_{p,+\infty}(G)} = \|f\|_{h^r_p(G)} = \sum_{j=1}^{n} \sup_{0<|h|\le 1} |h_j|^{s_j - r_j} \|\Delta_{h_j}^{(k_j)}(G) f_{x_j^{s_j}}^{(s_j)}\|_{L_p(G)}, \tag{4.6}$$

(for the notation $\Delta_{h_j}^{(k_j)}(G)$ see (3.15) and (3.16)) makes sense and is finite.

The ambiguity in the choice of the pairs k_j, s_j for each given j is not essential if $G = \mathbb{R}^n$ or if the geometric properties of G are adjusted to the index r: different admissible pairs k_j, s_j correspond to equivalent norms.

If $p = q$ the space $B^r_{p,q}$ will be denoted B^r_p:

$$B^r_p(G) = B^r_{p,p}(G). \tag{4.7}$$

As in the case of the H-classes, one can for sufficiently smooth domains in the definition of the norm (4.4) work with differences of orders 1 and 2 only. If $G = \mathbb{R}^n$ one can define the norm of a function $f \in B^r_{p,q}(\mathbb{R}^n)$ in terms of best approximations (so that it becomes equivalent to the norm (4.4)). This norm is defined by the formula

$$\|f\|^{*}_{B^r_{p,q}} \overset{\text{def}}{=} \|f\|_{L_p(\mathbb{R}^n)} + \left\{ \sum_{k=0}^{\infty} a^{kq} E_{a^{k/r_1},\dots,a^{k/r_n}}(f)_p \right\}^{1/q}, \tag{4.8}$$

where $a > 1$ and $E_\nu(f)_p = E_{\nu_1,\dots,\nu_n}(f)_p$ is the best approximation in $L_p(\mathbb{R}^n)$ of f by functions of exponential type $\nu = (\nu_1, \dots, \nu_n)$.

The best approximation $E_{a^{k/r_1},\dots,a^{k/r_n}}(f)_p$ in (4.8) can be replaced by the approximation of f by entire functions $g_{a^{k/r_1},\dots,a^{k/r_n}}$ which depend on f in a linear manner and are given as convolutions of f by special entire functions.

§ 3. Imbedding Theorems for Nikol'skiĭ-Besov Spaces. Solution of the Problem of Traces of Functions in Sobolev Spaces

The *imbedding theorems for the classes* $B^r_{p,q}(G)$ in the case $G = \mathbb{R}^n$ were proved by O.V. Besov with the aid of method from approximation theory, i.e. on the basis of the representation in the form of the series (3.23) and an application of the norm (4.8). These theorems generalize S.M. Nikol'skiĭ's theorems for H-classes and coincide with the latter if $q = +\infty$. Let us give their formulation.

Theorem 1.

$$B^r_{p,q}(\mathbb{R}^n) \hookrightarrow B^\varrho_{p_0,q}(\mathbb{R}^n), \quad 1 \le p < p_0 \le +\infty, \tag{4.9}$$

$$\varrho = \kappa_1 r, \quad \kappa_1 = 1 - (\frac{1}{p} - \frac{1}{p_0}) \sum_{j=1}^n \frac{1}{r_j} > 0. \tag{4.10}$$

Theorem 2.

$$B^r_{p,q}(\mathbb{R}^n) \overset{\hookrightarrow}{\leftarrow} B^\varrho_{p,q}(\mathbb{R}^m), \quad 0 \le m < n, \tag{4.11}$$

$$\varrho = \kappa_2 r, \quad \kappa_2 = 1 - 1/p \sum_{j=m+1}^n \frac{1}{r_j} > 0, \quad 1 \le p \le +\infty. \tag{4.12}$$

Theorem 3. *If* $f \in B^r_{p,q}(\mathbb{R}^n)$, $\alpha = (\alpha_1, \dots, \alpha_n) \in \mathbb{N}^n_0$, $1 \le p < +\infty$, $\kappa_3 = 1 - \sum_{j=1}^n \alpha_j / r_j > 0$, $\varrho = \kappa_3 r$, *then*

$$\|f^{(\alpha)}\|_{B^\varrho_{p,q}(\mathbb{R}^n)} \le c \|f\|_{B^r_{p,q}(\mathbb{R}^n)} \tag{4.13}$$

where the constant $c > 0$ *does not depend on* f. *If* $D^\alpha B^r_{p,q}$ *denotes the space of derivatives of functions in* $B^r_{p,q}$, *then (4.13) can be formulated in terms of imbeddings as follows*

$$D^\alpha B^r_{p,q}(\mathbb{R}^n) \hookrightarrow B^\varrho_{p,q}(\mathbb{R}^n). \tag{4.14}$$

For $q = +\infty$ Theorems 1, 2 and 3 reduce to the theorems expressed respectively by the formulae (3.17), (3.18)-(3.19) and (3.21)-(3.23).

In the case of limit indices ($\kappa_j = 0$, $j = 1, 2, 3$) one has the following results.

If $\kappa_1 = 0$ then

$$B^r_{p,q}(\mathbb{R}^n) \hookrightarrow L_{p_0}(\mathbb{R}^n), \quad 1 \le p < p_0 < +\infty, \quad 1 \le q \le p_0,$$

where if $q > p_0$ there is no imbedding (the isotropic case was considered by Grisvard, Peetre and K.K. Golovkin, the one dimensional periodic case by P.L. Ul'yanov and the anisotropic case in the monograph Besov-Il'in-Nikol'skiĭ [1975]).

In Theorem 2 the corresponding extension operator from $\mathbb{R}^m$ to $\mathbb{R}^n$ is linear. If $\kappa_2 < 0$ or $\kappa_2 = 0$, $q > 1$, then the trace on $\mathbb{R}^m$ of a function $f \in B^r_{p,q}(\mathbb{R}^n)$, generally speaking, does not exist. If $\kappa_2 = 0$, $q = 1$, then

$$B^r_{p,1}(\mathbb{R}^n) \overset{\hookrightarrow}{\leftarrow} L_p(\mathbb{R}^m), \quad 1 \le p < +\infty,$$

(the isotropic case is due to Peetre [1975] and the anisotropic case to V.I. Burenkov and M.L. Gol'dman [1979]). The corresponding extension operator from $\mathbb{R}^m$ to $\mathbb{R}^n$ here is nonlinear and V.I. Burenkov and M.L. Gol'dman [1979] showed that there exists no bounded linear extension operator (but for each $q > 1$ there exists a linear extension operator to the larger space $B^r_{p,q}(\mathbb{R}^n)$).

Finally, if $\kappa_3 = 0$ then

$$D^\alpha B^r_{p,1}(\mathbb{R}^n) \hookrightarrow L_p(\mathbb{R}^n), \quad 1 \le p \le +\infty,$$

(cf. Besov-Il'in-Nikol'skiĭ [1975]).

If one applies one after the other the imbeddings $B^r_{p,q}(\mathbb{R}^n) \hookrightarrow B^\varrho_{p_0,q}(\mathbb{R}^n)$ (see (4.9)) and $B^\varrho_{p_0,q}(\mathbb{R}^n) \hookrightarrow B^{\varrho_0}_{p_0,q}(\mathbb{R}^m)$ (see (4.11)), one gets as a result an imbedding from $B^r_{p,q}(\mathbb{R}^n)$ into $B^{\varrho_0}_{p_0,q}(\mathbb{R}^m)$, where the conditions on the parameters of these spaces, under which this imbedding holds true, follow from (4.10) and (4.12) (cf. Nikol'skiĭ [1977], 7.1).

Thus the spaces $H^r_p(\mathbb{R}^n)$ are contained in the larger family of spaces $B^r_{p,q}(\mathbb{R}^n)$:

$$B^r_{p,+\infty}(\mathbb{R}^n) = H^r_p(\mathbb{R}^n)$$

and

$$H^{r+\varepsilon}_p(\mathbb{R}^n) \hookrightarrow B^r_{p,q}(\mathbb{R}^n) \hookrightarrow H^r_p(\mathbb{R}^n), \quad 1 \le q < +\infty, \tag{4.15}$$
$$\varepsilon = (\varepsilon_1, \ldots, \varepsilon_n), \quad \varepsilon_j > 0, \quad j = 1, 2, \ldots, n,$$

with preservation of the formulations of the imbedding theorems and their converses.

Moreover, as we have already noted, with the aid of the spaces B^r_p one can obtain a complete solution of the problem of traces of functions in Sobolev spaces W^l_p (recall that l is vectorial): it turned out that it was possible to give an exact description of the space of the traces in question in terms of the spaces B^r_p (cf. (4.7)); more precisely, one has

$$W^l_p(\mathbb{R}^n) \overset{\hookrightarrow}{\leftarrow} B^r_p(\mathbb{R}^m), \quad l = (l_1, \ldots, l_n) \in \mathbb{N}^n,$$

$$r = (r_1, \ldots, r_n),\ r = \kappa l,\ \kappa = 1 - \frac{1}{p}\sum_{j=1}^n \frac{1}{l_j},\ 1 \le m < n,\ 1 < p < +\infty. \tag{4.16}$$

It is interesting to note that for the same parameters p, r, m and l holds

$$B^l_{p,q}(\mathbb{R}^n) \overset{\hookrightarrow}{\leftarrow} B^r_{p,q}(\mathbb{R}^m),$$

regardless of the fact that for $p \neq 2$

$$B^l_p\mathbb{R}^m \neq W^l_p\mathbb{R}^n.$$

In the isotropic case Theorem (4.16) was obtained by O.V. Besov and in the anisotropic case by V.P. Il'in and V.A. Solonnikov.

Along with property (4.16), the connection betweeen the space $W^l_p(\mathbb{R}^n)$ and $B^l_{p,q}(\mathbb{R}^n)$, $l = (l_1, \ldots, l_n)$, is characterized by the following properties:

$$W^l_2(\mathbb{R}^n) = B^l_2(\mathbb{R}^n);$$

$$B_p^l(\mathbb{R}^n) \hookrightarrow W_p^l(\mathbb{R}^n) \hookrightarrow B_{p,2}^l(\mathbb{R}^n), \quad 1 < p \le 2;$$

$$B_{p,2}^l(\mathbb{R}^n) \hookrightarrow W_p^l(\mathbb{R}^n) \hookrightarrow B_p^l(\mathbb{R}^n), \quad 2 \le p < +\infty;$$

$$B_{p,1}^l(\mathbb{R}^n) \hookrightarrow W_p^l(\mathbb{R}^n) \hookrightarrow B_{p,+\infty}^l(\mathbb{R}^n) = H_p^l(\mathbb{R}^n), \quad 1 \le p \le +\infty. \tag{4.17}$$

The fact that Theorem 2 (cf. (4.11)-(4.12)) cannot be improved in terms of B-spaces (in particular, not the imbedding (3.9) for H-classes) follows at once from the fact that these imbeddings are reversible. That the imbeddings (4.9) and (3.17) for the spaces $B_{p,q}^r$ and H_p^r respectively cannot be improved follows from the existence of so-called marginal functions in B_p^r and H_p^r. A function f is said to be a *marginal function* in the space $H_p^r(\mathbb{R}^n)$ if it belongs to $H_p^r(\mathbb{R}^n)$ and if it does not belong to $H_p^{p+\varepsilon}(\mathbb{R}^n)$ for any $\varepsilon = (\varepsilon_1, \dots, \varepsilon_n)$, $\varepsilon_j > 0$, $j = 1, 2, \dots, n$, $|\varepsilon| > 0$. S.M. Nikol'skiĭ proved that there exist marginal functions in the spaces $H_p^r(\mathbb{R}^n)$ which under the imbedding (3.17) are mapped into marginal functions in the spaces $H_p^\varrho(\mathbb{R}^m)$, so that the index ϱ in (3.17) cannot be increased by any $\varepsilon = (\varepsilon_1, \dots, \varepsilon_n)$, $\varepsilon_j \ge 0$, $j = 1, 2, \dots, n$, $|\varepsilon| > 0$. In view of the imbedding (4.15) this shows also that (4.9) is unimprovable.

Along with marginal functions there exists another simpler and more effective method for proving that the imbedding theorems are sharp in the given terms. This may be called the *"dimension" method*: one studies for a fixed function $f = f(x)$ the behavior of

$$\lambda(\varepsilon) = \|f_\varepsilon\|_{B_{p,q}^r(\mathbb{R}^n)}$$

for $\varepsilon = (\varepsilon_1, \dots, \varepsilon_n)$, $f_\varepsilon(x) = f(\varepsilon_1 x_1, \dots, \varepsilon_n x_n)$, which is very simple and amounts only to a change of variable. If we compare $\lambda(\varepsilon)$ for various p, q, r, n and let the ε_i tend to zero or infinity, one can derive information showing the impossibility of certain inequalities between norms. This method is described in the monograph [1977] by S. M. Nikol'skiĭ.

Chapter 5

Sobolev-Liouville Spaces

§ 1. The Spaces L_p^r

The original definition of Sobolev-Liouville spaces, which we denote by $L_p^{(r)}(\mathbb{R}^n)$, $r = (r_1, \dots, r_n)$, $1 \le p \le +\infty$, was formulated in analogy with the definition of the Sobolev spaces (cf. § 3 of Chap. 3) using fractional derivatives (cf. § 5 of Chap. 2). As this definition requires some cumbersome auxiliary constructions, we give another equivalent definition (cf. P.I. Lizorkin [1969] and the monograph [1977] by S.M. Nikol'skiĭ) based on the representation of the functions as potentials with the kernel

$$G_r(x) = \frac{1}{(2\pi)^{n/2}} \int_{\mathbb{R}^n} \frac{e^{iyx}}{[1+\varrho^2(y)]^{r^*/2}} dy, \quad r = (r_1, \dots, r_n), \tag{5.1}$$

where $1/r^* = 1/n \sum_{j=1}^n 1/r_j$ and $\varrho(y)$ is the solution of the equation

$$\sum_{j=1}^n y_j^2 \varrho^{-2a_j} = 1,$$

$a_j = r^*/r_j$, $x = (x_1, \dots, x_n)$, $y = (y_1, \dots, y_n)$, $x_j \in \mathbb{R}$, $y_j \in \mathbb{R}$, $k = (k_1, \dots, k_n)$, $k_j \in \mathbb{N}_0$, $j = 1, 2, \dots, n$.

The kernel $G_r(x)$ is infinitely differentiable for $x \neq 0$ and summable over $\mathbb{R}^n$.

The space $L_p^r(\mathbb{R}^n)$, $1 < p < +\infty$, consists of all functions $f \in L_p(\mathbb{R}^n)$ admitting the representation

$$f(x) = \int_{\mathbb{R}^n} G_r(x-y) g(y) dy, \quad g(y) \in L_p(\mathbb{R}^n). \tag{5.2}$$

The norm of a function $f \in L_p^r(\mathbb{R}^n)$ can be defined as the norm in $L_p(\mathbb{R}^n)$ of the function g appearing to the right in (5.2).

The classes $L_p^{(r,\dots,r)}(\mathbb{R}^n)$ will be written $L_p^{(r)}(\mathbb{R}^n)$ and are called the *isotropic Sobolev-Liouville spaces.*

For nonnegative integer valued vectors $r = (r_1, \dots, r_n)$, i.e. vectors having nonnegative entries, the Sobolev-Liouville classes coincide with the Sobolev classes:

$$L_p^r(\mathbb{R}^n) = W_p^r(\mathbb{R}^n). \tag{5.3}$$

In view of results by V. G. Maz'ya and Nagel [1978] the norm in L_2^r, $r = (r_1, \dots, r_n)$, $0 < r_j < 1$, is equivalent to the norm

$$\left(\int_{\mathbb{R}^n} \int_{\mathbb{R}^n} \frac{|u(x+h) - u(x)|^2}{(\sum_{j=1}^n |h_i|^{r_i})^{2+\sum_{j=1}^n 1/r_j}} dx dh \right)^{1/2} + \left(\int_{\mathbb{R}^n} |u(x)|^2 dx \right)^{1/2}.$$

This result is contained in an analogous more general statement regarding a space with the norm

$$\left(\int_{\mathbb{R}^n} |Fu(\xi)|^2 \sum_{j=1}^n \mu_i(|\xi_i|) d\xi \right)^{1/2}, \xi = (\xi_1, \dots \xi_n),$$

where the μ_i are increasing functions subject to appropraite assumptions.

The following *imbedding theorems* hold true for Sobolev-Liouville spaces:
I)

$$L_p^r(\mathbb{R}^n) \hookrightarrow L_q^\varrho(\mathbb{R}^m), \tag{5.4}$$

$$1 \le m \le n, \quad 1 < p < q < +\infty, \quad \varrho_j = \kappa r, \quad j = 1, 2, \ldots, m$$

$$\kappa = 1 - \left(\frac{1}{p} - \frac{1}{q}\right) \sum_{j=1}^{m} \frac{1}{r_j} - \frac{1}{p} \sum_{j=m+1}^{n} \frac{1}{r_j} \ge 0.$$

II)

$$L_p^r(\mathbb{R}^n) \overset{\hookrightarrow}{\leftarrow} B_p^\varrho(\mathbb{R}^m), \tag{5.5}$$

$$1 \le m < n, \quad 1 < p < +\infty, \quad \varrho_j = \kappa r_j > 0, \quad j = 1, 2, \ldots, m$$

$$\kappa = 1 - \frac{1}{p} \sum_{j=m+1}^{n} \frac{1}{r_j}.$$

$$B_{p,q}^r(\mathbb{R}^n) \hookrightarrow L_p^r(\mathbb{R}^n), \quad 1 \le q \le \min\{p, 2\}, \tag{5.6}$$

III)

$$L_p^r(\mathbb{R}^n) \hookrightarrow B_{p,q}^r(\mathbb{R}^n), \quad q \ge \max\{p, 2\}. \tag{5.7}$$

For $m = n$ the first theorem takes in the isotropic case the form

$$L_p^{(r)}(\mathbb{R}^n) \hookrightarrow L_q^{(\varrho)}(\mathbb{R}^n), \tag{5.8}$$

$$1 < p < q < +\infty, \quad 0 \le \varrho = r - n(1/p - 1/q).$$

If for an integer r it turns out that ϱ is a nonnegative integer, then formula (5.8) reduces to the corresponding formula in S.L. Sobolev's theorem (cf. § 3 of Chap. 2) and in the remaining cases it generalizes or sharpens this theorem.

It follows from (5.2) the convolution operator $G_r * f$ of a function f in $L_p(\mathbb{R}^n)$ with kernel G_r maps $L_p(\mathbb{R}^n)$ isomorphically onto the anisotropic (in general) space $L_p^r(\mathbb{R}^n)$. P.I. Lizorkin [1972] likewise obtained a description of anisotropic spaces $L_p^r(\mathbb{R}^n)$ with the aid of truncated differentiation and in terms of differences of singular integrals.

For isotropic spaces $L_p^{(r)}(\mathbb{R}^n)$ the kernel $G_r(x)$ reduces to the *Bessel-Macdonald kernel*

$$G_r(x) = F^{-1}[(1 + |\lambda|^2)^{-r/2}], \quad r > 0 \tag{5.9}$$

(F is the Fourier transform) and the space $L_p^{(r)}(\mathbb{R}^n)$, also called the *Bessel potential space*. It has been studied in work of Calderón, Aronszajn, Smith, P.I. Lizorkin and others. In this case the convolution operators $G_r * f$ can be written in the form

$$I_r f(x) \overset{\text{def}}{=} G * f = F^{-1}(F(G)F(f)) = F^{-1}[(1 + |\lambda|^2)^{-r/2} F(f)], \quad r > 0. \tag{5.10}$$

With the aid of the operator I_r one can show that each Nikol'skiĭ-Besov space $B^{(r)}_{p,q}(\mathbb{R}^n)$ is isomorphic to the zero class $B^{(0)}_{p,q}(\mathbb{R}^n)$ (S.M. Nikol'skiĭ-Lions-P.I. Lizorkin [1965]), which is a subspace of the Schwartz space of tempered distributions S', and the isomorphism is given by the operator (5.10):

$$I_r B^{(0)}_{p,q}(\mathbb{R}^n) = B^{(r)}_{p,q}(\mathbb{R}^n). \tag{5.11}$$

It follows that the spaces $B^{(r)}_{p,q}$ for different r all are isomorphic: if $r_2 > r_1$ then one has

$$I_{r_2-r_1} B^{(r_1)}_{p,q}(\mathbb{R}^n) = B^{(r_2)}_{p,q}(\mathbb{R}^n). \tag{5.12}$$

Note that it is natural to set

$$I_{-r} \overset{\text{def}}{=} I_r^{-1}, \quad r > 0, \tag{5.13}$$

where I_r^{-1} is the inverse of the operator I_r.

The dual $(B^{(r)}_{p',q'})^*$ of $B^{(r)}_{p',q'}$, $1/p + 1/p' = 1/q + 1/q' = 1$, can be obtained by the convolution operator I_r with negative r:

$$(B^{(r)}_{p',q'})^* = I_{-r} B^{(0)}_{p,q}, \quad r > 0.$$

Therefore $(B^{(r)}_{p',q'})^*$ will also be written $B^{(-r)}_{p,q}$, $r > 0$.

The Sobolev-Liouville spaces can be described in terms of hypersingular integrals. A necessary and sufficient condition for a function $f \in L_p(\mathbb{R}^n)$, $1 < p < n$, to belong to the space $L^r_p(\mathbb{R}^n)$ is that the following hypersingular integral exists in the sense of convergence in L_p:

$$(T^r f)(x) = \int_{\mathbb{R}^n} \frac{\Delta_h^{2l} f(x)}{\varrho^{n+r^*}(h)} dh \overset{\text{def}}{=} \lim_{\varepsilon \to 0} \int_{\varrho(h) > \varepsilon} \frac{\Delta_h^{2l} f(x)}{\varrho^{n+r^*}(h)} dh,$$

where $2l > \max_j r_j$ and $\varrho(h)$ and r^* have the same significance as in formula (5.1). Moreover

$$c_1 \|f\|_{L^r_p(\mathbb{R}^n)} \leq \|f\|_{L_p(\mathbb{R}^n)} + \|T^r f\|_{L_p(\mathbb{R}^n)} \leq c_2 \|f\|_{L^r_p(\mathbb{R}^n)},$$

where the positive constants c_1, c_2 do not depend on$f$ (P.I. Lizorkin [1972]). The theory of hypersingular integrals has been developed in the work of S.G. Samko [1984]. In particular, he constructed for the *generalized Riesz potential*

$$(K^\alpha \varphi)(x) = \int_{\mathbb{R}^n} \frac{\omega\left(\dfrac{x-t}{|x-t|}\right)}{|x-t|^{n-\alpha}} \varphi(t) dt \tag{5.14}$$

with homogeneous characteristic ω, in the case of a nondegenerate symbol of this potential (i.e. the Fourier transform of the kernel)

$$(\mathscr{K}^\alpha)(x) = \int_{\mathbb{R}^n} \frac{\omega\left(\dfrac{t}{|t|}\right)}{|t|^{n-\alpha}} e^{ixt} dt$$

on the unit sphere S^{n-1} of $\mathbb{R}^n$, the inverse of the operator (5.14) in the form of a hypersingular integral

$$(D^\alpha f)(x) \stackrel{\text{def}}{=} \int_{\mathbb{R}^n} \frac{(\Delta_t^{(l)} f)(x)}{|t|^{n+\alpha}} \Omega\left(\frac{t}{|t|}\right) dt,$$

which is a multidimensional analogue of the so-called *Marchaud derivatives* (in general of fractional order).

The space

$$J^\alpha(L_p) \stackrel{\text{def}}{=} \{f : f = K^\alpha \varphi, \varphi \in L_p(\mathbb{R}^n)\}$$

is called the *space of Riesz potentials*. If $1 < p < n/\alpha$ it reduces to a space of ordinary functions belonging to $L_{np/(n-\alpha p)}(\mathbb{R}^n)$ but for $p \geq n/\alpha$ it consists in general of generalized functions. One has the following connection between the Riesz potential spaces $J^\alpha(L_p)$ and the spaces $L^\alpha_{p,q}$ of functions $f : \mathbb{R}^n \to \mathbb{R}$ with the norm $\|f\|_q + \|D^\alpha f\|_p$:

$$J^\alpha(L_p) \cap L_q = L^\alpha_{p,q}.$$

If $p = q$ this result follows from P.I. Lizorkin's results. S.G. Samko likewise obtained imbedding theorems for the spaces $L^\alpha_{p,q}$ when the indices α, p, q are varied and if $1 < p < n/\alpha$, $p \leq q \leq np/(n - \alpha p)$ he established the density of infinitely differentiable functions in $L^\alpha_{p,q}$.

The theory of multidimensional singular operators, to the extent it involves this section, is set forth in Samko's book [1984].

§ 2. Study of Function Spaces with the Aid of Methods of Harmonic Analysis

In the study of the spaces $H^r_p(\mathbb{R}^n)$ and $B^r_{p,q}(\mathbb{R}^n)$ the method of expanding functions in a series of entire functions of exponential type has been very expedient. This method combined with methods of harmonic analysis (Fourier transform, multipliers, maximal functions etc.) also turned out to be very productive and has led to many new results.

Set

$$Q_k \stackrel{\text{def}}{=} \{y = (y_1, \dots, y_n) : -2^k \leq y_j \leq 2^k, j = 1, 2, \dots, n\},\ k = 0, 1, \dots,\ Q_{-1} = \emptyset, \tag{5.15}$$

and $\Gamma_k \stackrel{\text{def}}{=} Q_k \backslash Q_{k-1}$ (a "*corridor*"), $k = 0, 1, 2, \dots$. We denote by f_k, $k = 0, 1, 2, \dots$, entire functions such that the support of the Fourier transform Ff_k is contained in the corridor Γ_k:

$$\operatorname{supp} Ff_k \subset \Gamma_k, \quad k = 0, 1, \dots$$

The function $f \in L_p$ will be written as a series

$$f = \sum_{k=0}^{\infty} f_k. \tag{5.16}$$

If f is given then the functions f_k can be obtained in the following way. Let χ_k be the characteristic function of Γ_k. As $\mathbb{R}^n = \bigcup_{k=0}^{\infty} \Gamma_k$ and $\Gamma_k \cap \Gamma_{k'} = \emptyset$ for $k \neq k'$, the functions χ_k make up a partition of unity on $\mathbb{R}^n$:

$$1 = \sum_{k=0}^{\infty} \chi_k, \tag{5.17}$$

and for the functions f_k holds

$$f_k = F^{-1}\chi_k F f, \quad k = 0, 1, 2, \ldots. \tag{5.18}$$

If $f \in B_{p,q}^{(r)}(\mathbb{R}^n)$, $1 < p < +\infty$, $1 \le q \le +\infty$, then its norm is equivalent to the functional

$$\left(\sum_{k=0}^{\infty} 2^{krq} \|f_k\|_p^q \right)^{1/q}, \quad r \in \mathbb{R}. \tag{5.19}$$

This result is due to S.M. Nikol'skiĭ [1977]. It shows also that the norms (5.19) can be used for the definition of $B_{p,q}^{(r)}$ for negative r and that it is easy to establish with the aid of these norms the isomorphism of the classes $B_{p,q}^{(r)}$ for different r.

Let us introduce the *space* l_q of sequences $x = (x_0, x_1, \ldots, x_k, \ldots)$ with the norm

$$\|x\|_{l_q} = \|\{x_k\}\|_{l_q} \stackrel{\text{def}}{=} \begin{cases} \left(\sum_{k=0}^{\infty} |x_k|^q\right)^{1/q} & \text{for } 1 \le q < +\infty, \\ \sup_k |x_k| & \text{for } q = +\infty. \end{cases}$$

Then (5.19) can be written as

$$\|\{2^{kr} \|f_k\|_{L_p}\}\|_{l_q}. \tag{5.20}$$

We emphasize that the expression (5.19) is equivalent to the norm in $B_{p,q}^{(r)}$ also for $r < 0$.

In the case $p = 1$ and $p = +\infty$ (Nikol'skiĭ [1977]) an analogous result holds true under somewhat different assumptions about the expansion (5.16). It is also possible to carry over this result to the case of anisotropic spaces $B_{p,q}^{r}$. Let us further remark that in the study of similar questions it turns out to be useful to use other decompositions of $\mathbb{R}^n$ (instead of the "corridors").

§ 3. Lizorkin-Triebel Spaces

By the *Paley-Littlewood theorem* the norm of a function f in L_p is equivalent to the functional

$$\left\| \left(\sum_{k=0}^{\infty} |f_k(x)|^2 \right)^{1/2} \right\|_{L_p(\mathbb{R}^n)} = \| \, \|\{f_k\}\|_{l_2} \|_{L_p(\mathbb{R}^n)}. \tag{5.21}$$

Comparison of the norms (5.20) and (5.21) leads to the idea to consider the function spaces $L_{p,q}^{(r)} = L_{p,q}^{(r)}(\mathbb{R}^n)$ with the norm

$$\|f\|_{L_{p,q}^{(r)}(\mathbb{R}^n)} = \| \, \|\{2^{kr}|f_k(x)|\}\|_{l_q} \|_{L_p(\mathbb{R}^n)}, \tag{5.22}$$

where the functions f_k are defined by the formula (5.18). A systematic study of these spaces was accomplished by P.I. Lizorkin and later by Triebel. Therefore the spaces $L_{p,q}^{(r)}$ are called *Lizorkin-Triebel spaces*. It turned out that up to equivalence of norm the spaces $L_{p,2}^{(r)}$ coincide with the Sobolev-Liouville space $L_p^{(r)}(\mathbb{R}^n)$. In the work of L.I. Lizorkin and Triebel analogues of the spaces $L_{p,q}^{(r)}$ are considered which correspond to expansions of functions in series generalizing the expansion (5.16) and further spaces which generalize the spaces $L_{p,q}^{(r)}$ to the anisotropic case. In the periodic case and for $p = 2$ analogous norms, equivalent to norms of type (5.22), were earlier treated by N.S. Nikol'skaya [1973].

We have already remarked that the functions (5.20) are norms equivalent to the norms previously introduced for functions $f \in B_{p,q}^r(\mathbb{R}^n)$, where r may have arbitrary sign. Analogous results were established by Triebel [1977] for the functionals (5.22), i.e. for the norms of functions in $L_{p,q}^{(r)}(\mathbb{R}^n)$. He further established the following *imbedding theorems*:

$$B_{p,q}^{(r)}(\mathbb{R}^n) \hookrightarrow L_{p,q}^{(r)}(\mathbb{R}^n) \hookrightarrow B_{p,p}^{(r)}(\mathbb{R}^n), \quad 1 < q \leq p < +\infty, \tag{5.23}$$

$$B_{p,p}^{(r)}(\mathbb{R}^n) \hookrightarrow L_{p,q}^{(r)}(\mathbb{R}^n) \hookrightarrow B_{p,q}^{(r)}(\mathbb{R}^n), \quad 1 < p \leq q < +\infty, \tag{5.24}$$

$$B_{p,p}^{(r)}(\mathbb{R}^n) = L_{p,p}^{(r)}(\mathbb{R}^n), \tag{5.25}$$

worked out the interpolation theory for these spaces (cf. Sect. 8.3) and with the aid of this proved imbedding theorems with different p-metrics and different dimensions (Triebel [1978c]).

For the Nikol'skiĭ-Besov spaces $B_{p,q}^{(r)}(\mathbb{R}^n)$ one has found various equivalent norms, for example, in terms of difference functions, in terms of approximations and in terms of Fourier transforms. This allows one to understand better the structure of these spaces. Therefore there arose naturally the question to describe norms in Lizorkin-Triebel spaces without the use of the Fourier transform. The solution to this problem, which to some extent is the converse of the problem which is solved by the Littlewood-Paley theorems, was obtained by G.A. Kal'yabin [1977b], [1980].

The spaces $L^{(r)}_{p,q}$ and their generalizations have also been studied by M.L. Gol'dman [1980], [1984a], [1984b], Peetre [1975] and others.

§ 4. Generalizations of Nikol'skiĭ-Besov and Lizorkin-Triebel Spaces

Let $Q = \{Q_k\}_{k=0}^{\infty}$ be a sequence of sets decomposing $\mathbb{R}^n$ into parallepipeds with edges parallel to the coordinates axis, $\bigcup_{k=0}^{\infty} Q_k = \mathbb{R}^n$. Let $s = \{s_k\}_{k=0}^{\infty}$, $s_k < 0$. We denote by $B^{(Q,s)}_{p,q}$ the set of all functions $f \in L_p(\mathbb{R}^n)$ such that there exists for them a series expansion (denoted by (q)) in L_p with respect to entire functions q_k:

$$f = \sum_{k=0}^{\infty} q_k, \tag{5.26}$$

with

$$\operatorname{supp} F q_k \subset Q_k, \quad \|f\|_{(q)} = \| \, \|\{s_k q_k(x)\}\|_{L_p} \|_{l_q} < +\infty \tag{5.27}$$

The norm in $B^{(Q,s)}_{p,q}$ is defined by the formula

$$\|f\|_{B^{(Q,s)}_{p,q}} \stackrel{\text{def}}{=} \inf_{(q)} \|f\|_{(q)}, \tag{5.28}$$

where the infimum is taken over all expansions (q) of f as a series (5.26) subject to the condition (5.27).

Along with the spaces $B^{(Q,s)}_{p,q}$ generalizing the Nikol'skiĭ-Besov spaces considered with the norm (5.20) one has also studied spaces $B^{\omega(\cdot)}_{p,q}$ obtained with the aid of a generalization of the norm (4.4) (P.L. Ul'yanov [1967], [1968]), A.S. Dzhafarov [1963], M.L. Gol'dman [1984b], G.A. Kalyabin [1977b], [1980] etc.).

Let $\omega_j(\delta)$ be functions such that $\omega_j(0) = 0$, $\omega_j(\delta) \uparrow$, $\omega_j(\delta)\delta^{-\beta_j} \downarrow$, $0 < \beta_j \le k_j$, $k_j \in \mathbb{N}$, $\omega(\delta) = (\omega_1(\delta), \dots, \omega_n(\delta))$. Let $f \in L_p$ with the moduli of continuity $\omega_p^{(k_j)}(f;\delta)$ of order k_j in x_j, $j = 1, 2, \dots, n$. Set

$$\|f\|_{B^{\omega(\cdot)}_{p,q}} \stackrel{\text{def}}{=} \|f\|_p + \left\{ \sum_{j=1}^{n} \int_0^1 \left[\frac{\omega_p^{(k_j)}(f;\delta)}{\omega_j(\delta)} \right]^q \frac{d\omega_j(\delta)}{\omega_j(\delta)} \right\}^{1/q}. \tag{5.29}$$

The *space* $B^{\omega(\cdot)}_{p,q}(\mathbb{R}^n)$ consists of functions for which the functional (5.29) is finite; the latter is taken as the norm in this space. If $\omega_j(\delta) = \omega^{r_j}$, $0 < r_j < k_j$, $r = (r_1, \dots, r_n)$, then

$$B^{\omega(\cdot)}_{p,q} = B^{(r)}_{p,q}.$$

In the periodic case with $q = \infty$ and $k = n = 1$ P.L. Ul'yanov [1967], [1968] gave necessary and sufficient conditions for the imbedding

$$H_p^{\omega(\cdot)} \hookrightarrow L_{p_0}, \quad 1 \leq p < p_0 \leq +\infty,$$

of the form

$$\left(\int_0^1 \left[\frac{\omega(\delta)}{\delta^{1/p-1/p_0}} \right]^{\varrho} \frac{d\delta}{\delta} \right)^{1/\varrho} < +\infty,$$

where $\varrho = p_0$ if $p_0 < +\infty$ and $\varrho = 1$ for $p_0 = +\infty$ (cf. also § 3 of Chap. 4).

M.L. Gol'dman [1984b] established the exact conditions for the imbedding

$$B_{p,q}^{\omega(\cdot)}(\mathbb{R}^n) \hookrightarrow L_{p_0}(\mathbb{R}^n), \quad 1 \leq p < p_0 \leq +\infty,$$

and further the general imbedding with different metrics:

$$B_{p,q}^{\omega(\cdot)}(\mathbb{R}^n) \hookrightarrow B_{p_0,q_0}^{\psi(\cdot)}(\mathbb{R}^n), \quad 1 \leq p < p_0 \leq +\infty, \quad 0 < q, q_0 \leq +\infty.$$

In the isotropic case ($k_j = k$, $\omega_j(\delta) = \omega(\delta)$, $j = 1, 2, \ldots, n$) the condition for imbedding into L_{p_0} is formulated in terms of the function

$$\omega_q(\delta) = \delta^k \left(\int_0^\delta \left[\frac{t^k}{\omega(t)} \right]^q \frac{dt}{t} \right)^{1/q}.$$

A necessary and sufficient condition for imbedding into L_{p_0} in this case is

$$\left(\int_0^1 \left[\frac{\omega_q(t)}{\delta^{n(1/p-1/p_0)}} \right]^{\varrho} \frac{d\delta}{\delta} \right)^{1/\varrho} < +\infty,$$

where $\varrho = +\infty$ for $q \leq p_0^*$, $\varrho = qp_0^*/(q - p_0)$ for $q > p_0$, while $p_0^* = p_0$ for $1 \leq p < +\infty$ and $p_0^* = 1$ for $p_0 = +\infty$. He also obtained generalizations of these results to the case of imbedding into Lorentz, Orlicz and other spaces.

In an appropriate choice of the system Q and s the spaces $B_{p,q}^{(Q,s)}(\mathbb{R}^n)$ and $B_{p,q}^{\omega(\cdot)}(\mathbb{R}^n)$ coincide with each other, up to equivalence of norm. In the work of M.L. Gol'dman and G.A. Kalyabin imbedding theorems for the spaces $B_{p,q}^{(Q,s)}(\mathbb{R}^n)$ were obtained for different metrics and different dimensions. In particular, necessary and sufficient conditions were given for the existence of traces on $\mathbb{R}^m$, $0 \leq m < n$ (the case $m = 0$ corresponding to imbedding into $C(\mathbb{R}^m)$) and further an exact description of the spaces of traces, which coincides with an appropriate class $B_{p,q}^{Q',s'}(\mathbb{R}^n)$, where the system (Q', s') is derived from (Q, s).

In the paper M.L. Gol'dman [1980] a theory of more general spaces is developed on the basis of a covering method generalizing the series expansions (5.16) and (5.26) and allowing to include in the discussion also spaces with a dominating mixed derivative (cf. § 4 of Chap. 6), zero classes and classes with negative smoothness parameter.

Next let us consider *generalized Lizorkin-Triebel spaces*. Let $N = \{N(k)\}_{k=0}^{\infty}$, $N(k) = (N_1(k), \dots, N_n(k))$, $N_j(k+1) \geq \gamma_j N_j(k)$, $\gamma_j > 1$, $j = 1, 2, \dots, n$, $Q_k = \{y : |y_j| \leq N_j(k), j = 1, 2, \dots, n\}$, χ_k being the characteristic function of the set $Q_k \backslash Q_{k-1}$, $k = 1, 2, \dots$, χ_0 the one for Q_0, $s = \{s_k\}_{k=0}^{\infty}$, $s_k > 0$, $f \in L_p(\mathbb{R}^n)$, $f_k = F^{-1}\chi_k F f$, $k = 0, 1, 2, \dots$.

We denote by $L_{p,q}^{(N,s)}(\mathbb{R}^n)$ the set of functions f in $L_p(\mathbb{R}^n)$, $1 < p < +\infty$, such that the functional

$$\|f\| = \| \, \|\{s_k f_k(x)\}\|_{l_q} \|_{L_p(\mathbb{R}^n)} \tag{5.30}$$

is finite and, hence, a norm. For $N_j(k) = 2^k$, $s_k = 2^{kr}$, the space $L_{p,q}^{(N,s)}(\mathbb{R}^n)$ reduces to the Lizorkin-Triebel spaces. The spaces $L_{p,q}^{(N,s)}$ are also studied in papers by G.A. Kalyabin and M.L. Gol'dman (the latter considers also spaces with auxiliary decompositions of the corridors into "packages"). They obtained conditions for the existence of traces on planes $\mathbb{R}^m$ of dimension m, $0 \leq m < n$, and a description of the trace spaces is given. It turns out that in the case of decompositions into corridors the trace space does not depend on q and coincides with an appropriate space $B_{p,q}^{(Q',s')}(\mathbb{R}^m)$. In the case when there is an auxiliary decomposition into packages the trace space depends on q and for its descriptions one requires some new estimates for entire functions. Let us state them.

For each locally summable function f on the real axis the *Hardy-Littlewood maximal function* is defined as

$$(Mf)(x) = \sup_{v>0} \frac{1}{2v} \int_{x-v}^{x+v} |f(y)| dy.$$

G.A. Kalyabin [1979] proved that if g is any entire function of exponential type $\lambda > 0$ then

$$|g(t)| \leq c(1 + \lambda|t - x|)(Mg)(x), \tag{5.31}$$

and M.L. Gol'dman [1979] showed that if the continuous function $G(x^{(n-1)}, x_n)$, $x^{(n-1)} = (x_1, \dots, x_{n-1})$, for each $x^{(n-1)}$ is an exponential function of type $\lambda > 0$ and if $g(x^{(n-1)}) = G(x^{(n-1)}, 0)$ then for each $r = 1, 2, \dots$ holds the inequality

$$\|g\|_{L_p(\mathbb{R}^{n-1})} \leq c_r \left\{ \lambda \int_{|x_n| \leq 1/\lambda} \|G(\cdot, x_n)\|_{L_p(\mathbb{R}^{n-1})} dx_n + \lambda^{1-r} \int_{|x_n| > 1/\lambda} \frac{\|G(\cdot, x_n)\|_{L_p(\mathbb{R}^{n-1})}}{|x_n|^r} dx_n \right\}. \tag{5.32}$$

In view of Hölder's inequality the right hand side of (5.32) for $r \geq 2$ does not exceed the quantity

$$c\lambda^{1/p} \left(\int_{-\infty}^{+\infty} \|G(\cdot, x_n)\|_{L_p(\mathbb{R}^{n-1})}^p dx_n \right)^{1/p} = c\lambda^{1/p} \|G\|_{L_p(\mathbb{R}^n)}.$$

Therefore the estimate (5.32) generalizes the aforementioned inequality of S.M. Nikol'skiĭ for the norm of entire functions in a different number of variables (cf. § 1 of Chap. 3); here we do not assume the finiteness of the norm $\|G\|_{L_p(\mathbb{R}^n)}$. The inequalities (5.31) and (5.32) can be generalized to the anisotropic case and functions of the form $G(x^{(m)}, \tilde{x}^{(m)})$, $1 \leq m \leq n-1$.

These estimates in conjunction with the maximal inequalities of Stein-Fefferman and P.I. Lizorkin's estimates for the Hilbert transform in Lizorkin-Triebel spaces play an essential role in the study of spaces of this type.

By analogy with Nikol'skiĭ-Besov spaces G.A. Kalyabin discovered a description of generalized spaces of Liouville type and further Lizorkin-Triebel spaces without using the Fourier transform but in terms of difference functions, more exactly in terms of local averaged moduli of continuity.

The function

$$\omega_{x_j}^{(k_j)}(f\,;\delta\,;x) = \sup_{o<t\leq\delta} \frac{1}{t}\int_{-t}^{t} |\Delta_{h,x_j}^{(k_j)} f(x)|dh$$

is called the *local averaged modulus of continuity*. If $0 < \lambda < \Lambda$ and $\lambda s_k \leq s_{k+1} \leq \Lambda s_k$, $k = 0, 1, \ldots$, then (5.30) is equivalent to the norm

$$\|f\|_{L_p(\mathbb{R}^n)} + \sum_{j=1}^{n} \| \, \|\{s_k \omega_{x_j}^{(k_j)}(f\,;N_j(k)^{-1}\,;x)\}\|_{l_q}\|_{L_p(\mathbb{R}^n)}. \tag{5.33}$$

G.A. Kalyabin gave also a description of the spaces $L_{p,q}^{(N,s)}$ in terms of truncated differentiation operators and in the isotropic case even in terms of harmonic extensions to $\mathbb{R}^{n+1}$. Analogous results with the norm $\| \, \| \cdot \|_{l_q}\|_{L_p}$ replaced by $\| \, \| \cdot \|_{L_p}\|_{l_q}$ hold for Nikol'skiĭ-Besov spaces. Finally, G.A. Kalyabin obtained criteria for the spaces $L_{p,q}^{(N,s)}$ to be Banach algebras with respect to pointwise multiplication of functions (this coincides with the condition for imbedding in $C(\mathbb{R}^n)$).

In his work G.A. Kalyabin makes wide use of the method of Hardy-Littlewood maximal functions. If f is a locally summable function on $\mathbb{R}^n$ and $Q(x;r)$ denotes the ball of radius r with center at the point x then the function

$$(Mf)(x) = \sup_{r>0} \frac{1}{\mu Q(x;r)} \int_{Q(x;r)} |f(y)|dy$$

is called the *maximal function of* f.

G.A. Kalyabin proved that if f is an entire function of exponential type ν_j in the variable x_j, $j = 1, 2, \ldots, n$, then for arbitrary $x, \tilde{x} \in \mathbb{R}^n$, holds

$$|f(\tilde{x})| \leq c_n \cdot \prod_{j=1}^{n} (1+\nu_j)|\tilde{x}_j - x_j| \cdot Mf(x), \quad x = (x_1, \ldots, x_n), \tilde{x} = (\tilde{x}_1, \ldots, \tilde{x}_n)$$

(the constant c_n depends on n and on $\nu_1, \nu_2, \ldots, \nu_n$).

With the aid of this inequality, using the expansion of functions in Lizorkin-Triebel spaces in a series of entire functions, he obtained a description of the traces of generalized Besov spaces.

In conclusion, let us remark that with the aid of general theorems by V.I. Burenkov and M.L. Gol'dman [1979] on the connection of norms of operators in spaces of periodic and nonperiodic functions a large portion of the results covered here can be extended to the case of Nikol'skiĭ-Besov and Lizorkin-Triebel spaces of periodic functions.

Chapter 6

Spaces of Functions Defined in Domains

§ 1. Integral Representations of Functions

As was already remarked above one of the most powerful methods in the study of properties of functions in several variables is the method of integral representations. With the aid of this method S.L. Sobolev [1962] obtained his very first imbedding theorem for the spaces $W_p^{(l)}(G)$ in the case of domains subject to a cone condition. In his investigations S.L. Sobolev constructed and applied representations of functions in terms of their derivatives of given order m in the form of a sum of a polynomial and an integral of potential type (Sobolev [1962]).

Sobolev's integral representation of a function $f \in W_p^{(l)}(G)$ has the form

$$f(x) = \int_B f(y) \sum_{|\alpha|<l} \frac{(-1)^{|\alpha|}}{\alpha!} D_y^\alpha[(x-y)^\alpha \varphi(y)]dy +$$

$$+ l \sum_{|\alpha|=l} \frac{1}{\alpha!} \int_{V_x} \frac{(x-y)^\alpha}{|x-y|^n} w(x,y)(D^\alpha f)(y)dy$$

for almost all $x \in G$ where G is a domain starshaped with respect to all balls B,

$$V_x = \bigcup_{y \in B} \{z : z = \alpha x + (1-\alpha)y, 0 < \alpha < 1\} \quad \text{a cone,}$$

$$\varphi \in \overset{o}{C}{}^\infty(B), \quad \int_B \varphi(x)dx = 1,$$

$$w(x,y) = \int_{|x-y|}^{+\infty} \varphi(x + \varrho \frac{y-x}{|y-x|}) \varrho^{n-1} d\varrho.$$

The first term of the sum to the right of this identity is, as is readily seen, a polynomial and the remaining terms are integrals of potential type (cf. S.L. Sobolev [1974], Yu.G. Reshetnyak [1971], [1980], V.I. Burenkov [1969]).

V.I. Il'in (see Besov-Il'in-Nikol'skiĭ [1977]) has elaborated the machinery of S.L. Sobolev extending it to the anisotropic case. He also constructed integral representations for functions with differences of derivatives, with different families of derivatives etc. This machinery made it possible to study, in particular, properties of the space $W_p^{(l)}(G)$ and $B_{p,q}^{(l)}(G)$ of functions defined in a domain $G \subset \mathbb{R}^n$ with suitable geometric properties and to clarify the dependence of the imbedding theorems on the geometry of the domain.

Let us give an example of an integral representation with pure derivatives (Besov-Il'in-Nikol'skiĭ [1975]):

$$f(x) = f_{h^\lambda}(x) + \int_0^h \sum_{j=1}^n v^{-1-|\lambda|-l_j\lambda_j} dv \int_{\mathbb{R}^n} \frac{\partial^{l_j} f(x+y)}{\partial x_j^{l_j}} \varphi_j(v^{-\lambda}y)dy, \tag{6.1}$$

where $l_j \in \mathbb{N}$, $\lambda = (\lambda_1, \ldots, \lambda_n) \in (0, +\infty)^n$,
$|\lambda| = \sum_{j=1}^n \lambda_j$, $v^{-\lambda}y = (v^{-\lambda_1}y_1, \ldots, v^{-\lambda_n}y_n)$,

$$f_{h^\lambda}(x) = h^{-|\lambda|} \int_{\mathbb{R}^n} f(x+y)\varphi_0(v^{-\lambda}y)dy,$$

the supports of the functions φ_j, usually termed kernels, being contained in a given ball and $\varphi_j \in C_0^\infty(\mathbb{R}^n)$.

The identity (6.1) holds for almost all x for which the right hand side makes sense. It is constructed from the restriction of f to the *λ-horn*

$$x + \bigcup_{0<v\leq h} [av^\lambda + v^\lambda\delta^\lambda(-1,1)^n], \tag{6.2}$$

where $\delta > 0$, $a \in \mathbb{R}^n$, $av^\lambda = (a_1v^{\lambda_1}, \ldots, a_nv^{\lambda_n})$, $v^\lambda\delta^\lambda(-1,1)^n = \{x : x_j = v^{\lambda_j}\delta^{\lambda_j}y_j, |y_j| < 1, j = 1, 2, \ldots, n\}$, and the parameters a and δ may be chosen arbitrarily at the expense of a suitable choice of the support of the kernels φ_j. Also, the representation (6.1) makes sense for all domains G such that each point $x \in G$ is contained in a λ-horn of the form (6.2). If moreover one can in (6.2) restrict oneself to a finite number of different a's we say that G satisfies a *λ-horn condition.* A domain with a λ-horn condition may also contain degenerate (zero) angles whose degeneracy character is determined by the ratio $\lambda_1 : \lambda_2 : \cdots : \lambda_n$. If $\lambda_1 = \cdots = \lambda_n$ then the λ-horn condition reduces to the cone condition.

Here are some examples of domains satisfying a λ-horn condition. Let $n = 2$, $\lambda = (1, 2)$, $0 < \delta < 1$, $a_1 = a_2 = 1$. Then the horn (6.2) with vertex at the origin is given by the conditions

$$x_1 = v(1 + \delta y_1), \quad x_2 = v^2(1 + \delta^2 y_2),$$

$$0 \le v \le h, \quad -1 < y_1 < 1, \quad -1 < y_2 < 1,$$

or, equivalently, by

$$x_2 = \frac{1+\delta^2 y_2}{1+\delta y_1} x_1^2, \quad -1 < y_1 < 1, \quad -1 < y_2 < 1,$$

where moreover

$$0 < x_1 < h(1+\delta y_1), \quad 0 < x_2 < (1+\delta^2).$$

This horn is depicted in Fig. 1.

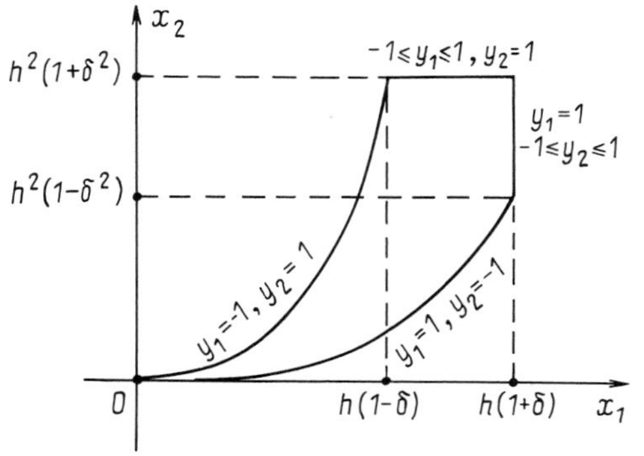

Fig. 1

By variation of the parameters $a = (a_1, a_2) \in \mathbb{R}^2$ and $\delta > 0$ one can obtain horns of the form drawn in Fig. 2.

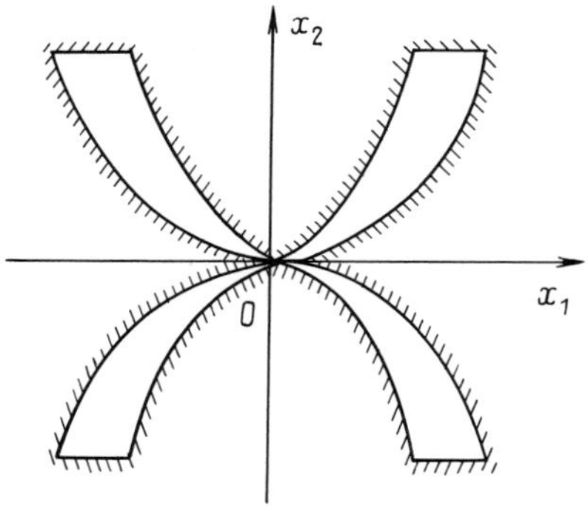

Fig. 2

If in a given domain G situated in the x_1, x_2-plane one can choose a finite number of values for the parameters a and δ and, thus, a finite number of horns of the type considered (their number can be larger than four at the expense of different parameters a and δ but all curves bounding them having tangency of order two with the x_1-axis at the origin) in such a way that each point $x \in G$ is contained in a horn with vertex at x, which is congruent to one of the horns in a given finite set of horns, then we say that G, in accordance with the previous definition, satisfies the $(1,2)$-*horn condition*. The boundaries of such domains have singularities of the type indicated. For example, the boundary may comprise zero angles and then at their vertices one has tangency of order two with the horisontal direction (the point A in Fig. 3). One may also have cusps of the first kind with tangency of any order (i.e. angles equal to 2π) and, finally, many other types of singularities, as indicated in Fig. 3.

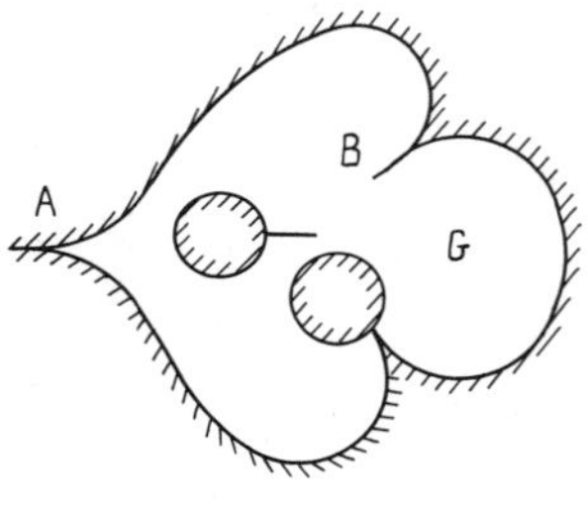

Fig. 3

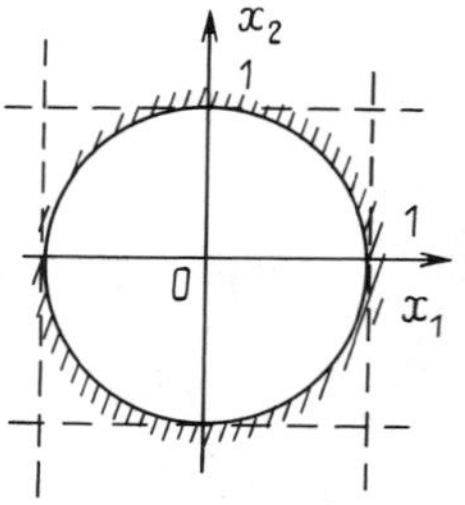

Fig. 4

Note that the disk $x_1^2 + x_2^2 < 1$ (Fig. 4) satisfies a (λ_1, λ_2)-horn condition only if

$$1/2 \leq \frac{\lambda_2}{\lambda_1} \leq 2,$$

because for $\lambda_2/\lambda_1 > 2$ if one tries to approach the points $(0,1)$ and $(0,-1)$ with the vertex of a λ-horn congruent to one of the horns of the type depicted in Fig. 2 (with the λ's considered here) it begins to slip out from the boundaries of the disk, while for $0 < \lambda_2/\lambda_1 < 1/2$ an analogous phenomenon takes place at the points $(-1,0)$ and $(1,0)$.

In Fig. 5 we have drawn a domain (the unit disk with a radius removed) which satisfies an appropriate horn condition but its boundary is not locally Lipschitz, because in no neighborhood of the point $(1,0)$ it can be represented as the graph of a singlevalued function.

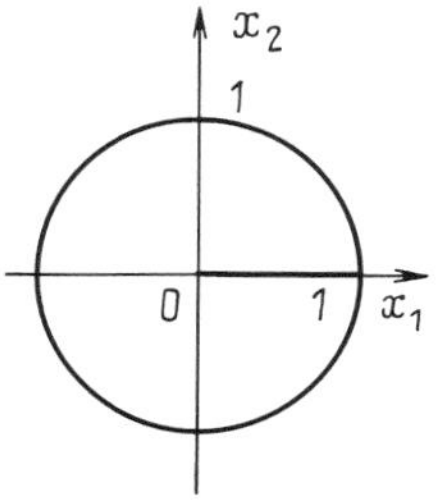

Fig. 5

§ 2. Imbedding Theorems for Domains

The proof of imbedding theorems with the aid of integral representations of functions leads to estimates of integral operators given by these representations: integrals of potential type, singular integrals etc.

For isotropic Sobolev spaces imbedding theorems of the type $W_p^{(l)}(G) \hookrightarrow L_q(G)$ on domains with degenerate angles were first obtained by I.G. Globenko [1962] and V.G. Maz'ya [1981]. The general picture of the dependence of the imbedding theorems on the geometry of a domain degenerating in terms of λ-horns was elucidated by V.P. Il'in [1981].

We give examples of imbedding theorems with a λ-horn condition, from which one can see this dependence.

Let $\alpha = (\alpha_1, \ldots, \alpha_n) \in \mathbb{N}^n$. If X is any function space, let $D^\alpha X$ denote the space of all derivatives $f^{(\alpha)}$ of functions $f \in X$ (cf. (4.14)).

One has the imbedding

$$D^\alpha W_p^l(G) \hookrightarrow L_q(G) \tag{6.3}$$

provided $1 \le p \le q \le +\infty$, $\kappa = \min_j \lambda_j l_j - |\lambda|(1/p - 1/q) - (\lambda, \alpha) \ge 0$, or, if $\kappa = 0$, either $1 < p < q < +\infty$ or $1 = p < q = +\infty$ or $1 < p = q < +\infty$.

One has further the imbedding

$$D^\alpha B_{p,q}^l(G) \hookrightarrow L_{p_0}(G) \tag{6.4}$$

for $1 \le p \le p_0 \le +\infty$, $\kappa \ge 0$, and either $q = 1$ or $1 \le q \le p_0$, $1 \le p < p_0 < +\infty$.

This and other imbedding theorems for domains satisfying a λ-horn condition were established by O.V. Besov and V.P. Il'in and are set forth in the monograph Besov-Il'in-Nikol'skiĭ [1977]. The imbedding theorems (6.3) and (6.4) have the same form as for $G = \mathbb{R}^n$ only when $\lambda_1 l_1 = \cdots = \lambda_n l_n$, i.e. for a domain G with a $1/l$-horn condition, or, otherwise put, only in the case when the geometry of the domain agrees with the differentiability properties of the functions.

The imbedding (6.3) for $1 < p = q < +\infty$, $\kappa = 0$, contains, in particular, an estimate of the L_p-norm of a mixed derivative in terms of the L_p-norms of the function itself for a domain with a cone condition.

For domains with sufficiently smooth boundary such estimates were first obtained by Yu. K. Solntsev in 1964. We also remark that in this estimate the assumption $1 < p < +\infty$ is essential: it is only in the integral metric $p > 1$ that it is possible to estimate mixed derivatives in terms of pure derivatives. That it is not so in the uniform metric is shown by an example by V.I. Yudovich:

$$f(x, y) = \begin{cases} xy \log\log \dfrac{1}{x^2 + y^2}, & \text{for } x^2 + y^2 > 0, \\ 0 & \text{for } x = y = 0. \end{cases}$$

This function has in the disk $x^2 + y^2 < 1$ bounded pure second derivatives but the second order mixed derivatives are unbounded in the disk with the origin removed.

From the point of view of imbedding theorems one has also studied generalizations of the spaces $W_p^l(G)$, $B_{p,q}^l(G)$, which are distinguished, first, by the fact that they are constructed with mixed L_p-norms, $p = (p_1, \ldots, p_n)$, (cf. Sect. 1.10 of Chap. 1) and, second, that in the various terms to the right in (3.13), (3.14), (4.5), (4.6) stand different L_p-norms. S.M. Nikol'skiĭ [1977], V.P. Il'in (see Besov-Il'in-Nikol'skiĭ [1975], also regarding the former two authors), V.K. Golovkin [1963], V.A. Solonnikov [1963] generalized many imbedding theorems in this direction. The integral representations of functions in terms of derivatives or differences were generalized by O.V. Besov [1984], [1985] to the case when the "core of the representation" is a "flexible λ-horn", which provided a possibility to generalize the imbedding theorems (6.3), (6.4) for the spaces $B_{p,q}^l(G)$, $W_p^l(G)$, as well as a number of other theorems, to the case of a flexible λ-horn condition. Let us state this more precisely.

A domain $G \subset \mathbb{R}^n$ is said to satisfy *a flexible λ-horn condition*, $\lambda \in (0,+\infty)^n$, (the *flexible cone condition* if $\lambda_1 = \cdots = \lambda_n$) if for some $\delta \in (0,1]$, $T \in (0,+\infty)$ there exist for each $x \in G$ a curve $\varrho(t^\lambda) = (\varrho_1(t^{\lambda_1}),\dots,\varrho_n(t^{\lambda_n})) = \varrho(t^\lambda, x)$, $0 \le t \le T$, satisfying the following conditions:

(a) for each $j \in \overline{1,n}$ the functions $\varrho_j(t)$ are absolutely continuous on $[0, T^{\lambda_j}]$; for almost all $u \in [0, T^{\lambda_j}]$ holds $|\varrho_j(u)| \le 1$;

(b) $\varrho(0) = 0$,

$$x + V(\lambda, x, \delta) \overset{\text{def}}{=} x + \bigcup_{0<t<T} [\varrho(t^\lambda, x) + t^\lambda \delta^\lambda (-1,1)^n] \subset G.$$

In the isotropic case the integral representations for a flexible cone were first obtained by Yu. G. Reshetnyak [1971], [1981].

The question on the "best" formulation of the imbedding theorems in domains satisfying a flexible λ-horn condition was studied by V.P. Il'in [1981]. It was reduced to the question of finding, for a domain satisfying a flexible λ-horn condition for λ in a certain set $\{\lambda\}$, a vector λ^* which is extremal in a certain sense. This vector is found as the solution of a certain linear programming problem and it defines the form of the inequalities among the parameters of the spaces which guarantee an imbedding with a "best" formulation.

§ 3. Traces of Functions on Manifolds

The properties of traces of functions on sufficiently smooth manifolds Γ^m of dimension m has been studied by O.V. Besov, V.P. Il'in, V.A. Solonnikov and S.V. Uspenskiĭ. They studied functions in the classes $W_p^r(G)$, $H_p^r(G)$, $B_{p,q}^r(G)$, where G is an open set in $\mathbb{R}^n$ and $\Gamma^m \subset \bar{G}$, $1 \le m < n$. In the isotropic case and for sufficiently nice domains the following reversible imbeddings were obtained

$$B_{p,q}^{(r)}(G) \overset{\hookrightarrow}{\leftarrow} B_{p,q}^{(r-(n-m)/p)}(\Gamma^m),$$

$$W_p^{(r)}(G) \overset{\hookrightarrow}{\leftarrow} B_{p,p}^{(r-(n-m)/p)}(\Gamma^m),$$

where $0 \le m < n$, $1 \le p \le +\infty$, $r - (n-m)/p > 0$.

These imbeddings clearly generalize the imbeddings (3.19) and (4.11).

A special case of similar imbeddings was encountered long ago: Douglas [1931] proved for the Dirichlet integral of harmonic functions in the disk $\{z = x + iy : |z| < 1\}$ the representation

$$\int\int_{|z|<1} |\nabla u(z)|^2 dxdy = \frac{1}{8\pi} \int_0^{2\pi} \int_0^{2\pi} \frac{|u(e^{i\theta}) - u(ei^{i\varphi})|^2}{\sin \dfrac{\theta - \varphi}{2}} d\theta d\varphi.$$

This formula was utilized by Beurling [1940]. In contemporary language it means that the space $W_2^{(1/2)}(|z| = 1)$ is the space of traces for $W_2^{(1)}(|z| < 1)$.

The properties of traces of anisotropic classes on planes of various dimensions were investigated by V.P. Il'in and V.A. Solonnikov and on sufficiently smooth manifolds by Ya.S. Bugrov and S.V. Uspenskiĭ. For domains whose boundary has corners the study of such questions was inaugurated by S.M. Nikol'skiĭ [1956/58] for the space $H_p^{(r)}$ and continued by G.N. Yakovlev [1967] and several other authors.

An inversible characterization of the traces of functions in isotropic spaces $W_p^{(r)}$, $1 < p < +\infty$, $B_{p,q}^{(r)}$, $1 \leq p \leq +\infty$, $1 \leq q \leq +\infty$, in the case when the number $r - (n-m)/p$ is positive and noninteger on m-dimensional manifolds Γ^m was obtained by O.V. Besov (see Besov-Il'in-Nikol'skiĭ [1975]).

The Nikol'skiĭ-Besov spaces for functions defined on domains in n-dimensional space and the imbedding theorems for them at once found applications in the theory of partial differential equations. In terms of these space it became possible to obtain necessary and sufficient conditions which one has to impose on the boundary data in order to have a solution of Dirichlet's problem for elliptic equations (S.L. Sobolev [1953], [1958], [1970], V.M. Babich and L.N. Slobodetskiĭ [1956], L.N. Slobodetskiĭ [1961], [1962] and others). O.V. Besov [1967] has in an essential way extended the sufficient conditions for the existence of a classical solution to the wave equation.

§ 4. Extension of Functions from Domains to the Whole Space with Preservation of Class

Let us first consider the Sobolev spaces $W_p^{(l)}(G)$ where $1 \leq p \leq +\infty$ and G is a domain in $\mathbb{R}^n$. The question alluded to in the title consists of founding a bounded, if possible, linear operator $A : W_p^{(l)}(G) \to W_p^{(l)}(\mathbb{R}^n)$ such that $(Af)(x) = f(x)$ for $x \in G$, $f \in W_p^{(l)}(G)$.

The possibility of extending functions in spaces $W_p^{(l)}(G)$, $H_p^{(l)}(G)$ with $1 \leq p \leq +\infty$ for domains G with sufficiently smooth boundary was established by S.M. Nikol'skiĭ (see Besov-Il'in-Nikol'skiĭ [1975]). In the construction of the extension operator A he used the method of Hestenes, where the operator is gotten as a linear combination of reflexions and dilations. Calderón [1961] established the possibility of extending functions in $W_p^{(l)}(G)$ for $1 < p < +\infty$ for domains G with Lipschitz boundary using a method of continuation based on integral representations of functions (as the extension one takes the expression in the right hand side of the integral representation (cf. § 1 of Chap. 6) defined for all $x \in \mathbb{R}^n$ and coinciding for all $x \in G$ with the given function).

Stein [1970] showed that the extension theorem for domains G with boundary of class Lip 1 is valid for arbitrary $1 \leq p \leq +\infty$.

V.I. Burenkov [1974c] completed these results by constructing for domains G with boundary of class Lip 1 an extension operator $A : W_p^{(l)}(G) \to W_p^{(l)}(\mathbb{R}^n)$

such that $(Af)(x) \in C^\infty(\mathbb{R}^n \backslash G)$ and for each α, $|\alpha| > l$

$$\|\varrho^{|\alpha|-l} D^{|\alpha|}(Af)\|_{L_p(\mathbb{R}^n \backslash \overline{G})} < +\infty, \quad \varrho = \varrho(x, \partial G); \tag{6.5}$$

he showed also that there does not exist such an operator for which this inequality holds with the exponent $|\alpha| - l - \varepsilon$, where ε. Furthermore, the operator A is optimal in the sense that the derivatives of the extension have in a determined sense minimal growth when one approaches the boundary of G. In the case of the extension from $\mathbb{R}^m$ to $\mathbb{R}^n$ "best" extension operators were constructed earlier in the papers L.D. Kudryavtsev [1959], Ya.S. Bugrov [1957b], [1963b] and S.V. Uspenskiĭ [1966].

We say that a domain G satisfies an *ε-δ-condition* (in brief: $G \in A_{\varepsilon,\delta}$) if any two points $x, y \in G$ whose distance is less than δ there exists a smooth curve $\Gamma \subset G$ connecting x and y and satisfying the conditions

1) the length of the curve satisfies $l(\Gamma) \leq \dfrac{1}{\varepsilon}|x-y|$;
2) for each point $z \in \Gamma$ holds

$$\varrho(z, \partial G) \geq \varepsilon \frac{|x-z||y-z|}{|x-y|}.$$

Note that any domain with Lip 1 boundary is in $A_{\varepsilon,\delta}$ and that the condition $G \in A_{\varepsilon,\delta}$ is close to the flexible cone condition in § 2 of Chap. 6.

S.K. Vodopyanov and V.M. Gol'dshteĭn (cf. Gol'dshteĭn-Reshetnyak [1983]) showed that in the case of a simply connected domain $G \subset \mathbb{R}^n$ a necessary and sufficient condition for the existence of an extension operator $A : W_2^{(1)}(G) \to W_2^{(1)}(\mathbb{R}^n)$ is that G for some $\varepsilon > 0$ and $\delta > 0$ satisfies an ε-δ-condition. Jones [1980] showed that for arbitrary $n, l, 1 \leq p \leq +\infty$ the imbedding theorems for the spaces $W_p^{(l)}(G)$ hold for domains $G \in A_{\varepsilon,\delta}$. V.M. Goldshteĭn [1981] proved that for simply connected domains $G \subset \mathbb{R}^2$ and $1 \leq p \leq +\infty$ a necessary and sufficient condition for the existence of extension operators

$$A_1 : W_p^{(1)}(G) \to W_p^{(1)}(\mathbb{R}^2), \quad A_2 : W_p^{(1)}(\mathbb{R}^2 \backslash \bar{G}) \to W_p^{(1)}(\mathbb{R}^2),$$

is that $G \in A_{\varepsilon,\delta}$ for some $\varepsilon > 0$ and $\delta > 0$.

Note that for each $0 < \gamma < 1$ there exist domains with Lipγ boundary such that the theorem on extension with preservation of differentiability properties does not hold true. In this hypothesis there arises the question on extension of functions in such domains with minimally deteriorated differentiability properties. V.I. Burenkov [1974c] proved for domains G with boundary of class Lip γ, $0 < \gamma < 1$, the imbedding theorem

$$W_p^{(l)}(G) \to W_p^{([\gamma l])}(\mathbb{R}^n), \quad 1 \leq p \leq +\infty,$$

(the exponent $[\gamma l]$ here is exact) and obtained estimates of type (6.5).

Let us also turn to the question of extension with preservation of differentiability properties for the seminormed space $w_p^{(l)}(G)$ where G is an open set in $\mathbb{R}^n$. For unbounded domains satisfying the so-called *infinite cone condition* the possibility of such an extension from G to $\mathbb{R}^n$ was established by O.V. Besov [1969] (cf. also Besov-Il'in-Nikol'skiĭ [1975]). However, in the general case extension from G to $\mathbb{R}^n$ is not possible, even if G has a sufficiently smooth boundary ∂G (for example, for a set G consisting of two balls with nonintersecting closures). V.I. Burenkov [1976a] proved that for $1 \le p \le +\infty$ for open sets with Lip 1 boundary there exists a bounded linear operator A such that

$$A : w_p^{(l)}(G) \to w_p^{(l)}(G^\delta),$$

where $G^{(\delta)}$ is the δ-neighborhood of G. If G is bounded then G^δ can be replaced by the whole space.

Consider $G = G_m \times \mathbb{R}^{n-m}$, $1 \le m < n$, where G_m is a bounded domain in $\mathbb{R}^m$ with boundary of class Lip 1. In this case it is not possible to replace G^δ by $\mathbb{R}^n$. V.I. Burenkov and, simultaneously, B.L. Faĭn found in a sense optimal conditions on the weight function $\varphi(x_1, \dots, x_n)$ under which there exists a bounded linear operator into the weighted space $W_{p,\varphi}^{(l)}(\mathbb{R}^n)$ (§ 4 of Chap. 7):

$$A : w_p^{(l)}(G) \to w_{p,\varphi}^{(l)}(\mathbb{R}^n)$$

(V.I. Burenkov [1976a], B.L. Faĭn [1970]).

The imbedding theorems discussed in detail above hold also for the spaces $B_{p,q}^{(l)}(G)$. In particular, P.A. Shvartsman proved in 1983 an extension theorem with preservation of differentiability properties when $G \in A_{\varepsilon,\delta}$.

Let us also mention the following result on extension with preservation of modulus of continuity. For a bounded domain G satisfying the strong cone condition O.V. Besov for $1 < p < +\infty$ and Yu.A. Brudnyĭ for $1 \le p \le +\infty$ have constructed an extension operator A such that

$$\omega_p^{(k)}(Af; \delta) \le c\omega_p^{(k)}(f; \delta), \quad 0 < \delta \le \delta_0.$$

Next consider the question of describing all extensions with preservation of class for a function $f \in B_{p,q}^{(l)}(G)$. This is equivalent to the *problem of pasting functions*, i.e. of finding necessary and sufficient conditions on two functions $f_1 \in B_{p,q}^{(l)}(G)$ and $f_2 \in B_{p,q}^{(l)}(\mathbb{R}^n \backslash \overline{G})$ for which the *pasting*

$$f(x) = \begin{cases} f_1(x), & x \in G, \\ f_2(x), & x \in \mathbb{R}^n \backslash \bar{G} \end{cases}$$

belongs to $B_{p,q}^{(l)}(\mathbb{R}^n)$.

For domains G with sufficiently smooth boundary and the spaces $H_p^{(l)}(G)$ with $l \neq m + 1/p$, $m \in \mathbb{N}$, this problem was solved by S.M. Nikol'skiĭ [1961a]; for the spaces $B_{p,q}^{(l)}(G)$ and $n = 1$, $q = p$, $0 < l < 1$, $l \neq p$ by G.N. Yakovlev

[1967]; for arbitrary n and l and $q = p$ by V.I. Burenkov [1976a]; for $q \neq p$ by Yu.V. Kuznetsov [1983a], [1983b]. Closely connected with these questions there is the problem of additivity of the corresponding function spaces (cf. S.M. Nikol'skiĭ [1960/61], V.I. Burenkov [1976b]).

The majority of the extension theorems set forth above carry over to the case of the anisotropic spaces $W_p^l(G)$, $B_{p,q}^l(G)$, where $l = (l_1, \dots, l_n)$, and even more general spaces. Let us point out the principal differences compared to the isotropic case appearing here. For anisotropic spaces a decisive role is played not by the smoothness of the boundary but the requirement that the geometry of the domain should agree with the smoothness parameters. Thus, if $n = 2$ and G is a disk then the imbedding

$$W_p^{(l_1,l_2)}(G) \to W_p^{(l_1,l_2)}(\mathbb{R}^n)$$

holds iff $1/2 \leq l_2/l_1 \leq 2$. O.V. Besov and V.P. Il'in (see further their book with S.M. Nikol'skiĭ [1975]) proved that if a domain satisfies the strong $1/l$-condition then the extension theorem for extension from G to $\mathbb{R}^n$ with preservation of differentiablity properties holds for the spaces $W_p^l(G)$, $1 < p < +\infty$. B.L. Faĭn and P.A. Shvartsman established for $1 \leq p \leq +\infty$ an imbedding theorem for the spaces $W_p^l(G)$ for domains of class $A_{\varepsilon,\delta}^l$, the anisotropic analogue of the class $A_{\varepsilon,\delta}$ (the condition $G \in A_{\varepsilon,\delta}^l$ is close to the flexible $1/l$-horn condition in § 3 of Chap. 6). One has likewise considered corresponding spaces with mixed norm (see the book Besov-Il'in-Nikol'skiĭ [1975]) and furthermore anisotropic Sobolev spaces whose definition comprises derivatives summable with different powers (V.I. Burenkov and B.L. Faĭn [1979], B.L. Faĭn [1970]).

In the one dimensional case V.I. Burenkov obtained exact estimates for the minimal possible norm of the extension operator

$$A : W_p^{(l)}(a,b) \to W_p^{(l)}(\mathbb{R}^1),$$

of the form

$$c_1^l \left(1 + \frac{l^l}{(b-a)^l}\right) \leq \inf_A \|A\| \leq c_2^l \left(1 + \frac{l^l}{(b-a)^l}\right),$$

where c_1 and c_2 are constants indendent of l.

§ 5. Density of Infinitely Differentiable Functions

In many questions it is useful to have results concerning the density of various smooth functions in spaces $W_p^{(l)}(G)$, $B_{p,q}^{(l)}(G)$ and in other function spaces considered here on domains $G \subset \mathbb{R}^n$.

If $G = \mathbb{R}^n$ the issue of the density of the set $C^\infty(\mathbb{R}^n)$ in $W_p^{(l)}(\mathbb{R}^n)$ $(1 \leq p < +\infty)$, $B_{p,q}^{(l)}(\mathbb{R}^n)$ $(1 \leq p, q < +\infty)$ is solved with the aid of averaging (cf. § 3 of Chap. 2). In the paper Deny-Lions [1953/54] (and later in Meyers-Serrin [1964]) it was shown that for any domain $G \subset \mathbb{R}^n$ the set $C^\infty(G)$ is dense in $W_p^{(l)}(G)$. Below we state a more general result due to V.I. Burenkov on the density of $C^\infty(G)$ from which follows the statement concerning its density in various concrete function spaces. Let $Z(G)$ be a normed (or seminormed) space satisfying the following conditions:

1) $\overset{\circ}{C}{}^\infty(G) \subset Z(G) \subset \operatorname{loc} L(G)$;

2) for each measurable function $\Phi(x,y)$ on $G \times E$, where E is an arbitrary measurable subset of a Euclidean space, holds Minkowski's inequality

$$\left\| \int_E \Phi(x,y)dy \right\|_{Z(G)} \le \int_E \|\Phi(x,y)\|_{Z(G)}\, dy;$$

3) if $\varphi \in \overset{\circ}{C}{}^\infty(G)$, $f \in Z(G)$ then $f\varphi \in Z(G)$.

A sufficient (and, if some auxiliary conditions on $Z(G)$ are fulfilled, also necessary) condition for $C^\infty(G)$ to be dense in $Z(G)$ is that for every finite function $f \in Z(G)$ holds

$$\lim_{h \to 0} \|\overset{\circ}{f}(x+h) - f(x)\|_{Z(G)} = 0,$$

where

$$\overset{\circ}{f}(x) = \begin{cases} f(x), & x \in G \\ 0, & x \in \mathbb{R}^n \setminus G, \end{cases} \tag{6.6}$$

in other words, that the extension by zero off G of a function f on G enjoys the property of continuity in norm.

The corresponding sequence $\varphi_k \in C^\infty(G)$ ($\varphi_k \to f$ in $Z(G)$ for $k \to \infty$) is obtained with the aid of (nonlinear) averaging with variable step of the form

$$\sum_{m=1}^{\infty} (f\psi_m)_{h_m}(x),$$

where $\psi_m(x)$ is a suitable partition of unity and $(f\psi_m)_{h_m}$ is an average with sufficiently small step h_m (whose choice depends on f).

If $G = \mathbb{R}^n \setminus \mathbb{R}^m$ then averaging with variable step of a different type was introduced, and studied and applied to the construction of in a sense best extension from a hyperplane to the whole space by L.D. Kudryavtsev [1983] (for more details see § 3 of Chap. 7).

In the supplementary hypothesis that

$$\|f\varphi\|_{Z(G)} \le c_\varphi \|f\|_{Z(G)}, \quad \varphi \in \overset{\circ}{C}{}^\infty(G), f \in Z(G),$$

one can construct the sequence $\{\varphi_k\}$ in such a way that it has in a sense the same boundary values as f itself: namely for each function $\mu \in C^\infty(G)$

(with arbitrary fast growth when the boundary ∂G is approached) holds the condition

$$\|(\varphi_k - f)\mu\|_{Z(G)} < +\infty.$$

In the case of arbitrary domains $G \subset \mathbb{R}^n$ one often considers Sobolev type spaces $\tilde{W}_p^{(l)}(G)$ consisting of functions f with the finite norm

$$\|f\|_{\tilde{W}_p^{(l)}(G)} = \sum_{|s| \le l} \|f^{(s)}\|_{L_p(G)},$$

where one in contrast to (2.13) takes not only the top derivatives but all derivatives of orders s, $0 \le |s| \le l$ (if G is a domain with locally continuous boundary then $\tilde{W}_p^{(l)}(G) = W_p^{(l)}(G)$; cf. the book Nečas [1967]).

In the work of V.I. Burenkov the question on the existence of approximating functions φ_k preserving the boundary values of f for functions $f \in \tilde{W}_p^{(l)}(G)$, which are optimal as regards to the growth of the derivatives $\varphi_k^{(s)}$ for $|s| > l$. It is shown that there exists functions $\varphi_k \in C^\infty(G)$ such that $\varphi_k \to f$ in $\tilde{W}_p^{(l)}(G)$ $(1 \le p < +\infty)$ and

$$\|(f^{(s)} - \varphi_k^{(s)})\varrho^{|s|-l}\|_{L_p(G)} < +\infty, \quad |s| \le l,$$

$$\|\varphi_k^{(s)}\varrho^{|s|-l}\|_{L_p(G)} < +\infty, \quad |s| > l,$$

where $\varrho = \varrho(x)$ is the distance from the point $x \in G$ to the boundary ∂G; in the last inequality it is not possible to replace the exponent $|s| - l$ by $|s| - l - \varepsilon$, where $\varepsilon > 0$. Using a result by B.V. Tandit [1972] on the trace of functions in general weighted Sobolev spaces (cf. §4 of Chap. 7), E.A. Popova proved that the factor $\varrho^{|s|-l}$ cannot be replaced by $\varrho^{|s|-l}\varphi(\varrho)$, where φ is any continuous function such that $\lim_{t\to+\infty} \varphi(t) = +\infty$. V.I. Burenkov obtained a more general result with the function $\varphi(t)$ replaced by an arbitrary positive continuous function on G.

The functions φ_k are constructed with the aid of an averaging opeartor E_δ with variable step of the form

$$(E_\delta f)(x) = \sum_{m=-\infty}^{\infty} \psi_m(x)(f)_{\delta v_m}(x),$$

where $\psi_m(x)$ is a partition of unity corresponding to the decomposition of G into layers

$$G_m = \{x \in G : a^{-m} < \varrho(x) \le a^{-m-1}\}, \quad a > 1,$$

and v_m does not depend on f, $m = 0, \pm 1, \pm 2, \dots$.

Denote by $\check{C}^\infty(G)$ the set of function φ defined on G such that there exists a neighborhood G_φ of the set $\bar{G}$ and a function $\psi \in C^\infty(G_\varphi)$ which coincides with φ on G. The density of the space $\check{C}^\infty(G)$ in spaces $W_p^{(l)}(G)$, $B_{p,q}^{(l)}(G)$ etc. has been established by many authors (S.M. Nikol'skiĭ, L.N. Slobodetskiĭ, Gagliardo,

V.P. Il'in etc.) for domains with sufficiently smooth or Lipschitz domains. In the monograph Besov-Il'in-Nikol'skiĭ [1975] the density of $\check{C}^\infty(G)$ in the spaces $\tilde{W}_p^{(l)}(G)$, $\tilde{B}_{p,q}^{(l)}(G)$ (defined by analogy with $\tilde{W}_p^{(l)}(G)$) and generalizations is proved for domains G satisfying the condition of local shifts, in particular, for domains with locally continuous boundary. This result was established with the aid of "averages with shift", as introduced by Gagliardo.

Note that in contradistinction to the case of $C^\infty(G)$ the set $\check{C}^\infty(G)$ is not dense in $W_p^{(l)}(G)$ for arbitrary G (for example, this is not the case for a disk with one radius removed).

Let $V_p^{(l)}(G)$ be the space obtained as the closure in the norm of the space $W_p^{(l)}(G)$ of the set $\check{C}^\infty(G) \cap W_p^{(l)}(G)$ (cf. e.g. Kufner's monograph [1980]). Then $V_p^{(l)}(G) \hookrightarrow W_p^{(l)}(G)$ and the reversible imbedding

$$V_p^{(l)}(G) \overset{\hookrightarrow}{\hookleftarrow} W_p^{(l)}(G)$$

holds true iff $\check{C}^\infty(G)$ is dense in $W_p^{(l)}(G)$. The space $V_\infty^{(l)}(G)$ is isomorphic to $C^{(l)}(G)$ and is therefore always separable (in opposition to $W_\infty^{(l)}(G)$) but, as well as, $V_1^{(l)}(G)$, nonreflexive. The spaces $V_p^{(l)}(G)$, $1 < p < +\infty$, are separable and reflexive.

Let us pass to the question of approximation of functions in the spaces $W_p^{(l)}(G)$ by functions in $\overset{\circ}{C}{}^\infty(G)$. If $G = \mathbb{R}^n$ then $\overset{\circ}{C}{}^\infty(\mathbb{R}^n)$ is dense in $W_p^{(l)}(\mathbb{R}^n)$ $(1 \le p < +\infty)$. If $G \ne \mathbb{R}^n$, for a function $f \in W_p^{(l)}(G)$ to be arbitrarily well approximable by functions in $\overset{\circ}{C}{}^\infty(G)$ we must assume that f as well as appropriate derivatives of f vanish at the boundary ∂G. We denote by $\overset{\circ}{W}{}_p^{(l)}(G)$ the closure of $\overset{\circ}{C}{}^\infty(G)$ in the norm of the space $W_p^{(l)}(G)$.

If G is a domain with sufficiently smooth boundary then a necessary and sufficient condition for a function f to be in $\overset{\circ}{W}{}_p^{(l)}(G)$ is that $f \in W_p^{(l)}(G)$ and that

$$f^{(s)}|_{\partial G = 0}, \quad |s| \le l - 1,$$

(cf. Sobolev's books [1962], [1974]).

For an arbitrary domain $G \subset \mathbb{R}^n$, V.I. Burenkov [1974b] proved that for $n < p < +\infty$ a necessary and sufficient condition for f to be in $\overset{\circ}{\tilde{W}}{}_p^{(l)}(G)$ (i.e. the closure of $\overset{\circ}{C}{}^\infty(G)$ in the norm of the space $\tilde{W}_p^{(l)}(G)$) is that $f \in \tilde{W}_p^{(l)}(G)$ and that for each $x \in \partial G$ holds

$$\lim_{r \to 0} \left\{ \frac{1}{\mu(Q_r(x) \cap G)} \int_{Q_r(x) \cap G} |f^{(s)}(y)|^q dy \right\}^{1/q} = 0, \quad |s| \le l - 1,$$

for some q, $1 \le q \le +\infty$ (for two different q's the conditions being equivalent).

The question when a function f belongs to $\overset{\circ}{\tilde{W}}{}_p^{(l)}(G)$ is closely connected to the question when the function $\overset{\circ}{f}$ (cf. (6.6)) belongs to $\tilde{W}_p^{(l)}(\mathbb{R}^n)$. (Note that if $f \in \mathring{\tilde{W}}_p^{(l)}(G)$ then $\mathring{f}$ is in $\tilde{W}_p^{(l)}(G)$, this for any domain G.) G.G. Kazaryan (see Besov-Il'in-Nikol'skiĭ [1975]) proved that if the domain G satisfies the condition of exterior local shifts and $\mathring{f} \in \tilde{W}_p^{(l)}(\mathbb{R}^n)$ then $f \in \tilde{W}_p^{(l)}(G)$. V.I. Burenkov [1974b] found necessary and sufficient conditions on a domain G for the equivalence of $f \in \mathring{\tilde{W}}_p^{(l)}(G)$ and $\mathring{f} \in \tilde{W}_p^{(l)}(\mathbb{R}^n)$. These conditions have the form: for each neighborhood U of any point $x \in \partial G$ holds $\mu(U \backslash G) > 0$.

As we have already mentioned, $\mathring{C}^\infty(\mathbb{R}^n)$ is dense in $W_p^{(l)}(\mathbb{R}^n)$. It is much harder to solve the problem when $\mathring{C}^\infty(\mathbb{R}^n)$ is dense in the larger space $w_p^{(l)}(\mathbb{R}^n)$ with the finite seminorm $\|\cdot\|_{w_p^{(l)}(\mathbb{R}^n)}$ (cf. §3 of Chap. 2). S.L. Sobolev [1963] proved that $\mathring{C}^\infty(\mathbb{R}^n)$ is dense in $w_p^{(l)}(\mathbb{R}^n)$. V.R. Portnov [1964], [1965] generalized this to the case of weighted spaces of functions defined on a domain in suitable assumptions on the domain and the weight (for details cf. §4 of Chap. 7). O.V. Besov proved that $\mathring{C}^\infty(\mathbb{R}^n)$ is dense in the space $b_{p,q}^{(l)}(\mathbb{R}^n)$ of all functions with the finite seminorm (4.5) and (4.6). Sect. 14 of the monograph Besov-Il'in-Nikol'skiĭ [1975] is devoted to a similar question.

Many of the results mentioned in this section extend to the case of the various generalizations of the spaces $W_p^{(l)}(G)$ and $B_{p,q}^{(l)}(G)$.

§6. Spaces with a Dominating Mixed Derivative and Other Spaces

For all the spaces considered below in Chap. 2-6 a decisive role was played by a certain family of partial derivatives which in case at hand involved all pure partial derivatives. However, in many problems of mathematical physics a basic role can be played by a mixed derivative. In this connection S.M. Nikol'skiĭ began around 1963 to study function *spaces* $S_p^r H$, where the definition of the norm contains a certain mixed derivative playing a dominating part, and constructive characterizations and imbedding theorems were obtained for them. In particular, the functions $f \in S_p^r H(\mathbb{R}^n)$, $r = (r_1, \ldots, r_n) \in \mathbb{N}^n$, were characterized by the existence of a representation as a series

$$f(x) = \sum_{|k| \geq 0} q_k(x), \quad k = (k_1, \ldots, k_n) \in \mathbb{N}_0^n,$$

where

$$q_k(x) = q_{2^{k_1 r_1}, \ldots, 2^{k_n r_n}}(x)$$

are entire functions of exponential type 2^{k_j} in the variable x_j and

$$\|q_k\|_{L_p(\mathbb{R}^n)} \leq c 2^{-kr}, \quad kr = \sum_{j=1}^n k_j r_j,$$

with c independent of k.

These results were later generalized to other *spaces* $S_p^r W, S_p^r L, S_p^r B$, where in the definition of the norm a dominating role is played by a mixed Liouville derivative or multiple differences of mixed derivatives. For example, the norm in $S_p^r W(\mathbb{R}^n)$ is defined by the formula

$$\|f\|_{S_p^r W(\mathbb{R}^n)} \stackrel{\text{def}}{=} \sum_{\alpha_j \le r_j,\ j=1,2,\ldots,n} \|f^{(\alpha)}\|_{L_p(\mathbb{R}^n)},$$

$$\alpha = (\alpha_1, \ldots, \alpha_n) \in \mathbb{N}_0^n, \quad r = (r_1, \ldots, r_n) \in \mathbb{N}_0^n.$$

These and related themes are also treated in papers by N.S. Bakhvalov [1963], O.V. Besov (see the monograph Besov-Il'in-Nikol'skiĭ [1975]), M. K. Potapov [1969], A. D. Dzhabrailov [1972], T. I. Amanov [1976] and others.

The problem of estimating the derivative $f^{(\beta)}$,

$$\|f^{(\beta)}\|_{L_q(G)} \le c \sum_{\alpha \in \mathscr{A}} \|f^{(\alpha)}\|_{L_p(G)}, \quad q \ge p, \tag{6.7}$$

where $\mathscr{A}$ is a given family of multi-indices α, was considered by V.P. Il'in (see the book Besov-Il'in-Nikol'skiĭ [1975]): he found necessary and sufficient conditions for the validity of (6.7). The result is formulated in terms of the mutual position of the multi-index β and the set $\mathscr{A}$.

An L_p-estimate of the derivative $f^{(\beta)}$ in terms of general homogeneous differential operators $P_j(D)$ was obtained by O.V. Besov (see Besov-Il'in-Nikol'skiĭ [1975]) and L_p-estimates in terms of "nonregular" differential polynomials by G.G. Kalyabin [1974], [1976], [1977], [1979a], [1979b], extending the results of V.P. Mikhaĭlov [1967] for "regular" differential polynomials.

From the point of view of imbedding theorems V.P. Il'in ([1980] and the book Besov-Il'in-Nikol'skiĭ [1975]) considered spaces $W_{p,a,\kappa}^r(G)$ defined, in contrast to the spaces $W_p^r(G)$, starting with the metric $L_{p,a,\kappa}$, generalizing the L_p-metric; in the isotropic case it is defined by the formula

$$\|f\|_{L_{p,a,\kappa}(G)} \stackrel{\text{def}}{=} \sup_{\varrho > 0} \sup_{y} (1 + \varrho^{-na/p}) \|\chi_{y,\varrho} f\|_{L_p(G)}, \quad 1 \le p \le +\infty,$$

where $\chi_{y,\varrho}$ is the characteristic function of the ball $\{x : |x - y| < \varrho\}$. In the isotropic case similar function spaces for special values of the parameters were studied by Morrey (cf. § 11 of Chap. 1), who showed, in particular, that the functions in such spaces are Hölder continuous.

Various spaces characterized by integral estimates of the local deviation from polynomials were studied by Zygmund and Calderón, John and Nirenberg, Campanato. Generalizations of these spaces have been considered by V.P. Il'in, who clarified the differential-difference properties for their functions and the connection with other spaces.

§ 7. Multiplicative Estimates for Derivatives and Differences

Until now we have considered estimates of derivatives and differences of functions in terms of quantities written as sums of certain terms. It is of some interest to have estimates of derivatives and differences of functions in terms of a product of suitable derivatives: in this case it is clear that the vanishing of one of these quanitities entails also the vanishing of the quantity we are estimating. Multiplicative estimates, connecting various L_p-norms of various derivatives of functions, were established by V.P. Il'in, Nirenberg, Gagliardo. In the general anisotropic case one has the estimate

$$\|f^{(\alpha)}\|_q \leq c\|f\|_{p^{(0)}}^{\mu_0} \prod_{j=1}^{n} \|f_{x_j^{l_j}}^{(l_j)}\|_{p^{(j)}}^{\mu_j}, \quad f \in L_{p^{(0)}}(\mathbb{R}^n), \tag{6.8}$$

in the following relations between the parameters:

$$1 < p^{(j)} < +\infty, \quad 1 < q < +\infty, \quad j = 0, 1, \ldots, n,$$

$$0 \leq \mu_0 < 1, \quad 0 \leq \mu_k \leq 1, \quad k = 1, 2, \ldots, n, \quad \sum_{j=1}^{n} \mu_j = 1,$$

$$\alpha \in \mathbb{N}_0^n, \quad \frac{1}{q} \leq \sum_{j=0}^{n} \frac{\mu_j}{p^{(j)}}, \quad \alpha_k - \frac{1}{q} = l_k - \sum_{j=0}^{n} \frac{\mu_j}{p^{(j)}}, \quad k = 1, 2, \ldots, n.$$

The estimate (6.8) was for $1/q < \sum_{j=0}^{n} \mu_j/p^{(j)}$ obtained by A.D. Dzhabrailov [1972] and for $p^{(1)} = \cdots = p^{(n)} = n$, $\mu_0 > 0$ by V.A. Solonnikov [1972] in the hypothesis that

$$|\alpha : l| + \left(\frac{1}{p} - \frac{1}{q}\right)\left|\frac{1}{l}\right| < 1, \quad p \leq q,$$

where

$$|\alpha : l| = \frac{\alpha_1}{l_1} + \cdots + \frac{\alpha_n}{l_n};$$

here p, q and $p^{(0)}$ are allowed to take the value 1. In the above form (6.8) was established by O.V. Besov and generalized to the case of mixed norms (cf. Besov-Il'in-Nikol'skiĭ [1975]). He further obtained multiplicative estimates for integral moduli of continuity. Finally, A.P. Terekhin [1979] has obtained exhaustive estimates connecting various norms with mixed differences.

§ 8. Stabilizing Functions at Infinity to Polynomials

For functions f with the finite seminorm $\|\nabla f\|_{L_p(\mathbb{R}^n)}$ (or, in the notation (2.12), with the finite seminorm $\|f\|_{w_p^{(1)}(\mathbb{R}^n)}$) S.V. Uspenskiĭ has for $p < n$

established the existence of a constant $c = c(f)$ to which a suitable function equivalent to the given function converges along any ray issuing from the origin as the argument tends to infinity. He obtained also estimates for the difference $f - c(f)$ in the norm $\|f\|_{L_p(\mathbb{R}^n)}$.

S.L. Sobolev [1963], [1974] studied the stabilization at infinity to a polynomial when f has a finite seminorm $\|f\|_{W_p^{(l)}(\mathbb{R}^n)}$ (cf. (2.12)).

Stabilization of functions to polynomials is closely connected to approximation with finite infinitely differentiable functions: usually these two questions are studied simultaneously (cf. § 5 of Chap. 6). For anisotropic seminormed Sobolev spaces with mixed L_p-norms, $p = (p_1, \dots, p_n)$, O.V. Besov ([1969] and the book Besov-Il'in-Nikol'skiĭ [1975]) has studied the question of stability at infinity of functions to polynomials. More general questions of stabilization at infinity of functions to, in particular, trigonometric polynomials were treated by L.D. Kudryavtsev (for more details regarding this question see § 4 of Chap. 7).

§ 9. Compact Imbeddings

In the study of boundary problems in mathematical physics compactness of the operators considered often plays an important role. In this connection one has in the study of function spaces, along with the boundedness of the imbedding operator, also considered conditions assuring that it is, moreover, compact (completely continuous). The question of compactness of imbeddings of the spaces W_p^l, H_p^l, $B_{p,q}^l$ into spaces into the same type but with different parameters has been treated by S.M. Nikol'skiĭ, O.V. Besov, V.P. Il'in and P.I. Lizorkin (see Besov-Il'in-Nikol'skiĭ [1975], Lizorkin-Nikol'skiĭ [1969]). Here is a typical result of this type: if

$$B_{p,q}^l(G) \hookrightarrow B_{p_0,q}^r(G),$$

then the imbedding operator

$$B_{p,q}^l(G) \hookrightarrow B_{p_0,q}^{r'}(G^*), \quad r' < r,$$

is completely continuous, where $G^* \subset \bar{G}$, for instance, $G^* = G$ or G^* a submanifold of lower dimension.

Necessary and sufficient conditions for the compactness of sets in spaces $W_p^l(G)$, $H_p^l(G)$ and $B_{p,q}^l(G)$ have been given for $G = \mathbb{R}^n$ by O.V. Besov, P.I. Lizorkin and S.M. Nikol'skiĭ. (This was carried over to the general case by A.F. Kocharli; see Besov-Il'in-Nikol'skiĭ [1975].) In their work, in particular, an *analogue of the Riesz compactness criterion* is established (cf. Besov-Il'in-Nikol'skiĭ [1975]): a necessary and sufficient condition for a set in $W_p^l(G)$ or $B_{p,q}^l(G)$, $1 < p < +\infty$, $1 < p < +\infty$, where G is a domain in a suitable class, to be compact in $L_p(G)$ is that it be equicontinuous with respect to shifts in $W_p^l(G)$ and $B_{p,q}^l(G)$ respectively. An analogous statement holds also for subspaces of $H_p^l(G)$ consisting of functions which are continuous with respect to the shift in $H_p^l(G)$.

Chapter 7

Weighted Function Spaces

§ 1. The Role of Weighted Spaces in the General Theory

In the solution of many problems for the usual function spaces there arise also in a natural way *weighted function spaces*, i.e. spaces consisting of functions such that not the absolute values of appropriate derivatives are summable on the domain but rather the result of their multiplication by a suitable nonnegative functions, called the *weight*, which is equivalent to saying that the derivatives in question are summable with respect to a specially chosen measure. Let us illustrate this at the hand of the following example. In the theory of the usual function spaces and their applications a great role is played by *Steklov's inequality*: if $f : G \to \mathbb{R}$, G being a bounded domain in $\mathbb{R}^n$, $\nabla f \in L_p(G)$, $f|_{\partial G} = 0$ (as usual, ∂G is the boundary of G), then there exists a constant $c > 0$, not depending on f in the given class, such that

$$\|f\|_{L_p(G)} \le c\|\nabla f\|_{L_p(G)}. \tag{7.1}$$

In the case of functions defined in unbounded domains, for example, in the halfspace $\overset{+}{\mathbb{R}}{}^n$ in $\mathbb{R}^n$ and vanishing on the hyperplane $\mathbb{R}^{n-1}$, restricting oneself to this halfspace, the analogue of inequality (7.1) is not true. On the other hand, for each f such that $\nabla f \in L_p(\overset{+}{\mathbb{R}}{}^n)$ and $f|_{\mathbb{R}^{n-1}} = 0$ holds for each $\varepsilon > 0$ the "weighted" inequality

$$\left\| \frac{f(x)}{(1+|x|)^{n/p+\varepsilon}} \right\|_{L_p(\overset{+}{\mathbb{R}}{}^n)} \le c\|\nabla f\|_{L_p(\overset{+}{\mathbb{R}}{}^n)}, \quad n \ge p > 1.$$

Thus in this case $(1+|x|)^{n/p+\varepsilon}$ is the *weight function*.

Weighted functions arise also in connection with the problem of extension of functions with given smoothness on manifolds contained in an Euclidean space to, for example, infinitely differentiable functions outside the given manifold. In fact, if the given functions and the manifolds have given smoothness, we expect to have on the whole space exactly the same smoothness, i.e. the extended functions do not have in general derivatives of arbitrary high order belonging to a given function space but derivatives belonging to it only after multiplication by a suitable function.

Weighted spaces appear in a natural way likewise in the study of boundary problems for partial differential equations with variable coefficients. For example, if we consider the equation

$$\sum_{|\alpha|\le l}(-1)^{|\nu|}(a_\nu(x)u^{(\nu)}(x))^{(\nu)} = 0, \quad x \in G, \tag{7.2}$$

as the Euler equation of the functional

$$\int_G \sum_{|\nu|=l} a_\nu(x)[u^{(\nu)}(x)]^2 dx, \quad \nu \in \mathbb{N}_0^n, \tag{7.3}$$

then we must seek the solution in the class of functions for which the functional (7.3) (the "*energy integral*") is finite. The set of all functions for which (7.3) is finite is a weighted space, namely a space of functions whose partial derivatives are not just simply square integrable over G but only after multiplication with the factors $a_\nu(x)$.

Finally, weighted spaces are also very expedient in the study of ordinary unweighted function spaces. Thus some theorems on extension of functions to the entire space from manifolds of lower dimension can be proved by prolonging the given functions to functions in a weighted space and afterwards applying imbedding theorems from weighted spaces into ordinary ones and, hence, as a result it follows that the extension lies in an ordinary unweighted space.

§ 2. Weighted Spaces with a Power Weight

The study of weighted spaces began with spaces with a weight of the type of a power. Let G be an open set in n-dimensional Euclidean space $\mathbb{R}^n$ and let $\varrho(x)$ be the distance from $x \in G$ to the boundary of G. For a function $u : G \to \mathbb{R}$ set

$$\|u\|_p = \|u\|_{L_p(G)}, \quad \|u\|_{p,\alpha} = \|\rho^\alpha u\|_p.$$

Denote by $w_{p,\alpha}^{(r)} = w_{p,\alpha}^{(r)}(G)$ the seminormed space consisting of all functions $u : G \to \mathbb{R}$ having on G partial derivatives $f^{(k)}$, $k \in \mathbb{N}_0^n$, up to order r with

$$\|u\|_{w_{p,\alpha}^{(r)}} \overset{\text{def}}{=} \sum_{|k|=r} \|f^{(k)}\|_{p,\alpha} < +\infty. \tag{7.4}$$

For sufficiently small exponents α the weighted spaces $w_{p,\alpha}^{(r)}$ preserve many properties of the unweighted spaces $w_p^{(r)} = w_{p,0}^{(r)}$ (in the first place the imbeddings connected with them) and the methods for the proof of these properties.

In this case we say that the $w_{p,\alpha}^{(r)}$ are spaces with *weak degeneracy*. Thus, for example, if $\alpha < r-(n-m)/p$ then on each portion of the boundary ∂G which is a sufficiently smooth manifold of dimension m, situated at positive distance from the remainder of the boundary, each $u \in w_{p,\alpha}^{(r)}$ has a trace in the sense of mean convergence. If however α is sufficiently large then functions $u \in w_{p,\alpha}^{(r)}$ need not have traces on any portion of the boundary of G. Therefore imbedding theorems for the spaces $w_{p,\alpha}^{(r)}$ give, in particular, in the solution in such spaces of boundary problems of partial differential equations a possibility to determine which pieces of the boundary should be free from boundary conditions.

For the weighted seminormed spaces $w_{p,\alpha}^{(r)}$ one has the following imbeddings (Yu.S. Nikol'skiĭ [1979], L.D. Kudryavtsev [1983])

$$w_{p,\beta}^{(r)} \subset w_{p,\alpha}^{(r)}, \quad \beta < \alpha, \tag{7.5}$$

$$w_{p,\alpha}^{(r)} \subset w_{p,\alpha-1}^{(r-1)}, \quad r \geq 1, \alpha > 1-1/p, \tag{7.6}$$

If $\alpha > 1-1/p$ then the partial derivatives of order $r-1$ in $w_{p,\alpha}^{(r)}$ do not, in general, posses traces on the boundary of G. If however $\alpha < 1-1/p$ then for each m-dimensional portion of the boundary the derivative of order $r-1$ do have traces provided also $\alpha < 1-(n-m)/p$. In particular, the traces exist always if the boundary is $(n-1)$-dimensional, and in this case $f^{(r-1)}|_{\partial G} \in L_p(\partial G)$.

Moreover, if $\alpha < 1-1/p$ then the imbedding (7.6) does not hold true but one has the imbedding

$$w_{p,\alpha}^{(r)} \subset w_{p}^{(r-1)}. \tag{7.7}$$

Denote by $W_{p,\alpha}^{(r)} = W_{p,\alpha}^{(r)}(G)$ the complete normed linear space with the norm

$$\|f\|_{W_{p,\alpha}^{(r)}} \stackrel{\text{def}}{=} \|f\|_p + \|f\|_{w_{p,\alpha}^{(r)}}.$$

For the normed spaces $W_{p,\alpha}^{(r)}$ the inclusions (7.6) and (7.7) correspond to

$$W_{p,\alpha}^{(r)} \hookrightarrow W_{p,\alpha-l}^{(r-l)}, \quad \alpha > l-1/p, 1 \leq l \leq r, \tag{7.8}$$

$$W_{p,\alpha}^{(r)} \hookrightarrow W_{p}^{(r-l)}, \quad 0 < \alpha < l \leq r. \tag{7.9}$$

If there are suitable relations between p, r and α the weighted spaces $W_{p,\alpha}^{(r)}$ imbed in ordinary spaces:

$$W_{p,\alpha}^{(r)} \hookrightarrow B_{p}^{(r-\alpha)}, \quad r > \alpha > 0. \tag{7.10}$$

It follows from (7.10) that (cf. § 3 of Chap. 4) that the trace of a function $f \in W_{p,\alpha}^{(r)}$ exists on an m-dimensional piece Γ^m of the boundary ∂G of G, provided $r-(n-m/p)-\alpha > 0$ and we have the imbedding

$$W_{p,\alpha}^{(r)}(G) \hookrightarrow B_{p}^{(r-\frac{n-m}{p}-\alpha)}(\Gamma^m), \quad \alpha > 0. \tag{7.11}$$

With the aid of the imbedding (7.10) we can establish not only the existence of traces and describe their properties for the functions $f \in W_{p,\alpha}^{(r)}$ themselves but we can also do the same thing for their derivatives. Let us restrict attention to the case when the boundary of the open set has dimension $n-1$. Let n be the exterior unit normal to the boundary ∂G of G and let $\partial^s f/\partial n^s$ be the normal derivative of f of order s. If $f \in W_{p,\alpha}^{(r)}$,

$$r-\alpha-\frac{1}{p}>0, \quad \alpha>-\frac{1}{p}, \tag{7.12}$$

and s_0 is the least integer greater or equal to $r-\alpha-1/p$, i.e.

$$r-\alpha-\frac{1}{p} \leq s_0 < r-\alpha-\frac{1}{p}+1, \tag{7.13}$$

then the derivative $\partial^s f/\partial n^s$, $s=0,1,\ldots,s_0-1$, admits on ∂G the trace

$$\frac{\partial^s f}{\partial n^s}\Big|_{\partial G} \in B_p^{(r+\alpha-1/p-s)}(\partial G)$$

and one has the inequality

$$\left\| \frac{\partial^s f}{\partial n^s}\Big|_{\partial G} \right\|_{B_p^{(r+\alpha-1/p-s)}(\partial G)} \leq c\|f\|_{W_{p,\alpha}^{(r)}(G)},$$

where c does not depend on f in $W_{p,\alpha}^{(r)}(G)$ (S.V. Uspenskiĭ [1966], S.M. Nikol'skiĭ [1977])

Also the converse holds true: if the inequalities (7.12) and (7.13) are fulfilled and on the boundary ∂G of G functions

$$\varphi_s \in B_p^{(r-\alpha-1/p-s)}(\partial G), \quad s=0,1,\ldots,s_0-1, \tag{7.14}$$

are given, then there exists a function $f \in W_{p,\alpha}^{(r)}(G)$ such that

$$\frac{\partial^s f}{\partial n^s}\Big|_{\partial G} = \varphi_s, \quad s=0,1,\ldots,s_0-1, \tag{7.15}$$

and there exists a constant $c>0$ not depending on the functions φ_s such that

$$\|f\|_{W_{p,\alpha}^{(r)}(G)} \leq c\sum_{s=0}^{s_0-1} \|\varphi_s\|_{B_p^{(r-\alpha-1/p-s)}(\partial G)}. \tag{7.16}$$

The first result of the type of an imbedding for weighted spaces of functions of several variables was obtained in a paper by V.I. Kondrashov [1938]. The systematic study of weighted spaces with the weight equal to the distance from the boundary of the domain raised to a positive power was undertaken by

L.D. Kudryavtsev [1959]. He obtained an imbedding of the form (7.10) from a weighted space into the corresponding H-space of Nikol'skiĭ (the B-spaces were at the time yet not invented in mathematics). The imbedding (7.10) was in the case $p = 2$, $r = 1$ proved by A.A. Vasharin [1959] and in the general case by P.I. Lizorkin [1960] and S.V. Uspenskiĭ [1966].

L.D. Kudryavtsev found a necessary and sufficient condition, namely the inequality $r - (n - m)/p - \alpha > 0$, for the existence of a trace for each function in $W_{p,\alpha}^{(r)}(G)$ on an m-dimensional piece of the boundary ∂G of the open set G. He also obtained an imbedding of the type (7.11) when in the right hand side of this relation there is a corresponding H-space and further solved the problem of type (7.14)-(7.16) when the properties of the given functions are described in terms of H-spaces (Kudryavtsev [1959]). The more exact imbedding (7.11) as well as a complete solution of problem (7.14)-(7.16) was obtained later on in the paper [1966] by S.V. Uspenskiĭ. For a positive $\alpha > 0$ L.D. Kudryavtsev too obtained the imbedding (7.9) but (7.8) only up to an arbitrary $\varepsilon > 0$ in the exponent of the power in the weight function. In this form the imbedding (7.8) was proved by P.I. Lizorkin [1960] in some isolated cases, and by S.V. Uspenskiĭ [1966] in its final form. The theory of weighted spaces with a powerlike weight has been further developed by Poulsen [1962], Nečas [1967], Kufner [1980], Triebel [1978].

§ 3. Applications of the Theory of Weighted Spaces

As an application of the theory of weighted spaces L.D. Kudryavtsev [1959] considered the best extension of functions in the class of infinitely differentiable functions from an m-dimensional manifold Γ^m in n-dimensional space $\mathbb{R}^n$, $0 \leq m \leq n$, to the set $\mathbb{R}^n \setminus \Gamma^m$. Here the word "best" means that the derivatives of the extended function have in some sense minimal growth when the argument approaches the given manifold Γ^m. In order to solve this problem one constructed an averaging operator with radius of averaging tending to zero when the center of averaging approaches the boundary of the domain of definition of the functions to be averaged, and the properties of this operator were investigated (Kudryavtsev [1959]). Similar operators have also been used by Deny and Lions [1953/54]. In the case of a halfspace it was shown that the operator leaves the Nikol'skiĭ spaces invariant, preserves the boundary values of functions and maps locally summable functions into infinitely differentiable ones on the halfspace. With the aid of this operator and imbedding theorems for functions and systems of functions belonging to H-classes on a hyperplane one found infinitely differentiable extensions of the latter to a halfspace and for each derivative of a given order of the extension one found a weight with the aid of which its modulus becomes summable, and such that the exponent of the power differs from the minimal among all possible exponents by an arbitrary $\varepsilon > 0$ (Kudryavtsev [1959]).

These investigations were pursued by Ya.S. Bugrov [1963] and S.V. Uspenskiĭ [1966], pushing the result down to the exact power. They obtained the best infinitely differentiable extensions as solutions of certain differential equations of a special kind. In particular, they studied harmonic and polyharmonic extensions from the point of view of membership in weighted classes.

The problem of best extensions of a function from a domain $G \subset \mathbb{R}^n$ with Lipschitz boundary to an infinitely differentiable function on $\mathbb{R}^n$ was solved by V.I. Burenkov [1974c] (for more details see § 4 of Chap. 6).

Imbedding theorems for weighted spaces, as well as for ordinary unweighted ones, have been applied to the solution of boundary problems for partial differential equations; indeed, demands in the solution of such problems have proved to be one of the factors which stimulated the development of weighted function spaces. L.D. Kudryavtsev [1959] obtained estimates for the stability of the solution of Dirichlet's problem in the disk under deformation and, in a similarly way as in the unweighted case, he studied in detail the variational method for solving the first boundary problem for self-adjoint elliptic partial differential equations in the weighted spaces $w_{2,\alpha}^{(r)}$ with weak degeneracy. For model elliptic equations of the second order singular along the whole boundary sufficient conditions were found for the solvability of the first boundary problem in terms of membership of the boundary functions in H-classes and it was shown that this conditions coincide "up to an arbitrary $\varepsilon > 0$" with the necessary ones. Necessary and, at the same time, sufficient conditions for the solvability of this problem by variational methods were in the case of a bounded domain with sufficiently smooth boundary obtained (in terms of membership of the boundary functions in B-classes) by A.A. Vasharin [1959]. The study of boundary problems for degenerate elliptic equations with the aid of imbedding theorems for weighted spaces with powerlike weights is the object of the papers Lizorkin-Nikol'skiĭ [1981], [1983], Sedov [1972a,b] as well as many others.

S.M. Nikol'skiĭ and P.I. Lizorkin [1981], [1983] have considered the solution of the first boundary problem in the case of strong degeneracy (we speak of *strong degeneracy* if the number of boundary conditions appearing is less than l, where $2l$ is the order of the equation). They found that the solution of the boundary problem is correct if the strong degeneracy occurs along the whole boundary of the domain where we seek the solution. An essential role here is played by the inequality

$$\|f\|_{L_p(G)} \leq c\left(\sum_{j=0}^{s-1}\left\|\frac{\partial^j f}{\partial n^j}\right\|_{L_p(\Gamma)} + \sum_{|k|=r}\|\varrho^\alpha f^{(k)}\|_{L_p(G)}\right),$$

where G is a domain in $\mathbb{R}^n$, $f \in W_{p,\alpha}^{(r)}(G)$, $s \in \mathbb{N}$ and

$$r/2 \leq s \leq r, \quad r - \frac{1}{p} - s < \alpha < r + 1 - \frac{1}{p} - s.$$

In suitable assumptions on the growth of the coefficients of the elliptic equation (including the assumption of strong degeneracy) they proved an existence and uniqueness theorem for the solution of the first boundary problem and further an isomorphism theorem between the corresponding weighted spaces and a certain space which, in general, consists of generalized functions. The isomorphism is induced by the differential operator in the left hand side of the equation under consideration (Lizorkin-Nikol'skiĭ [1981], [1983]).

The traces of function in weighted spaces with a powerlike weight in domains with piecewise smooth boundaries have been studied in Vasharin [1959], Dzhabrailov [1967a,b], Kudryavtsev [1959], Fokht [1982], Stein [1970], Triebel [1973a].

The method of integral representations of functions for obtaining imbedding theorems for weighted spaces has been worked out by V.P. Il'in and A.D. Dzhabrailov and the method of expanding in a series of exponential functions by A. S. Dzhafarov. The latter proved a series of imbedding theorems for weighted spaces with dominating mixed derivatives (cf. § 6 of Chap. 6). A.F. Kochahrli, V.G. Perepelkin and S.V. Uspenskiĭ obtained imbedding theorems for weighted spaces in domains with Hölder boundary. S.I. Agalarov proved weighted multiplicative estimates (cf. § 5 of Chap. 6) for the norms of derivatives in $\mathbb{R}^n$, generalizing the corresponding estimates in the unweighted case.

Multiplicative estimates of the norms of intermediate derivatives in the case of a bounded domain with sufficiently smooth boundary can be found in Troisi [1969] and A.S. Fokht [1982]. For such a domain G and the weight function $\psi(\varrho(x))$, $\varrho(x) = \varrho(x, \partial G)$, where $\psi(\varrho)$ is a continuous increasing function, $\psi(0) = 0$, with the finite or infinite limit $\lim_{\varrho\to+0} \psi(\varrho)/\varrho$. A.S. Fokht got weighted estimates for the norms of intermediate derivatives with a weight which is a power of ψ in terms of the corresponding norms of derivatives of higher orders and the norm of the function. Similar estimates were applied by him in the study of solutions of elliptic partial differential equations

Function spaces with a powerlike weight having a singularity at infinity are considered in L.D. Kudryavtsev [1966], [1967], Yu.S. Nikol'skiĭ [1969], [1971], [1974], T.S. Pigolkina [1969], S.L. Sobolev [1974], Triebel [1976], [1978]. L.D. Kudryavtsev [1966], [1967] studied the space $W^{(r)}_{p,\alpha}(\mathbb{R}^n|\infty)$ consisting of functions f with

$$\|f\|_{W^{(r)}_{p,\alpha}(\mathbb{R}^n|\infty)} \overset{\text{def}}{=} \|f\|_{w^{(r)}_{p,\alpha}(\mathbb{R}^n|\infty)} + \|f\|_{L_p(Q^n)} \tag{7.17}$$

where

$$\|f\|_{w^{(r)}_{p,\alpha}(\mathbb{R}^n|\infty)} \overset{\text{def}}{=} \sum_{|k|=r} \|(1+|x|)^{-\alpha} f^{(k)}\|_{L_p(\mathbb{R}^n)}$$

and

$$Q^n = \{x : |x| < 1\},$$

defining the space $W^{(r)}_{p,\alpha}(\overset{+}{\mathbb{R}}{}^n|\infty)$, where $\overset{+}{\mathbb{R}}{}^n = \{x = (x_1, \dots, x_n) : x_n > 0\}$, in an analogous manner. In particular, he established the imbedding

$$W^{(r)}_{p,\alpha}(\mathbb{R}^n|\infty) \hookrightarrow W^{(r-k)}_{p,\alpha+k}(\mathbb{R}^n|\infty),$$

$$0 \le k \le r, \quad 1 \le p \le +\infty, \quad \alpha > n/p - 1,$$

and proved the equivalence of the norm (7.17) with the norm

$$\|f\|_{W^{(r)}_{p,\alpha}(\mathbb{R}^n|\infty)} + \|(1+|x|)^{-\alpha-r} f\|_{L_p(\mathbb{R}^n)}, \quad \alpha > n/p - 1.$$

For the traces of functions of the class $W^{(r)}_{p,\alpha}(\overset{+}{\mathbb{R}}{}^n|\infty)$ to the hyperplane $\mathbb{R}^{n-1} = \{x : x_n = 0\}$ the relation

$$W^{(r)}_{p,\alpha}(\mathbb{R}^n|\infty) \hookrightarrow W^{(r-1)}_{p,\alpha+1-1/p}(\mathbb{R}^{n-1}|\infty), \quad \alpha > n/p - 1,$$

was found, an exact description of these traces in terms of correspondingly defined spaces $L^{(r)}_{p,\alpha}(\mathbb{R}^{n-1})$, in general with a fractional smoothness index $r > 0$, was given and, finally, an invertable extension theorem was proved (Kudryavtsev [1959]). In the sequel the spaces $W^{(r)}_{p,\alpha}(\mathbb{R}^n|\infty)$ were studied by S.L. Sobolev [1974], Triebel [1976], [1977], T.S. Pigolkina [1969]; generalizations were considered by Yu. S. Nikol'skiĭ [1969], [1971].

The imbedding theorems obtained for spaces of type $W^{(r)}_{p,\alpha}(\mathbb{R}^n|\infty)$ made it possible to develop the classical variational methods for the solution of boundary problems in the case of unbounded domains. Thus, for example, L.D. Kudryavtsev [1967] proved an existence and uniqueness theorem for the first boundary problem for weakly degenerating (in particular nondegenerating) elliptic equations of the second order in a halfspace in the sole assumption of the finiteness of the energy integral. It was likewise shown that the condition of finiteness of the energy integral contains all restrictions on the behavior of the solution at infinity which are necessary for the uniqueness of the solution of the problem under consideration.

The systematic study of imbedding theorems for anisotropic function spaces with powerlike weights began in the 60's. S.L. Uspenskiĭ [1961], [1966] obtained imbedding theorems for such spaces and studied the question of the existence of mixed derivatives for functions whose pure derivatives belong, generally speaking, to different weighted spaces. Later on similar estimates were obtained by A. F. Kocharli [1974], V.G. Perepelkin and, more exact and more general ones, by O.V. Besov [1983].

§ 4. General Weighted Spaces

In the 70's one began to study systematically imbeddings of *anisotropic weighted spaces* with a rather general weight, i.e. spaces consisting of functions such that the moduli of certain partial derivatives (of any order) are summable to appropriate powers with respect to given, in general, different weights. First of all one studied the properties of traces of functions belonging to such spaces. For estimting the differences of these traces an inequality was proved (Kudryavtsev [1974]) with an arbitrary parameter in the right hand side: let $\varphi_j(t)$ be nonnegative measurable functions defined for $t > 0$, $\overset{+}{\mathbb{R}}{}^n_b = \{x : 0 < x_n < b\}$, $b < +\infty$ and consider the space $L^{(r_j)}_{p,\varphi_j}$ of all function f such that

$$\|f\|_{L^{(r_j)}_{p,\varphi_j}} \overset{\text{def}}{=} \left\| \varphi_j \frac{\partial^{r_j} f}{\partial x_j^{r_j}} \right\|_{L_p(\overset{+}{\mathbb{R}}{}^n_b)} < +\infty, \quad j = 1, 2, \ldots, n.$$

Set

$$L^{(r_j,r_k)}_{p,\varphi_j,\varphi_k} = L^{(r_j)}_{p,\varphi_j} \cap L^{(r_k)}_{p,\varphi_k},$$

$$\|f\|_{L^{(r_j,r_k)}_{p,\varphi_j,\varphi_k}} = \|f\|_{L^{(r_j)}_{p,\varphi_j}} + \|f\|_{L^{(r_k)}_{p,\varphi_k}}, \quad j, k = 1, 2, \ldots, n.$$

If $r_j \geq 1$, $r_n \geq 1$,

$$\int_0^b [\varphi_j(t)]^{-p'} dt < +\infty, \qquad \int_0^b [\varphi_n(t)]^{-p'} dt < +\infty,$$

$$f \in L^{(r_j,r_n)}_{p,\varphi_j,\varphi_n}(\overset{+}{\mathbb{R}}{}^n_b)$$

(j fixed, $j \in \overline{1,n}$), $1/p + 1/p' = 1$, $0 \leq s < r_n$, $s \in \mathbb{N}_0$, $k \geq r_j$, $k \in \mathbb{N}$,

$$G_\nu(t) \overset{\text{def}}{=} \frac{1}{t} \left\{ \int_0^t [\varphi_\nu(t)]^{-p'} dt \right\}^{1/p'}, \quad \nu = 1, 2,$$

then for each $t \in (0, b)$ holds the inequality

$$\left\| \Delta^{(k)}_{x_i}(h) \frac{\partial^s f(x^{(n-1)}, x_n)}{\partial x_n^s} \right\|_{L_p(\mathbb{R}^{n-1})} \leq$$

$$\leq c\|f\|_{L^{(r_j,r_n)}_{p,\varphi_j,\varphi_n}} \left[|h|^{r_j} t^{-s} G_j(t) + t^{r_n - s} G_n(t) \right]. \tag{7.18}$$

If $t = t(h)$ is a function satisfying in a neighborhood of the origin the condition

$$|h|^{r_j} G_j(t) = t^{r_n} G_n(t), \tag{7.19}$$

then it follows from (7.18) with $s = 0$ that

$$\|\Delta^{(k)}_{x_j}(h) f\|_{L_p(\mathbb{R}^{n-1})} \leq c\|f\|_{L^{(r_j,r_n)}_{p,\varphi_j,\varphi_n}} |G_j(t(h))| \cdot |h_j|^{r_j}.$$

In the case when $s = 0$ and $\varphi_j(t) = t^{\alpha_j}$, $\varphi_n(t) = t^{\alpha_n}$, taking $t = h^\lambda$ in inequality (7.18), with λ determined by the equation (7.19), we obtain an earlier result due to S.V. Uspenskiĭ [1966],

$$\|\Delta_{x_j}^{(k)}(h) f\|_{L_p(\mathbb{R}^{n-1})} \leq c\|f\|_{L^{(r_j, r_n)}_{p, x_j^{\alpha_j}, x_n^{\alpha_n}}} |h|^{\mu_j},$$

where

$$\mu_j = \frac{r_j(r_n - \alpha_n - 1/p)}{r_n + \alpha_j - \alpha_n}.$$

From this result it is seen how the connection between the properties of functions in anisotropic weighted classes and the properties of their traces deteriorates compared to the isotropic case. The study of the properties of traces of functions in general weighted classes was continued in the research of G.A. Kalyabin [1977a] and B.V. Tandit [1980a,b,c]. But so far this problem is not completely solved. That is, not solved in the sense that one does not possess a complete description of the spaces of traces under minimal restrictions on the weight functions guaranteeing the existence of the traces of the functions, even in the case when these weight functions depend on one variable, namely the distance of the point to the boundary of definition of the functions under study. Such a description was obtained by G.A. Kalyabin and B.V. Tandit in certain supplementary restrictions on the weight function (an exact description was given only in the case when in the definition of the weighted space involves only first order derivatives). This problem, interesting in itself, has important applications to the theory of elliptic p.d.e: its solution would imply an exact description of the properties of the boundary data under which the boundary values exist.

One can get an idea of the difficulties which appear here and the possible approaches towards overcoming them from the above mentioned results by G.A. Kalyabin, with complements by B.V. Tandit, devoted to the model case when the weight functions depend on a "transversal" variable, otherwise only with minimal restrictions.

Denote by $L^{(r,1)}_{p,\varphi,\psi}$ the class of functions $f(x, y)$ which possess on the halfspace $\mathbb{R}^{n-1} = \{(x, y) : x \in \mathbb{R}^n, y > 0\}$ generalized derivatives $f_x^{(l)}$, $l \in \mathbb{N}_0^n$, $0 \leq |l| \leq r$, and f'_y such that the functional

$$\|f\|_{L^{(r,1)}_{p,\varphi,\psi}} = \left\{ \int_{\mathbb{R}^n} \left[\int_0^{+\infty} \left((\varphi(y) \sum_{0 \leq |l| \leq r} |f_x^{(l)}|)^p + |\psi(y) f'_y|^p \right) dy \right] dx \right\}^{1/p}, \quad (7.20)$$

where $1 < p < +\infty$ and the weight functions are positive and continuous for $y > 0$, is finite.

The functional (7.20) is a seminorm in $L^{(r,1)}_{p,\varphi,\psi}$. Concerning the behavior of the functions φ and ψ near $y = 0$ it is assumed that

$$\xi(y) = \int_0^y \psi^{-p'}(t) dt < +\infty, \quad 1/p + 1/p' = 1, \quad (7.21)$$

$$\eta(y) = \int_0^y \varphi(t)^p dt < +\infty. \tag{7.22}$$

These assumptions are natural in the study of the space of traces of functions in $L^{(r,1)}_{p,\varphi,\psi}$ (the traces are taken on the hyperplane $\mathbb{R}^n$). In fact, if for example for sufficiently small y holds $\eta(y) = +\infty$ then the trace of an arbitrary function in $L^{(r,1)}_{p,\varphi,\psi}$ (if it exists) by necessity must vanish for all $x \in \mathbb{R}^n$. If however $\eta(y) < +\infty$ but $\xi(y) = +\infty$ one can find a function $f \in L^{(r,1)}_{p,\varphi,\psi}$ which does not have a trace. Let us note that if (7.21) and (7.22) are fulfilled and the functional (7.20) is finite then the trace on $\mathbb{R}^n$ always exists and belongs to $L_p(\mathbb{R}^n)$.

The membership of a function $g(x)$ in the space of traces of functions in $L^{(r,1)}_{p,\varphi,\psi}$ will be expressed in terms of L_p-norms of segments of their expansions in a series of Dirichlet sums (cf. below (7.26)) whose order (and this is the key idea) is defined by the weight functions. Moreover, in some situations one has to consider, instead of norms of the sections themselves, the L_p-norms of their derivatives.

Let us pass to the exact formulations. Let $y_k \downarrow 0$ in such a way that for $k = 1, 2, \dots$ holds

$$0 < a \le \max\left\{\frac{\xi(y_k)}{\xi(y_{k-1})}, \frac{\eta(y_k)}{\eta(y_{k-1})}\right\} \le b < 1. \tag{7.23}$$

There exist always sequences with the property (7.23): for example, one can pick $y_0 > 0$ in an arbitrary manner and the put

$$y_k = \max\{y : \xi(y) \le \frac{1}{2}\xi(y_{k-1}), \eta(y_k) \le \frac{1}{2}\eta(y_{k-1})\}, \quad k = 1, 2, \dots.$$

With this definition of the sequence $\{y_k\}$ condition (7.23) holds with $a = b = 1/2$.

The set of all integers k such that $\xi(y_k) \ge a\xi(y_{k-1})$ will be denoted by $\mathcal{M}_1$ and the set of the remaining ones by $\mathcal{M}_2$. In view of (7.23) we have $\eta(y_k) \ge a\eta(y_{k-1})$ for all $k \in \mathcal{M}_2 = \mathbb{N} \setminus \mathcal{M}_1$.

With the aid of the sequence $\{y_k\}$ we further define the numbers

$$N_k = \left(\xi^{1/p'}(y_k)\eta^{1/p}(y_k)\right)^{-1/r} \tag{7.24}$$

and the operator

$$A_k g(x) = S_{N_k} g(x) - S_{N_{k-1}} g(x), \quad S_{N_0} g(x) \equiv 0, \tag{7.25}$$

where the *Dirichlet sum* $S_N g(x)$ of the function g is defined by the formula

$$(S_N g)(x) = (F^{-1}\chi_{Q_N} F g)(x) \tag{7.26}$$

(χ_{Q_N} is the characteristic function of the n-dimensional cube

$$Q_n = \{x = (x_1, \dots, x_n) : |x_j| < N, j = 1, 2, \dots, n\}$$

and F and F^{-1} are the direct and inverse Fourier transforms).

The following theorem holds true.

The function $g \in L_p(\mathbb{R}^n)$ is the trace on $\mathbb{R}^n$ of a function $f \in L^{(r,1)}_{p,\varphi,\psi}(\overset{+}{\mathbb{R}}{}^{n+1})$ iff the sums

$$\sigma_1(g) \overset{\text{def}}{=} \sum_{k \in \mathscr{M}_1} \left(\xi^{-1/p'}(y_k) \|A_k g(x)\|_{L_p(\mathbb{R}^n)}\right)^p,$$

$$\sigma_2(g) \overset{\text{def}}{=} \sum_{k \in \mathscr{M}_2} \left(\eta^{1/p}(y_k) \sum_{j=1}^{n} \left\| \frac{\partial^r}{\partial x_j^r} A_k g(x) \right\|_{L_p(\mathbb{R}^n)}\right)^p$$

are finite.

As an example of a corollary to this theorem we mention that a necessary and sufficient condition for the trace of a function $f \in L^{(r,1)}_{p,\varphi,\psi}$ be continuous on $\mathbb{R}^n$ is that the following condition holds true

$$\int_0^1 \left(\xi^{1/p'}(y) \eta^{1/p}(y)\right)^{-np'/pr} d\xi(y) < +\infty.$$

The above theorem simplifies considerably in additional assumptions, which imply that one of the sets $\mathscr{M}_1$ and $\mathscr{M}_2$ is empty. Thus, if for example the function $\xi^\alpha(y)\eta^{-1}(y)$ is increasing for some $\alpha > 0$ then the space of traces on $\mathbb{R}^n$ of functions in $L^{(r,1)}_{p,\varphi,\psi}(\overset{+}{\mathbb{R}}{}^{n+1})$ reduces to the generalized Nikol'skiĭ-Besov space with the norm

$$\left(\sum_{k=1}^{\infty} \left[2^k \|A_k g(x)\|_{L_p(\mathbb{R}^n)}\right]^p\right)^{1/p}, \tag{7.27}$$

where the sequence $\{y_k\}$, defining the operator A_k (cf. (7.25)), is obtained as the roots of the equation $\xi(y_k) = 2^{-kp'}$. It is known that the norm (7.27) is equivalent to the norm

$$\|g\|_{L_p(\mathbb{R}^n)} + \left(\sum_{k=1}^{\infty} [2^k \omega_p^{(r)}(g; N_k^{-1})]^p\right)^{1/p},$$

where $\omega_p^{(r)}(g; \delta)$ is the modulus of continuity of order r of g in the L_p-metric (G.A. Kalyabin [1980]).

Many properties of anisotropic weighted spaces were obtained in the work of D.F. Kalinichenko 1964-67 and Yu.S. Nikol'skiĭ 1965-1975. Conditions were found under which the traces of functions in a weighted class are trigonometric or algebraic polynomials (L.D. Kudryavtsev [1975a]). The conditions for the traces on a hyperplane to be algebraic polynomials were obtained without the assumption that the weight function is in some sense estimated by a function in one variable. Furthermore, the question of stabilization of functions in weighted spaces to algebraic and trigonometric polynomials when the argument tends to some hyperplane or to infinity was studied (L.D. Kudryavtsev [1975a,b], V.N. Sedov [1972], S.L. Sobolev [1963], [1974], S.V. Uspenskiĭ [1961] etc.).

We say that a *function* $f : \mathbb{R} \to \mathbb{R}$ *stabilizes of order* m *to a function* $g : \mathbb{R} \to \mathbb{R}$, for example, when the argument tends to $+\infty$ if

$$\lim_{x \to +\infty} [f(x) - g(x)]^{(k)} = 0, \quad k = 0, 1, \ldots, m.$$

Necessary and sufficient conditions were found for the stabilization of order $2n$ for $x \to +\infty$ of a function to a trigonometric polynomial $T_n(x)$ of degree n, formulae were obtained for the computation of the coefficients of $T_n(x)$ and the rate of the stabilization was determined. In the study of stabilization of functions to algebraic polynomials one introduced a representation of an n times differentiable function f on an interval of the form (L.D. Kudryavtsev [1975a])

$$f(x) = \sum_{k=0}^{n} A_k(x) x^k,$$

where

$$A_k(x) = \frac{1}{k!} Q_{n-k}\left(x \frac{d^k}{dx^k}\right) f(x), \quad Q_s(x) = \sum_{\nu=0}^{s} \frac{(-1)^\nu}{\nu!} x^\nu, \quad s = 0, 1, 2, \ldots.$$

The question of compactness of imbeddings of weighted function spaces has been studied in Triebel [1973], in P.I. Lizorkin and M. Otelbaev [1980] and in other places. Poulsen in 1962, V.R. Portnov in 1965, O.V. Besov, Kufner in 1966, L.D. Kudryavtsev in 1980, P.I. Lizorkin in 1978, O.V. Besov in 1983, as well as other authors solved the problem of the density of infinitely differentiable functions in weighted spaces in various assumptions on the weight functions (cf. the monograph Kufner [1980]).

The theory of multipliers and singular integral operators in weighted spaces was studied in papers by V.M. Kokilashvili [1978a,b], [1979], [1983], [1985a,b,c]. He established necessary and sufficient conditions for the weight functions under which operators defined by fractional maximal functions or Riesz potentials are bounded in Lebesgue spaces. He studied both the case of scalar as well as vector valued functions for isotropic and anisotropic spaces. He gave also a complete description of the class of weight functions for which the maximal function defines a bounded operator in the corresponding weighted Lorentz spaces (for the definition of Lorentz spaces see § 2 of Chap. 8). He further solved an analogous problem for operators of potential type.

A rather complete interpolation theory for weighted spaces is developed in the work of Triebel (see [1978]).

Estimates of the norm of intermediate derivatives in weighted spaces and applications of these estimates to the solution of variational problems were obtained in L.D. Kudryavtsev [1983], [1986]. Consider the space $L^{(r)}_{p,\varphi} = L^{(r)}_{p,\varphi}(0, T)$, $0 < T \leq +\infty$, of functions u which on each interval contained in $(0, T)$ have absolutely continuous derivatives of order $r-1$, $r \in \mathbb{N}$, and such that $\|\varphi u\|_{L_p(0,T)} < +\infty$ (φ is a given nonnegative measurable function on $(0, T)$). If $\varphi^{-1} \in L_q((0, T), \text{loc})$, $1/p + 1/q = 1$, then we can in $L^{(r)}_{p,\varphi}$ introduce a norm: the functional

$$\|u\|_p^{(1)} = \sum_{m=0}^{r-1} |u^{(m)}(t_0)| + \|\varphi u^{(r)}\|_{L_p(0,T)} \tag{7.28}$$

yields for different choices of the point $t_0 \in (0, T)$ norms equivalent to each other on this space.

Let $0 < T < +\infty$ and assume that one of the following three conditions are fulfilled:

$$1) \quad t^{r-m_0-1}(T-t)^{r-m_1-1}\varphi^{-1}(t) \in L_q(0, T), \quad m_0, m_1 \in \{0, 1, \dots, r-1\}, \tag{7.29}$$

$$2) \quad \varphi^{-1} \in L_q(0, T/2), \tag{7.30}$$

$$3) \quad \varphi^{-1} \in L_q(T/2, T). \tag{7.31}$$

Consider a system of indices

$$i_1, i_2, \dots, i_k, j_1, j_2, \dots, j_l, \tag{7.32}$$

where

$$0 \leq i_1 < i_2 < \dots < i_k \leq r-1, \quad 0 \leq j_1 < j_2 < \dots < j_l \leq r-1.$$

If $k + l = r$ we denote by $\{\bar{i}_\nu\}_{\nu=1}^{\nu=l}$ and $\{\bar{j}_\mu\}_{\mu=1}^{\mu=k}$ the complements in the set $\{0, 1, \dots, r-1\}$ of $\{i_m\}_{\mu=1}^{\mu=k}$ and $\{j_\nu\}_{\nu=1}^{\nu=l}$ respectively. We can always assume that $\bar{i}_1 < \bar{i}_2 < \dots < \bar{i}_l$, $\bar{j}_1 < \bar{j}_2 < \dots < \bar{j}_k$.

If $k + l = r$ and $j_1 \leq \bar{i}_1, j_2 \leq \bar{i}_2, \dots, j_l \leq \bar{i}_l$ then the system of indices (7.32) is called a system satisfying the *Pólya condition* (Pólya [1931]). This condition was obtained by Pólya as a necessary and sufficient condition for the existence and uniqueness of a polynomial $P(t)$ of degree at most $r-1$ taking on the endpoints of the interval $[0, 1]$ given values $P^{(i_\mu)}(0)$, $\mu = 1, 2, \dots, k$, $P^{(j_\nu)}(1)$, $\nu = 1, 2, \dots, l$, $k + l = r$.

The system of indices (7.32) is said to be *complete* if it contains a subsystem satisfying the Pólya condition.

A system of indices (7.32) is called *admissible* if in the assumption of (7.29) we have $i_\mu \leq m_0$, $\mu = 1, 2, \dots, k$, $j_\nu \leq m_1$, $\nu = 1, 2, \dots, l$; if only (7.30) holds we require $\{j_\nu\}_{\nu=1}^{\nu=l} = \emptyset$ and if (7.31) holds $\{i_m\}_{\mu=1}^{\mu=k} = \emptyset$.

If $u \in L^{(r)}_{p,\varphi}(0,T)$, $\varphi^{-1} \in L_q((0,T), \mathrm{loc})$, $0 < T < +\infty$, and complete, then 1) the functional

$$\|u\|^{(2)}_p = \sum_{\mu=1}^{k} |u^{(i_\mu)}(0)| + \sum_{\nu=1}^{l} |u^{(j_\nu)}(T)| + \|\varphi u^{(r)}\|_{L_p(0,T)}$$

is a norm in $L^{(r)}_{p,\varphi}(0,T)$ equivalent to the norm (7.28); 2) for each $\varepsilon > 0$ there exists a constant $c_\varepsilon > 0$ such that the inequality

$$|u(t)| \leq c_\varepsilon\Big(\sum_{\mu=1}^{k} |u^{(i_\mu)}(0)| + \sum_{\nu=1}^{l} |u^{(j_\nu)}(T)| + \|\varphi u^{(r)}\|_{L_p(0,T)}\Big)$$

is fulfilled for all $t \in [\varepsilon, T - \varepsilon]$ and, if (7.29) holds, the inequality

$$|u^{(s)}(t)| \leq c\Big(\sum_{\mu=1}^{k} |u^{(i_\mu)}(0)| + \sum_{\nu=1}^{l} |u^{(j_\nu)}(T)| + \|\varphi u^{(r)}\|_{L_p(0,T)}\Big), \qquad (7.33)$$

$$s = 0, 1, \ldots, m = \min\{m_0, m_1\}.$$

Analogous results hold also in the case of the infinite interval $(0, +\infty)$, provived one imposes on φ conditions such that every function $u \in L^{(r)}_{p,\varphi}(0,+\infty)$ stabilizes to a polynomial $P(t) = \sum_{s=0}^{r-1} a_s t^s$ of degree at most $r-1$ as the argument tends to infinity and its derivatives stabilize to derivatives of this polynomial. In this case, for example, the analogue of inequality (7.33) takes the form

$$|u^{(s)}(t)| \leq c\Big(\sum_{\mu=1}^{k} |u^{(i_\mu)}(0)| + \sum_{\nu=1}^{l} |a_{j_\nu}| + \|\varphi u^{(r)}\|_{L_p(0,+\infty)}\Big) \cdot (1+t)^{r-s-1}. \qquad (7.34)$$

With the aid of the inequalities (7.33) and (7.34) in corresponding weighted spaces one can prove the existence and uniqueness of the solution of the variational problem for the minimum of a quadratic functional ($p = 2$) when for $t = 0$ and, in the case $T < +\infty$, likewise for $t = T$ the values of the derivatives $u^{(i_\mu)}(0)$, $\mu = 1, 2, \ldots, k$, $u^{(j_\nu)}(T)$, $\nu = 1, 2, \ldots, l$, corresponding to an admissible and complete system (7.32), and for $T = +\infty$ the corresponding family $\{a_j\}_{\nu=1}^{\nu=l}$ of coefficients of the polynomial $P(t)$ are prescribed.

§ 5. The Weighted Spaces of Kipriyanov

In 1967 I.A. Kipriyanov considered a class constructed on the basis of the Fourier-Bessel transform.

Let $\mathbb{R}^{n+1}$ be a Euclidean space with points $x = (x^{(n)}, y)$, $x^{(n)} = (x_1, \dots, x_n) \in \mathbb{R}^n$, $y \in \mathbb{R}$, and set $\overset{+}{\mathbb{R}}{}^{n+1} = \{x : y > 0\}$. For each $\nu \geq -1/2$ we define the weighted space $L_{2,\nu}(\overset{+}{\mathbb{R}}{}^{n+1})$ as the set of Lebesgue measurable functions $f : \overset{+}{\mathbb{R}}{}^{n+1} \to \mathbb{R}$ for which the norm

$$\|f\|_{L_{2,\nu}(\overset{+}{\mathbb{R}}{}^{n+1})} = \left\{ \int_{\overset{+}{\mathbb{R}}{}^{n+1}} |f(x)|^2 y^{2\nu+1} dx \right\}^{1/2}$$

is finite.

If $J_\nu(t)$ is the Bessel function of the first kind and

$$j_\nu(t) = \frac{2^\nu \Gamma(\nu+1)}{t^\nu} J_\nu(t), \tag{7.35}$$

then the *Fourier-Bessel transform* F_ν is defined by the formula

$$(F_\nu f)(\xi) = c_{n,\nu} \int_{\overset{+}{\mathbb{R}}{}^{n+1}} e^{-i(\xi^{(n)}, x^{(n)})} J_\nu(y\eta) f(x) y^{2\nu+1} dx, \tag{7.36}$$

where

$$c_{n,\nu} = \frac{1}{2^{\nu+n/2}\pi^{n/2}\Gamma(\nu+1)}, \quad \xi = (\xi^{(n)}, \eta).$$

That F_ν is unitary on $L_{2,\nu}(\overset{+}{\mathbb{R}}{}^{n+1})$ is a consequence of a formula of Parseval type:

$$\|f\|_{L_{2,\nu}(\overset{+}{\mathbb{R}}{}^{n+1})} = \|F_\nu f\|_{L_{2,\nu}(\overset{+}{\mathbb{R}}{}^{n+1})}. \tag{7.37}$$

As in the case of the Fourier transform, the Fourier-Bessel transform reduces a certain operation of differentiation to multiplication by the corresponding arguments. Namely, if

$$B_y = \frac{\partial^2}{\partial y^2} + \frac{2\nu+1}{y}\frac{\partial}{\partial y},$$

then for each function f, even in the variable y, holds the formula

$$(F_\nu \frac{\partial}{\partial x_j} B_y f)(\xi) = -i\xi_j \eta^2 (F_\nu f)(\xi). \tag{7.38}$$

Denote by $W_{2,\nu}^{(s)}(\overset{+}{\mathbb{R}}{}^{n+1})$, $s \geq 0$, the space of measurable functions f on $\overset{+}{\mathbb{R}}{}^{n+1}$ such that the norm

$$\|f\|_{W_{2,\nu}^{(s)}(\overset{+}{\mathbb{R}}{}^{n+1})} \overset{\text{def}}{=} \|(1+|\xi|^2)^{s/2} F_\nu f\|_{L_{2,\nu}(\overset{+}{\mathbb{R}}{}^{n+1})} \tag{7.39}$$

In his paper [1967] I.A. Kipriyanov proved the equivalence of the norm (7.31) with corresponding norms written in terms of the Fourier-Bessel transform; for example, for every $s \geq 0$ one of these norms takes the form

$$\left\{ \int_{\overset{+}{\mathbb{R}}^{n+1}} \left[|f(x)|^2 + |B_y^{s/2} f(x)|^2 + \sum_{j=1}^{n} \left| \frac{\partial^s}{\partial x_j^s} f(x) \right|^2 \right] y^{2\nu+1} dx \right\}^{1/2} .$$

The spaces $W_{2,\nu}^{(s)}(\overset{+}{\mathbb{R}}^{n+1})$ allow us to develop a rather complete theory for the solvability of an elliptic *equation of Euler-Poisson-Darboux type*

$$\sum_{j=1}^{n} \frac{\partial^2 u}{\partial x_j^2} + B_y u - \lambda^2 u = f(x), \quad x \in \overset{+}{\mathbb{R}}^{n+1} \tag{7.40}$$

with the boundary condition

$$\left. \frac{\partial u}{\partial y} \right|_{y=0} = 0. \tag{7.41}$$

This theory served as the starting point for obtaining imbedding theorems for traces of functions in the spaces $W_{2,\nu}^{(s)}$ and the development of a spectral theory for singular elliptic equations.

The spaces $W_{2,\nu}^{(s)}$ and their generalizations (anisotropic spaces of similar type, spaces with a complex parameter ν) have further found applications in the theory of pseudo-differential operators, in the study of the poles of kernels of fractional powers of singular elliptic operators, the development of the theory of singular parabolic systems of equations, as well as in many other questions.

Chapter 8

Interpolation Theory of Nikol'skiĭ-Besov and Lizorkin-Triebel Spaces

§ 1. Interpolation Spaces

The *method of interpolation of linear operators* is widely used in the theory of imbedding of function spaces. The essence of this method consists of concluding from the boundedness of a linear operator on certain in a sense "extremal" Banach spaces its boundedness also on intermediate Banach spaces.

Two Banach spaces X and Y form a *Banach pair* if they are algebraically and topologically imbedded in a Hausdorff topological vector space $\mathscr{A}$. This means that X and Y are linear subspaces of $\mathscr{A}$ and that the topology induced on X and Y by the topology of $\mathscr{A}$ is weaker than the original topology on these spaces.

A Banach space Z is called an *intermediate* for the pair X, Y if one has the imbedding

$$X \cap Y \hookrightarrow Z \hookrightarrow X + Y,$$

where the norm in the intersection $X \cap Y$ is defined by the formula

$$\|x\|_{X\cap Y} \stackrel{\text{def}}{=} \max\{\|x\|_X, \|x\|_Y\},$$

and in the sum $X + Y$ by

$$\|u\|_{X+Y} \stackrel{\text{def}}{=} \inf[\|x\|_X + \|y\|_Y], \quad u \in X + Y,$$

where the infimum is taken over all elements $x \in X$, $y \in Y$ whose sum equals u.

The space $X \cap Y$, normed in this way, is called the *intersection* of X, Y and $X + Y$ the *sum* of X, Y.

For example, if $1 \le p_1 < p_2 \le +\infty$ then any space $Z = L_p(0,1)$, $p_1 < p < p_2$, is an intermediate space for the Banach pair $X = L_{p_1}(0,1)$, $Y = L_{p_2}(0,1)$. Here $Y \subset X$ and as the containing topological vector space (see the definition of Banach pairs) we may take $\mathscr{A} = X = L_{p_1}(0,1)$.

A linear operator $T : X_1 + Y_1 \to X_2 + Y_2$ is said to be a *bounded operator from the Banach pair* X_1, Y_1 *into the Banach pair* X_2, Y_2 if its restriction to X_1 (respectively to Y_1) is a bounded operator from X_1 into Y_1 (respectively from X_2 into Y_2).

Let X_1, Y_1 and X_2, Y_2 be two Banach spaces and Z_1 and Z_2 intermediate spaces for the pairs X_1, Y_1 respectively X_2, Y_2.

The triple ("troïka") (X_1, Y_1, Z_1) is said to have the *interpolation property* with respect to the triple (X_2, Y_2, Z_2) if for each bounded linear operator T from the pair X_1, Y_1 into the pair X_2, Y_2 is a bounded operator from Z_1 into Z_2.

If $X_1 = X_2 = X$, $Y_1 = Y_2 = Y$, $Z_1 = Z_2 = Z$ we say that Z has the *interpolation property* (and that Z is an *interpolation space*) with respect to X, Y.

The historically first interpolation theorem was obtained by M. Riesz in 1926. In 1939 Thorin proved a generalization which is now called the *Riesz-Thorin convexity theorem.*

Let $(X, \mathscr{F}_x, \mu_x)$ and $(Y, \mathscr{F}_y, \mu_y)$ be two spaces equipped with the measures μ_x and μ_y respectively ($\mathscr{F}_x$ and $\mathscr{F}_y$ being the corresponding σ-algebras of measurable sets) and let $L_p(X)$ and $L_p(Y)$, $1 \le p \le +\infty$, be the Lebesgue spaces corresponding to these measures.

The triple of Banach spaces $L_{p_1}(X)$, $L_{p_2}(X)$, $L_{p_\theta}(X)$ has the interpolation property with respect to the triple $L_{q_1}(Y)$, $L_{q_2}(Y)$, $L_{q_\theta}(Y)$ provided

$$1/p_\theta = (1-\theta)/p_1 + \theta/p_2, \quad 1/q_\theta = (1-\theta)/q_1 + \theta/q_2, \quad 0 \le \theta \le 1.$$

Moreover, if T is a linear operator from the pair $L_{p_1}(X)$, $L_{p_2}(Y)$ into the pair $L_{q_1}(X)$, $L_{q_2}(Y)$, T_1 being the restriction to $L_{p_1}(X)$, T_2 the restriction to $L_{p_2}(X)$, and T_θ the restriction to $L_{p_\theta}(X)$, then

$$\|T_\theta\| \le \|T_1\|^{1-\theta}\|T_2\|^\theta.$$

The function $\log\|T_\theta\|$ is convex in θ, which explains the name of the theorem.

If $X = Y = \mathbb{R}^n$, $\mu_x = \mu_y =$ Lebesgue measure, $T = F =$ the Fourier transform then, as is well-known, F is a bounded linear map from $L_1(\mathbb{R}^n)$ into $L_\infty(\mathbb{R}^n)$ and from $L_2(\mathbb{R}^n)$ into itself. Therefore, taking $p_1 = q_1 = 2$, $p_2 = 1$, $q_2 = \infty$ we obtain the following result

Hausdorff-Young Theorem. *If $1 \le p \le 2$, $1/p + 1/p' = 1$, then the Fourier transform is continuous operator from $L_p(\mathbb{R}^n)$ into $L_{p'}(\mathbb{R}^n)$.*

In the early 60's the theory of interpolation of Banach spaces arose as an independent branch of functional analysis thanks to the work of Lions, Gagliardo, Grisvard, Stein and Weiss, S.G. Kreĭn and others. In his monograph [1978b] Triebel gave the first comprehensive presentation of the theory of interpolation and developed the interpolation theory of Nikol'skiĭ-Besov and Lizorkin-Triebel spaces. With the aid of interpolation techniques one can describe also many other spaces; see the monographs Bergh-Löfström [1976], Butzer-Behrens [1967], S.G. Kreĭn-Yu.I. Petunin-E.M. Semenov [1978], Peetre [1976], Stein [1970], Stein and Weiss [1971], Triebel [1977], [1978a], [1978b], [1983].

As an example of one of the methods for obtaining interpolation spaces let us mention the so-called *K-method* developed by Peetre in 1963-68.

Let X, Y be an interpolation pair and set

$$K(t,u) = K(t,u;X,Y) \stackrel{\text{def}}{=} \inf_{u=x+y}(\|x\|_X + \|y\|_Y),$$

$$u \in X+Y, \quad 0 < t < +\infty.$$

For $0 < \theta < 1$, $1 \le q < +\infty$ let $(X,Y)_{\theta,q}$ be the set of all $u \in X+Y$ such that

$$\|u\|_{(X,Y)_{\theta,q}} \stackrel{\text{def}}{=} \left(\int_0^{+\infty} [t^{-\theta}K(t,u)]^q dt/t\right)^{1/q} < +\infty,$$

and for $0 < \theta < 1$, $q = +\infty$ let $(X,Y)_{\theta,\infty}$ be the set of all $u \in X+Y$ such that

$$\|u\|_{(X,Y)_{\theta,\infty}} \stackrel{\text{def}}{=} \sup_{t>0} t^{-\theta}K(t,u) < +\infty.$$

The space $(X, Y)_{\theta,q}$, $0 < \theta < 1$, $1 \le q \le +\infty$, is an interpolation space between X and Y and if T is a linear operator mapping X, Y into itself, $T_{\theta,q}$ denoting its restriction to $(X, Y)_{\theta,q}$, then

$$\| T_{\theta,q} \| \le \| T_X \|^{1-\theta} \| T_Y \|^{\theta}.$$

§ 2. Lorentz and Marcinkiewicz Spaces

Let $\{X, \mathcal{F}, \nu\}$ be a space equipped with a σ-finite measure.
For each ν-measurable function $f : X \to \mathbb{R}$

$$\lambda_f(\sigma) = \nu\{x \in X : |f(x)| > \sigma\}, \quad 0 \le \sigma < +\infty$$

is called the *distribution function* of f.
Set

$$f^*(t) = \inf\{\sigma : \lambda_f(\sigma) < t\}.$$

The function $f^*(t)$ of one variable t, called the *(decreasing) rearrangement* of f, is decreasing and left continuous on the halfaxis $\overset{+}{\mathbb{R}}$ and its distribution function coincides with the distribution function of f:

$$\lambda_f(\sigma) = \lambda_{f^*}(\sigma).$$

Functions with the same distribution function are termed *equimeasurable*.
If $1 \le p < +\infty$, $1 \le q < +\infty$, we denote by $L_{p,q}(X)$ the set of all functions $f \in L_1(X) + L_\infty(X)$ such that

$$\|f\|_{L_{p,q}(X)} \overset{\text{def}}{=} \left[\int_0^\infty (t^{1/p} f^*(t))^q dt/t \right]^{1/q} < +\infty \tag{8.1}$$

If $1 \le p \le +\infty, q = +\infty$ we denote by $L_{p,\infty}(X)$ the set of all functions $f \in L_1(X) + L_\infty(X)$ such that

$$\|f\|_{L_{p,\infty}(X)} \overset{\text{def}}{=} \sup_{t>0} t^{1/p} f^*(t) < +\infty \tag{8.2}$$

The spaces

$$M_p \overset{\text{def}}{=} L_{p,\infty} \tag{8.3}$$

are called the *Marcinkiewicz spaces*, while the spaces $L_{p,q}$, $1 < p < +\infty$, $1 \le q \le +\infty$, are called the *Lorentz spaces*.
Note that the functional $\|f\|_{L_{p,q}(X)}$ is not a norm, only a quasinorm (except for $q = 1$), as it does satisfy the triangle inequality. However, for $p > 1$ it is possible to define a norm in $L_{p,q}$ equivalent to this functional.

In the literature one also encounters *Lorentz spaces* $\Lambda_{\psi,q}(X)$ and the *Marcinkiewicz space* $M_\psi(X)$. Let us define them.

Let ψ be a nondecreasing continuous concave function on the halfaxis $(0,+\infty)$.

The Lorentz space $\Lambda_{\psi,q}(X)$ is defined to be the space of all measurable functions f such that the norm

$$\|f\|_{\Lambda_{\psi,q}} \stackrel{\text{def}}{=} \left(\int_0^{+\infty} (f^*(t))^q d\psi(t)\right)^{1/q}, \quad 1 \le q < +\infty,$$

$$\|f\|_{\Lambda_{\psi,+\infty}} \stackrel{\text{def}}{=} \int_0^{+\infty} f^*(t) d\psi(t), \quad q = +\infty, \tag{8.4}$$

is finite, while the Marcinkiewicz space $M_\psi(X)$ consists of all measurable functions with the finite norm

$$\|f\|_{M_\psi} \stackrel{\text{def}}{=} \sup_{h>0} \left\{\frac{1}{\psi(h)} \int_0^h f^*(t)dt\right\}. \tag{8.5}$$

The spaces $\Lambda_{\psi,q}$, $1 \le q \le +\infty$, and M_ψ are Banach spaces. The space $\Lambda_{\psi,+\infty}$ is often written Λ_ψ. It is easy to see that $\Lambda_{t^{p/q},q} = L_{p,q}$.

If $\psi(0) \ne 0$ and $\lim_{t\to+\infty} \psi(t) < +\infty$ then $\Lambda_\psi = L_\infty$.

In the general case one has the inclusions

$$L_1 \cap L_\infty \subset \Lambda_\psi \subset L_1 + L_\infty.$$

If $\psi(t) = t^{1/p}$ and $q = 1$ then

$$\Lambda_{t^{1/p},1} = L_{p,1}. \tag{8.6}$$

This space will be written Λ_p.

Note that the condition $1 \le q_1 < q_2 \le +\infty$ is equivalent to the condition

$$L_{p,q_1} \hookrightarrow L_{p,q_2}, \tag{8.7}$$

and that

$$L_{p,p} = L_p. \tag{8.8}$$

Thus, the smallest among the spaces $L_{p,q}$ for fixed $p \in (1,+\infty)$ is the Lorentz space $\Lambda_p = L_{p,1}$ and the largest the Marcinkiewicz space $M_p = L_{p,+\infty}$, and

$$\Lambda_p \hookrightarrow L_p \hookrightarrow M_p. \tag{8.9}$$

The following *interpolation theorem* holds for the Lorentz spaces $L_{p,q}$.

If $0 < \theta < 1$, $1 < p_1 < p_2 < +\infty$, $1 \le q_1 < +\infty$, $1 \le q_2 < +\infty$, $1 \le q < +\infty$ and $1/p = (1-\theta)/p_1 + \theta/p_2$, then

$$(L_{p_1,q_1}(X), L_{p_2,q_2}(X))_{\theta,q} = L_{p,q}(X),$$

and if $\theta = 1/p', 1/p + 1/p' = 1$ then

$$(L_1(X), L_\infty(X))_{\theta,q} = L_{p,q}(X).$$

The second formula means that the Lorentz spaces $L_{p,q}$ can be obtained with the aid of the K-interpolation method from the Lebesgue spaces $L_1(X)$ and $L_\infty(X)$. The problem of describing all interpolation spaces between L_1 and L_∞ has been studied by Calderón, B. S. Mityagin, A. A. Sedaev and E. M. Semenov, Sparr and others.

We remark that the Marcinkiewicz spaces (and, consequently, generally Lorentz spaces) have applications not only to the solution of internal problems in the theory of functions but also in the theory of partial differential equations.

In the case of constant or sufficiently regular coefficients for the linear operator

$$\sum_{|\nu|\le l} (-1)^{|\nu|} a_\nu(x) u^{(\nu)}(x) = f(x)$$

where $f \in L_p$, $1 < p < +\infty$, one can obtain estimates of solutions in the L_p-metric and further existence and uniqueness theorems for the solutions of corresponding boundary problems. If however $p = 1$ then, in order to obtain the corresponding results, it is convenient to invoke the metric of the Marcinkiewicz space (cf. Bénilan-Brezis-Crandall [1975], Gallouet-Morel [1983]). Consider the equation

$$-\Delta u + \gamma(x, u) = f(x) \tag{8.10}$$

in the whole space $\mathbb{R}^n$, $n \ge 3$, where we assume that the function $\gamma(x, \xi)$: $\mathbb{R}^n \times \mathbb{R} \to \mathbb{R}$ is measurable, and decreasing in ξ, $\gamma(x, 0) = 0$, $\gamma(\cdot, x) \in L_1(\mathbb{R}^n)$, and that the map $x \mapsto \gamma(x, \xi)$ belongs to $L_1(\mathbb{R}^n, \mathrm{loc})$ for fixed ξ. Then, for each function $f \in L_1(\mathbb{R}^n)$ the equation has a unique solution in the Marcinkiewicz space $M_{t^{2/n}}$.

This result for $\gamma(x, \xi) = a(x)\xi$, i.e. in the case of the linear operator,

$$-\Delta u + au = f$$

shows that it is expedient to use the metric of the Marcinkiewicz space also in the context of linear problems for differential equations.

§ 3. Interpolation of the Nikol'skiĭ-Besov and Lizorkin-Triebel Spaces

The Nikol'skiĭ-Besov spaces $B^{(r)}_{p,q}(\mathbb{R}^n)$ and the Lizorkin-Triebel spaces $L^{(r)}_{p,q}(\mathbb{R}^n)$ for intermediate smoothness indices can be obtained from extremal spaces of the same type using the K-method of Peetre (Triebel [1978b]).

1. If $-\infty < r_0 < +\infty$, $-\infty < r_1 < +\infty$, $r_0 \neq r_1$, $1 < p < +\infty$, $1 \leq q_0 \leq +\infty$, $1 \leq q_1 \leq +\infty$, $1 \leq q \leq +\infty$ and $0 < \theta < 1$ then

$$(B_{p,q_0}^{(r_0)}, B_{p,q_1}^{(r_1)})_{\theta,r} = B_{p,q}^{(r)},$$

where

$$r = (1-\theta)r_0 + \theta r_1.$$

2. If $-\infty < r_0 < +\infty$, $-\infty < r_1 < +\infty$, $1 < p_0 < +\infty$, $1 < p_1 < +\infty$, $1 < q_0 < +\infty$, $1 < q_1 < +\infty$, $0 < \theta < 1$ and

$$r = (1-\theta)r_0 + \theta r_1, \quad 1/p = (1-\theta)/p_0 + \theta/p_1,$$

then for $r_0 \neq r_1$:

$$(L_{p_0,q_0}^{(r_0)}, L_{p_1,q_1}^{(r_1)})_{\theta,p} = B_{p,p}^{(r)} = L_{p,p}^{(r)},$$

and for $r_0 = r_1 = r$ and $q_0 \neq q_1$:

$$(L_{p_0,q_0}^{(r)}, L_{p_1,q_1}^{(r)})_{\theta,p} = B_{p,p}^{(r)}.$$

As we have already remarked, Triebel [1978b] has with the aid of interpolation theorems obtained direct and converse imbedding theorems for the spaces $B_{p,q}^{(r)}(\mathbb{R}^n)$ and $L_{p,q}^{(r)}(\mathbb{R}^n)$ and this for all (positive and negative) values of r. Moreover, he showed that the spaces $B_{p,q}^{(r)}$ and $L_{p,q}^{(r)}$, defined for $r < 0$ with the aid of the norms (5.20) and (5.25) are dual to the corresponding spaces for $r > 0$. Namely, if $-\infty < r < +\infty$, $1/p + 1/p' = 1/q + 1/q' = 1$, $1 < p < +\infty$, $1 \leq q < +\infty$, then

$$B_{p,q}^{(r)}(\mathbb{R}^n)^* = B_{p',q'}^{(-r)}(\mathbb{R}^n), \quad \overset{o}{B}{}_{p,\infty}^{(r)}(\mathbb{R}^n)^* = B_{p',1}^{(-r)}(\mathbb{R}^n).$$

where $\overset{o}{B}{}_{p,\infty}^{(r)} = \overset{o}{B}{}_{p,\infty}^{(r)}(\mathbb{R}^n)$ is the closure of $C^\infty(\mathbb{R}^n)$ in the norm of the space $B_{p,\infty}^{(r)}(\mathbb{R}^n)$.

Analogously, for $1 < p < +\infty$, $1 < q < +\infty$

$$L_{p,q}^{(r)}(\mathbb{R}^n)^* = L_{p,q}^{(-r)}(\mathbb{R}^n).$$

Concluding this section let us remark that Triebel further has developed an interpolation theory when the space $B_{p,q}^{(r)}$ consists of functions defined on a domain $G \subset \mathbb{R}^n$ (and not by necessity in the whole of $\mathbb{R}^n$). In § 3 of Chap. 10 we will also consider interpolation questions for the Calderón spaces $\Lambda(X, E)$.

Chapter 9

Orlicz and Orlicz-Sobolev Spaces

§ 1. Orlicz Spaces

All the function spaces which we have considered have consisted of functions which were at least summable of power p. A natural generalization of this condition is local summability of the function $\Phi(|f(x)|)$, where Φ is a given function. On the basis of such considerations Orlicz and Birnbaum in 1931 introduced in mathematics spaces which were called Orlicz spaces. The need to introduce more general function spaces than the ones previously considered was dictated by internal (i.e. arising in the theory of functions itself) as well as external causes. The latter were connected, in particular, with the demands of the theory of partial differential equations. As an example let us remark that if G is a bounded domain in $\mathbb{R}^n$ satisfying a cone condition, $l \in \mathbb{N}$, $lp = n$ and $p > 1$, then according to Sobolev's imbedding theorem

$$W_p^{(l)}(G) \hookrightarrow L_q(G), \tag{9.1}$$

provided $p \leq q < +\infty$. Simple examples reveal, however, that $W_p^{(l)}(G)$ is not contained in $L_\infty(G)$. Thus in the scale $L_q(G)$ there does not exist a smallest space such that the imbedding (9.1) holds true. As we will see below, there exists such a minimal space among Orlicz spaces. Consequently, the introduction of more general spaces than spaces with an L_p-metric gives us a possibility for a finer classification of functions according to their smoothness.

Let us define Orlicz spaces.

Let X be a measurable set in $\mathbb{R}^n$ and Φ a nonnegative function defined on the halfaxis $(0, +\infty)$. Denote by $\tilde{L}_\Phi(X)$ the set of all (complex) Lebesgue measurable functions f such that

$$\varrho(f\,; \Phi) \overset{\text{def}}{=} \int_X \Phi(|f(x)|)dx < +\infty. \tag{9.2}$$

As in the case of Lebesgue spaces, we identify equivalent functions (with respect to Lebesgue measure) and consider them as elements of the set $\tilde{L}_\Phi(X)$. The set $\tilde{L}_\Phi(X)$ is called an *Orlicz class*. The example of the function $\Phi(t) = e^t$, $X = (0, 1)$, reveals that $\tilde{L}_\Phi(X)$ is in general not a vector space, as in this case $f(x) = -\frac{1}{2}\log x \in \tilde{L}_\Phi(X)$ but $2f(x) = -\log x \notin \tilde{L}_\Phi(X)$. For $\tilde{L}_\Phi(X)$ to be a linear space one has to impose an additional restriction on Φ.

A function $\Phi : (0, +\infty) \to \mathbb{R}$ is called *convex* if for any t_1 and t_2, $0 < t_1 < t_2 < +\infty$, and any $\alpha_1 > 0$, $\alpha_2 > 0$ holds

$$\Phi\left(\frac{\alpha_1 t_1 + \alpha_2 t_2}{\alpha_1 + \alpha_2}\right) \leq \frac{\alpha_1 \Phi(t_1) + \alpha_2 \Phi(t_2)}{\alpha_1 + \alpha_2}.$$

This means that each chord of the graph of Φ lies above the part of the graph enclosed between the endpoints of the chord.

Every continuous convex function $\Phi(t)$ is absolutely continuous, satisfies a Lipschitz condition on each finite interval (consequently, it is almost everywhere differentiable), has at each point a right derivative $\Phi'_+(t)$ and a left derivative $\Phi'_-(t)$ with $\Phi'_-(t) \le \Phi'_+(t)$. The right derivative is a nondecreasing function. It follows, in particular, that each convex continuous function, satisfying the condition $\Phi(a) = 0$, admits the representation

$$\Phi(t) = \int_a^t \varphi(s)ds,$$

where $\varphi(t)$ is a continuous nondecreasing function.

A function $\Phi : \mathbb{R} \to \mathbb{R}$ is called a *Young function* (induced by the function $\varphi : \mathbb{R} \to \mathbb{R}$) if it admits the representation

$$\Phi(t) = \int_0^{|t|} \varphi(t)dt,$$

where φ is a positive function defined for $t > 0$, continuous from the right and nondecreasing, satisfying the conditions

$$\varphi(0) = 0, \quad \varphi(+\infty) = +\infty.$$

Every Young function Φ is convex and

$$\lim_{t\to 0} \frac{\Phi(t)}{t} = 0, \quad \lim_{t\to +\infty} \frac{\Phi(t)}{t} = +\infty.$$

If Φ is a Young function then all bounded measurable functions belong to $\tilde{L}_\Phi(X)$, but not all summable functions. However, each summable function on X, $\mu(X) < +\infty$, belongs to some Orlicz class. Thus, the Lebesgue space $L(X)$, $\mu(X) < +\infty$, is the union of all Orlicz classes.

If Φ_1 and Φ_2 are Young functions and the inequality

$$\Phi_1(t) \le \Phi_2(t), \tag{9.3}$$

is fulfilled for all $t \ge 0$ then clearly

$$\tilde{L}_{\Phi_2}(X) \subset \tilde{L}_{\Phi_1}(X). \tag{9.4}$$

Moreover, if $\mu X < +\infty$, for the inclusion (9.4) to hold true it is sufficient that inequality (9.3) holds only for all sufficiently large t.

Let Φ be a Young function induced by the function φ. Set

$$\psi(t) = \sup_{\varphi(s)\le t} s, \quad t \ge 0,$$

$$\Psi(t) = \int_0^t \psi(s)ds. \tag{9.5}$$

The function Ψ is called the conjugate *conjugate function* of Φ. One sees that Ψ too is a Young function and that its conjugate function is Φ. Therefore Φ and Ψ are called *conjugate Young functions.* For them holds an inequality, called *Young's inequality*

$$ab \le \Phi(a) + \Psi(b), \quad 0 < a < +\infty, \quad 0 < b < +\infty, \tag{9.6}$$

The pair of functions $\Phi(t) = t^p/p$ and $\Psi(t) = t^{p'}/p'$ where $1/p + 1/p' = 1$ is an example of conjugate Young functions.

The following theorem is in force.

If $\mu X < +\infty$ and Φ is a Young function then the set $\tilde{L}_\Phi(X)$ is a vector space iff Φ satisfies the condition:

there exist $c > 0$ and $T \ge 0$ such that for all $t \ge T$

$$\Phi(2t) \le c\Phi(t). \tag{9.7}$$

This condition is usually called the Δ_2*-condition.*

If $\mu X = +\infty$ and the Young function Φ satisfies condition (9.7) then $\tilde{L}_\Phi(X)$ likewise is a linear space.

Next define the notion of Orlicz space.

Let X be a compact set in $\mathbb{R}^n$, Φ a Young function, Ψ its conjugate function and f a measurable function on X. Then the functional

$$\|f\|_\Phi = \sup_g \int_X |f(x)g(x)|dx, \tag{9.8}$$

where the supremum runs over all functions $g \in \tilde{L}_\Psi$ such that (see (9.2)) $\varrho(g;\Psi) \le 1$, is a norm, called the Orlicz norm of f. The linear space of all measurable functions on X with finite norm (9.8) is called an *Orlicz space* and is written $L_\Phi(X)$.

From Young's inequality (9.6) it is clear that for $\varrho(g;\Psi) \le 1$ holds

$$\int |f(x)g(x)|dx \le \varrho(f;\Phi) + \varrho(g;\Psi) \le \varrho(f;\Phi) + 1.$$

Therefore

$$\tilde{L}_\Phi(X) \subset L_\Phi(X).$$

Furthermore, $L_\Phi(X)$ is the linear hull of $\tilde{L}_\Phi(X)$. It turns out that $L_\Phi(X)$ is a Banach space. If the Young function Φ satisfies condition (9.7) then the Orlicz class coincides with the Orlicz space.

The Orlicz norm possesses the *monotonicity* property: if $|f_1(x)| \le |f_2(x)|$ almost everywhere on X then

$$\|f_1\|_\Phi \le \|f_2\|_\Phi.$$

If $\Phi(t) = t^p$ then the Orlicz space $L_\Phi(X)$ coincides with the Lebesgue space $L_p(X)$.

For Orlicz spaces one has an *analogue of Hölder's inequality*: if Φ and Ψ are conjugate Young functions then for each pair of functions $f \in L_\Phi(X)$ and $g \in L_\Psi(X)$ holds

$$\left| \int_X f(x)g(x)dx \right| \leq \|f\|_\Phi \|g\|_\Psi .$$

If for two Young functions Φ_1 and Φ_2 there exist constants $c_1 > 0$ and $\varepsilon > 0$ such that for all t we have

$$\Phi_1(c_1 t) \leq \Phi_2(t) \leq \Phi_1(c_2 t), \tag{9.9}$$

then Φ_1 and Φ_2 are said to be *equivalent*. The Orlicz spaces $L_{\Phi_1}(X)$ and $L_{\Phi_2}(X)$ coincide iff Φ_1 and Φ_2 are equivalent.

Along with convergence in norm in an Orlicz space L_Φ one considers also *convergence in the Φ-mean sense*: a sequence $\{f_n\}$ is said to converge to a function f in the Φ-Orlicz sense if (cf. (9.2))

$$\lim_{n\to\infty} \varrho(f_n - f; \Phi) = 0.$$

Convergence in norm always implies Φ-mean convergence but if Φ satisfies condition (9.7) then both notions are equivalent.

Concluding our discussion let us note that the functional

$$\|f\|^*_{L_\Phi(X)} = \inf\left\{r > 0 : \int_X \Phi\left(\frac{1}{r}|f(x)|\right) dx \leq 1\right\}$$

likewise is a norm in $L_\Phi(X)$, called the *Luxemburg norm* and we have the inequality

$$\|f\|^*_{L_\Phi(X)} \leq \|f\|_{L_\Phi(X)} \leq 2\|f\|^*_{L_\Phi(X)},$$

that is, the Luxemburg norm is equivalent to the norm (9.8).

The notion of Orlicz space generalizes to the case when the value of the functions under examination belong to a Banach space.

Orlicz spaces are useful in the study of imbedding of Sobolev spaces $W_p^{(l)}(G)$, $G \subset \mathbb{R}^n$, in the case of limit indices, i.e. when $lp = n$. In this case we have in view of (2.14) for each $q > 1$ the imbedding

$$W_p^{(l)}(G) \hookrightarrow L_q(G),$$

but $W_p^{(l)}(G)$ is not contained in $L_\infty(G)$, as the space $W_p^{(l)}(G)$ for $lp = n$ contains essentially unbounded functions. Otherwise put, it is not possible to find, in terms of Lebesgue spaces L_q, a minimal space containing the Sobolev space $W_p^{(l)}$ with limit index. However, there exists an Orlicz space, containing all L_q, $1 < q < +\infty$, which contains $W_p^{(l)}(G)$, $lp = n$. Namely, S.I. Pokhoshaev [1965] proved that the Sobolev space $W_p^{(l)}(G)$, $lp = n$, where G is a bounded domain with Lip1 boundary, contains the Orlicz space L_{Φ_q},

$$\Phi_q(G) \asymp e^{|t|^2}, \quad 1/q + 1/q' = 1, \quad t \to +\infty,$$

for $q = p$, but not for $q < p$, while for $q > p$ the imbedding

$$W_p^{(l)}(G) \hookrightarrow L_{\Phi_q}(G)$$

is completely continuous. Thus, in particular, we see that in the scale of Orlicz spaces L_{Φ_q} there exists a minimal one which contains a Sobolev space with limit smoothness parameter.

This theorem generalizes with obvious changes to the case of unbounded domains G and to the case of fractional l.

§ 2. The Spaces $E_\Phi(X)$

We define the space $E_\Phi(X)$, $\mu X < +\infty$, as the closure of the essentially bounded functions in the space $L_\Phi(X)$.

The set of essentially bounded functions is dense in the Orlicz class $\tilde{L}_\Phi(X)$ in the sense of convergence in Φ-mean. It follows that if condition (9.7) is fulfilled then the set of bounded functions is dense in $L_\Phi(X)$, so that in this case the spaces $E_\Phi(X)$ and $L_\Phi(X)$ coincide. In general, $E_\Phi(X)$ is a proper subset of $L_\Phi(X)$ but, in contrast to $L_\Phi(X)$, it is always separable.

Many properties of Lebesgue spaces extend to the case of the space $E_\Phi(X)$. For example, every function $f \in E_\Phi(X)$ is *average continuous with respect to Φ-means*, i.e. if $f \in E_\Phi(X)$ and is extended by 0 off $X \subset \mathbb{R}^n$ then (cf. (9.2))

$$\lim_{h \to 0} \varrho(f(x+h) - f(x); \Phi) = 0.$$

For the spaces $E_\Phi(X)$ we have a representation of bounded linear functionals F of the form

$$F(f) = \int_X f(x)g(x)dx, \quad g \in E_\Psi(X),$$

where Φ and Ψ are complementary Young functions.

For subsets of $E_\Phi(X)$ one has an analogue of the criteria of F. Riesz and A.N. Kolomogorov for the pre-compactness of subsets of the Lebesgue spaces L_p (cf. § 5 of Chap. 1).

Many questions of the theory of Orlicz spaces are treated in the monographs Sobolev [1962], Nečas [1967].

§ 3. Sobolev-Orlicz Spaces

We have mentioned that the Sobolev spaces $W_p^{(l)}$ were introduced in mathematics in connection with the study of boundary problems for linear partial differential equations by variational methods. A natural class in which it is expedient to seek solutions is the class of functions with finite energy integral, i.e. a Sobolev class.

In the study of equations of more general type, generally speaking, nonlinear, in particular, quasilinear,

$$\sum_{|\nu|\le l}(-1)^{|\nu|}(a_\nu(x, D^l u))^{(\nu)} = f(x), \quad x \in G, \tag{9.10}$$

where $D^l u = \{u^{(\nu)}\}_{|\nu|\le l}$, $\nu = (\nu_1, \dots, \nu_n) \in \mathbb{N}_0^n$, G being a domain in $\mathbb{R}^n$, it is natural to seek their solutions in Sobolev spaces $W_p^{(l)}$ only in the case when the functions $a_\nu(x, \xi)$ have *polynomial order of growth*:

$$|a_\nu(x, \xi)| \le c(1 + \sum_{|\mu|\le l} |\xi_\mu|^{p-1}), \quad 1 < p < +\infty$$

$\mu = (\mu_1, \dots, \mu_n) \in \mathbb{N}_0^n$. But even in this case one gets results which are not sharp. In the beginning of the 60's V. I. Vishik showed that more exact results can be obtained using the *Sobolev-Orlicz space* $W^{(l)}L_\Phi$. This space is defined in an analogous fashion as the corresponding Sobolev space with the single difference that the metric L_p is replaced by an Orlicz metric L_Φ. Subsequently M.I. Vishik's investigations were continued by F. E. Browder and his students. However, in the first place one considered only equations with coefficients of polynomial growth, as the technique used worked only for reflexive spaces L_Φ. By the same token, one obtained exact estimates only in the L_p-metric, while the case of nonlinearities of nonpolynomial type, for example, equations of the type

$$-\sum_{j=1}^{n} \frac{\partial}{\partial x_j}\left(\frac{\partial u}{\partial x_j} \exp \frac{\partial u}{\partial x_j}\right) + \exp u = f$$

were not covered at all. Gossez [1971] has developed an approach which allowed also to incorporate boundary problems for the equation (9.10) with strongly or slowly growing coefficients. He employed not only nonreflexive Orlicz spaces L_Φ but also nonseparable ones. Thus, it was found that if the nonlinearity in (9.10) is of nonpolynomial type then in their study it was natural to invoke spaces more general than ordinary Sobolev spaces, namely Sobolev-Orlicz spaces. Let us define these spaces in greater detail than was done above.

Denote by $\mathcal{N}_{n,k}$ the set of all multi-indices $\nu = (\nu_1, \dots, \nu_n) \in \mathbb{N}_0^n$ such that $|\nu| \le k$, where $k \in \mathbb{N}_0^k$, and by $\kappa = \kappa_n(k)$ the number of elements in $\mathcal{N}_{n,k}$.

Let G be an open set in $\mathbb{R}^n$, $l \in \mathbb{N}$, and $\Phi = \{\Phi_\nu\}_{|\nu| \le l}$ a family of $\kappa_n(k)$ Young functions Φ_ν. We denote by $W^{(l)}L_\Phi(G)$ (respectively by $W^{(l)}E_\Phi(G)$) the set of all functions $f : G \to \mathbb{R}$ such that $f^{(\nu)} \in L_{\Phi_\nu}(G)$ (respectively $f^{(\nu)} \in E_{\Phi_\nu}(G)$), $\nu \in \mathcal{N}_{n,k}$.

The Sobolev-Orlicz spaces $W^{(l)}L_\Phi(G)$ and $W^{(l)}E_\Phi(G)$ are Banach spaces with the norm

$$\|f\|_{k,\Phi} = \sum_{|\nu| \le k} \|f^{(\nu)}\|_{L_{\Phi_\nu}(G)}.$$

The spaces $W^{(l)}E_\Phi(G)$ are separable and if all the functions Φ_ν, $|\nu| \le k$, satisfy condition (9.7) then the spaces $W^{(l)}L_\Phi(G)$ are likewise separable. The set of infinitely differentiable finite functions is dense in $W^{(l)}E_\Phi(\mathbb{R}^n)$.

The set $C^\infty(G)$ of all infinitely differentiable functions on a bounded domain G is contained in $W^{(l)}E_\Phi(G)$ and is dense in this space; if the boundary of G is Lipschitz then the set of restrictions to G of all infinitely differentiable functions on $\mathbb{R}^n$ is dense in $W^{(l)}E_\Phi$.

Let us give an example of the application of an Orlicz-Sobolev space with a quite general Young function Φ.

Let φ be a continuous nondecreasing positive function defined for $t > 0$, with $\varphi(0) = 0$, $\varphi(+\infty) = +\infty$. Consider the equation

$$\sum_{|\nu| \le l} (-1)^{|\nu|} (\varphi(u^{(\nu)}))^{(\nu)} = f, \tag{9.11}$$

and introduce the Young function

$$\Phi(t) = \int_0^{|t|} \varphi(s) ds$$

(we saw above that a large class of Young functions can be put in this form). It is natural to seek solutions of (9.11) in the Sobolev-Orlicz class $W^{(l)}L_\Phi$. If we in place of the single function $\varphi(t)$ have a family of functions $\varphi_\nu(t)$, $|\nu| \le l$, in this equation (cf. (9.10)), then an anisotropic Sobolev-Orlicz space evolves in a natural way.

§ 4. Imbedding Theorems for Sobolev-Orlicz Spaces

Before we formulate the imbedding theorems for Sobolev-Orlicz spaces we need yet another definition.

Let $\Phi = \{\Phi_\alpha\}_{\alpha \in \mathcal{A}}$ be a class of equivalent Young functions (cf. (9.9)), $\mathcal{A}$ being an index set. Set

$$g_{\Phi_\alpha}(t) \overset{\text{def}}{=} \frac{\Phi_\alpha^{-1}(t)}{t^{1+1/n}}, \quad t \ge 0, \quad \Phi_\alpha \in \Phi,$$

and assume that Φ is such that for all $\alpha \in \mathscr{A}$ holds

$$\int_1^{+\infty} g_{\Phi_\alpha}(t)dt = +\infty.$$

For any class Φ there exits always a function $g_{\Phi_{\alpha_0}}(t)$ such that

$$\int_0^1 g_{\Phi_{\alpha_0}}(t)dt < +\infty.$$

A class $\Phi^* = \{\Phi_\alpha^*\}$ of equivalent Young functions is said to be conjugate in the sense of Sobolev to the class Φ if it contains a function $\Phi_{\beta_0}^*$ such that

$$(\Phi_{\beta_0}^*)^{-1} = \int_0^t g_{\Phi_\alpha}(s)ds.$$

Any pair of Young functions $\Phi_\alpha \in \Phi$ and $\Phi_\beta^* \in \Phi^*$ belonging to Sobolev conjugate classes Φ and Φ^* is called a *pair of conjugate functions.*

If, for example, $\Phi_\alpha = \alpha t^p$, $1 < p < n$, $\alpha > 0$, then $\Phi_\beta^*(t) = \beta t^\beta$, $\beta > 0$, $q = np/(n-p)$.

The following theorem (Sobolev [1962]) holds true.

If G is a domain in $\mathbb{R}^n$ with Lipschitz boundary, $n \geq 2$, and Φ is a Young function with

$$\int_0^{+\infty} g_\Phi(t)dt = +\infty,$$

Φ^* being its conjugate, then we have the imbedding

$$W^1L_\Phi(G) \hookrightarrow L_{\Phi^*}(G).$$

If

$$\int_0^{+\infty} g(t)dt < +\infty,$$

then

$$W^1L_\Phi(G) \hookrightarrow L_\infty(G).$$

One has also theorems concerning imbedding into the space of traces. For example, let G be a bounded domain with Lipschitz boundary, $G \subset \mathbb{R}^n$, and let S^m be the intersection of the hyperplane $\mathbb{R}^m$ with $\bar{G}$, i.e. $S^m = \bar{G} \cap \mathbb{R}^m$.

Then if Φ is a Young function such that

$$\int_1^{+\infty} g_\Phi(t)dt = +\infty,$$

and the function $\Phi(t^{1/2})$ likewise is a Young function, assuming $n > m > n-r$ for $r \geq 1$ and $m = n-1$ for $r = 1$, one has the imbedding

$$W^1L_\Phi(G) \hookrightarrow L_\Psi(S^m),$$

where

$$\Psi(t) = [\Phi^*(t)]^{m/n}$$

(Φ^* is the Young function conjugate to Φ).

Sobolev-Orlicz spaces for bounded domains with smooth boundary have been studied by Lacroix and V.M. Kokilashvili. In the case of certain Young functions Φ defining nonseparable Orlicz spaces L_Φ, Lacroix [1974] described the traces of functions in the Sobolev-Orlicz spaces (cf. § 1 of Chap. 4). This is particularly useful from the point of applications to nonlinear boundary problems.

Chapter 10

Symmetric and Nonsymmetric Banach Function Spaces

§ 1. Ideal Spaces and Symmetric Banach Spaces of Measurable Functions

Let $(X, \mathscr{F}, \mu)$ be a space X with a σ-algebra $\mathscr{F}$ of measurable functions and a σ-finite measure and denote by $S = S(X, \mathscr{F}, \mu)$ the linear space of almost everywhere finite measurable functions factored by the set of all functions equal to zero almost everywhere. A linear subspace E of S is called a *Banach ideal space* if E is a Banach space and if it has the following property: if for two functions $f_1 \in S$ and $f_2 \in S$ holds for almost all $x \in X$ the inequality $|f_1(x)| \le |f_2(x)|$ it follows that $f_1 \in E$ and $\|f_1\|_E \le \|f_2\|_E$, i.e. if E contains a function then it contains also all measurable functions less or equal to it in modulus. A norm with this property is called *monotone*.

A Banach ideal space E is called *complete* if its unit ball is closed with respect to convergence in measure. This means the following. If $\{f_n\}$ is a sequence such that $\|f_n\|_E \le 1$, $n = 1, 2, \ldots$, and $f_n \to f$ in measure μ, $n \to \infty$, then $f \in E$ and $\|f\|_E \le 1$.

In the early 60's E.M. Semenov introduced, in conncection with interpolation of operators, the notion of symmetric space. This amounts to the following. Let $(X, \mathscr{F}, \mu)$ be a space equipped with a continuous σ-finite measure. A Banach ideal space $E \subset S(X, \mathscr{F}, \mu)$ is called *symmetric*[1] if E together with each function f contains all functions $g \in S(X, \mathscr{F}, \mu)$ which are equimeasurable to f, i.e. have the same distribution function; it is further understood that

$$\|f\|_E = \|g\|_E.$$

[1] *Translators's note.* In English literature one usually says *rearrangement invariant (r.i.) space.*

Examples of symmetric spaces are the Lebesgue spaces L_p, $1 \leq p \leq +\infty$, the Orlicz spaces L_Φ, the Marcinkiewicz spaces M_p and the Lorentz spaces $L_{p,q}$.

§ 2. Spaces with Mixed Norms

Examples of nonsymmetric Banach ideal spaces are provided by the Lebesgue spaces $L_p(\mathbb{R}^n)$, $p = (p_1, \dots, p_n)$, with a mixed norm (cf. § 1 of this chapter). Function spaces defined on the basis of mixed p-norms have been studied by many authors, including Benedek and Panzone [1961], Ya.S. Bugrov [1971], [1973] (imbedding theorems), M.K. Potapov [1980] (imbedding theorems and approximation of functions by trigonometric polynomials in a mixed norm). We define now spaces with mixed norm in the general case, for simplicity restricting ourselves to the case of two variables (each of these may be a point in a multidimensional space).

Let E_1 be a Banach ideal space on $(X_1, \mathscr{F}_2, \mu_1)$ and E_2 one on $(X_2, \mathscr{F}_2, \mu_2)$; we require the assumption that from almost eveywhere convergence of any decreasing sequence of functions $f_n \in E_2$, $n = 1, 2, \dots$, to a function $f \in E_2$ follows that

$$\lim_{n \to \infty} \|f_n\|_{E_2} = \|f\|_{E_2}.$$

(this condition is fulfilled for all the above mentioned spaces). Then *the space* $E_2[E_1]$ *with mixed norm* is defined as the space of all measurable functions $f(x_1, x_2)$, $x_1 \in X_1$, $x_2 \in X_2$, on $X_1 \otimes X_2$ with the fine norm

$$\|f\| = \| \, \|f(x_1, x_2)\|_{E_1, x_1} \|_{E_2, x_2}. \tag{10.1}$$

In the case of L_p-norms with the same p it follows from Fubini's theorem that

$$L_p(X_2)[L_p(X_1)] = L_p(X_1 \times X_2) = L_p(X_1)[L_p(X_2)] \tag{10.2}$$

This relation, i.e. the possibility of changing the order of computation of p-norms in different variables, is often used in the applications of Lebesgue spaces and, generally, of integral calculus in many branches of mathematics. An esssential obstruction toward the generalization of the methods applicable in the case of L_p-metrics to more general metrics is the fact that for every Banach ideal space $E \neq L_p$ there is no analogue of (10.2), that is, it is not true that

$$E_1[E_2] = E_2[E_1]. \tag{10.3}$$

Indeed, it was shown in 1930 by A.N. Kolomogorov and Nagumo that in the case when E_1 and E_2 are (in contemporary language) symmetric spaces formula (10.3) is possible only for L_p-spaces. A.V. Bukhvalov [1979] found a generalization of this theorem to ideal spaces: if E_1 and E_2 are any two ideal spaces and if the norms

$$\| \, \|f(x_1, x_2)\|_{E_2, x_2} \|_{E_1, x_1} \text{ and } \| \, \|f(x_1, x_2)\|_{E_1, x_1} \|_{E_2, x_2}$$

are equivalent on the set of functions of the form

$$f(x_1, x_2) = \sum_{j=1}^{n} g_j(x_1)h_j(x_2), \quad g_j \in E_1, \quad h_j \in E_2,$$

then the norms in E_1 and E_2 must be equivalent to L_p-norms with appropriate weights.

From this result it follows that it is impossible to split ideal function spaces, consisting of functions of several variables, into spaces of the separate variables, provided we consider spaces different from the Lebesgue spaces L_p. In view of this it is, for example, not possible to apply in the proof of theorems about the properties of traces of functions in spaces with mixed norm the usual methods of distinguishing one of the variables.

§ 3. Sobolev and Nikol'skiĭ-Besov Spaces Induced by Metrics of General Type

Many authors have studied Sobolev and Nikol'skiĭ-Besov spaces whose construction involves, instead of L_p-norms, norms in general symmetric spaces (Calderón [1961], Golovkin [1969], M.Z. Berkolaĭko [1983], O.V. Besov [1984], [1985] (see also the book Besov-Il'in-Nikol'skiĭ [1975]), Yu.A. Brudnyĭ [1971], A.V. Bukhvalov [1981], V.S. Klimov [1976] and others).

Let G be an open set in $\mathbb{R}^n$ and $E(G)$ an ideal Banach space. If $l \in \mathbb{N}$ let $W^{(l)}E(G)$ be the Banach space of all functions $f \in E(G)$ which have generalized partial derivatives $f^{(\nu)} \in E(G)$, $|\nu| \leq l$, with the norm

$$\|f\| = \sum_{|\nu| \leq l} \|f^{(\nu)}\|_{E(G)}.$$

If $E = L_p$ it is clear that $W^{(l)}L_p = W_p^{(l)}$ so that the spaces $W^{(l)}E$ constitute a generalization of the Sobolev spaces $W_p^{(l)}$. Next let us define the analogue of Sobolev-Liouville spaces (cf. Chap. 5).

Let $G_l(x)$, $l > 0$, be the Bessel-Macdonald kernel (cf. (5.9)). Let us denote by $L^{(l)}E(\mathbb{R}^n)$ the Banach space of all measurable functions on $\mathbb{R}^n$ of the form

$$f = G_l * g, \quad g \in E(\mathbb{R}^n)$$

with the norm

$$\|f\| = \|g\|_{E(\mathbb{R}^n)}.$$

A.V. Bukhvalov [1981] has found conditions under which for l integer the Sobolev space $W^{(l)}E(\mathbb{R}^n)$ coincides with the Sobolev-Liouville space $L^{(l)}E(\mathbb{R}^n)$ (if $E = L_p$ cf. (5.3)). Namely, he showed that if the symmetric space $E(\mathbb{R}^n)$ is an interpolation space between suitable spaces $L_{p_0}(\mathbb{R}^n)$ and $L_{p_1}(\mathbb{R}^n)$, $1 < p_0 < p_1 < +\infty$, then

$$W^{(l)}E(\mathbb{R}^n) = L^{(l)}E(\mathbb{R}^n). \tag{10.4}$$

Conversely, if $n > 1$ and the symmetric space E is separable and complete then (10.4) implies that E is an interpolation space between suitable spaces $L_{p_0}(\mathbb{R}^n)$ and $L_{p_1}(\mathbb{R}^n)$.

A study of the traces of functions in the Sobolev space $W^{(r)}E(\mathbb{R}^n)$, where E is a Lorentz space (cf. §2 of Chap. 8), was undertaken by Berkolaĭko [1983].

According to (2.14) one has for $1 < p < q < +\infty$ and $\varrho = r-s-n(1/p-1/q) \geq 0$

$$W_p^{(r)}(\mathbb{R}^n) \hookrightarrow W_q^{(s)}(\mathbb{R}^n).$$

Using the scale of Lorentz spaces (cf. §2 of Chap. 8) M.Z. Berkolaĭko strengthened this result for $\varrho > 0$ proving that

$$W^{(r)}M_p(\mathbb{R}^n) \hookrightarrow W^{(s)}\Lambda_q(\mathbb{R}^n). \tag{10.5}$$

He further investigated the problem of traces for the space $L_{q\theta}[L_p]$ with the mixed norm

$$\| \, \|f(x^{(m)}, \tilde{x}^{(m)})\|_{L_p, x^{(m)}} \|_{L_{q,\theta}, \tilde{x}^{(m)}}$$

(cf. (10.1)) where $L_p = L_p(\mathbb{R}^n)$ and $L_{q,\theta} = L_{q,\theta}(\mathbb{R}^n)$ is a Lorentz space, $x = (x^{(m)}, \tilde{x}^{(m)})$, $x^{(m)} = (x_1, \ldots, x_m)$, $\tilde{x}^{(m)} = (x_{m+1}, \ldots, x_n)$. He showed that the condition

$$\varrho = r - \frac{n-m}{q} > 0$$

is necessary and sufficient for a function in $W^{(r)}L_{q,\theta}[L_p](\mathbb{R}^n)$, $\theta > 1$, to have a trace on $\mathbb{R}^m$. (Here by a Sobolev space $W^{(r)}E_1[E_2]$ we intend a space, defined in an analogous way as $W^{(l)}E$, but by replacing in this definition the norm in E by the norm in $E_1[E_2]$). Furthermore

$$W^{(r)}L_{q,\theta}[L_p](\mathbb{R}^n) \overset{\hookrightarrow}{\leftarrow} B_{p,\theta}^{(\varrho)}(\mathbb{R}^m). \tag{10.6}$$

If $q = \theta$ this result was obtained earlier by Ya. S. Bugrov [1973].

Comparing the imbedding (10.6) for $p = q$ with the imbedding

$$W_p^{(\varrho)}(\mathbb{R}^n) \hookrightarrow W_{p,p}^{(\varrho)}(\mathbb{R}^m) \tag{10.7}$$

it is seen that the coincidence of the lower indices for the Nikol'skiĭ-Lorentz space in the right hand side of (10.7) is a consequence of formula (8.8), i.e. fixing ϱ and changing θ in (10.6) we obtain as the space of traces the whole scale of spaces $B_{p,\theta}^{(\varrho)}$; moreover, for $\theta = +\infty$ the space of traces of functions in $W^{(r)}L_{q,\infty}[L_p](\mathbb{R}^n)$ is the Nikol'skiĭ space $H_p^{(\varrho)}$.

In the case $\theta = 1$ the condition $\varrho = r-(n-m)/p$ is necessary and sufficient for a function in $W^{(r)}\Lambda[L_p](\mathbb{R}^n)$ (for the notation Λ_q cf. §2 of Chap. 8) to possess a trace on $\mathbb{R}^m$ and the

$$W^{(r)}\Lambda_q[L_p](\mathbb{R}^n) \overset{\hookrightarrow}{\leftarrow} B_{p,1}^{(\varrho)}(\mathbb{R}^m). \tag{10.8}$$

(Here we intend by $B_{p,1}^{(0)}(\mathbb{R}^m)$ the space L_p.)

Thus we see that the space $B_{p,1}^{(\varrho)}$, $\varrho > 0$, (the smallest in the scale $B_{p,\theta}^{(\varrho)}$ with respect to θ) also is a spaces of traces of a Sobolev space with mixed norm.

Let us underline that in the case (10.7), as in the case (10.8), the converse imbedding of the space of traces of the Sobolev space is for $\varrho > 0$ realized with the aid of linear extension operators; if $\varrho = 0$ it is only possible to find a nonlinear extension operator (concerning linear and nonlinear extension operators cf. also §4 of Chap. 6).

In the case $p = q = (n-m)/r$ with $(n-m)/p$ integer it follows from (10.8) that

$$W^{(\frac{n-m}{p})}\Lambda_p[L_p] \overset{\hookrightarrow}{\leftarrow} L_p(\mathbb{R}^m) \tag{10.9}$$

(here too the extension operator is nonlinear).

It is well-known that not all functions in $W_p^{(\frac{n-m}{p})}(\mathbb{R}^n) = W^{(\frac{n-m}{p})}L_p[L_p](\mathbb{R}^n)$ have a trace on an m-dimensional plane $\mathbb{R}^m$. From (8.7) and (10.7) it is seen that this is the case also for the space $W^{(\frac{n-m}{p})}L_{p,\theta}[L_p]$, $\theta > 1$, and it is only for $\theta = 1$ that all functions in $W^{(\frac{n-m}{p})}L_{p,\theta}[L_p]$ have a trace. Thus, the space L_p is in the case at hand too wide, within the scale $L_{p,\theta}$, to allow one to solve with the aid of it the question which functions in the Sobolev space $W_p^{(\frac{n-m}{p})}$ have a trace on $\mathbb{R}^m$.

In 1961 Calderón introduced the spaces $\Lambda^{(k)}(X,E)$ generalizing the Sobolev spaces $W_p^{(k)}(G)$ and the Nikol'skiĭ-Besov spaces $B_{p,q}^{(r)}(G)$ in the following direction. First, in the definition of the norms of the functions their moduli of continuity of their derivatives the space $L_p(G)$, $G \subset \mathbb{R}^n$, is replaced by a more general Banach space $X = X(G)$, admitting a strongly continuous semigroup of shifts. Second, the differential properties of functions in $\Lambda^{(k)}(X,E)$ are characterized by membership of the moduli of continuity (in the norm X) of the functions in some auxiliary ideal space $E = E(0,+\infty)$.

The study of the spaces $\Lambda^{(k)}(X,E)$ was begun by Calderón and the spaces are now called *Calderón spaces* or *Lipschitz spaces*. Let us give a more detailed definition in the case $G = \mathbb{R}^n$.

Let $X = X(\mathbb{R}^n)$ be a Banach space of functions defined on $\mathbb{R}^n$, which is invariant with respect to the shift, the shift being assumed to be continuous in X (if X is a symmetric space this is the case if X is separable). Denote by $\omega_X^{(k)}(f;\delta)$ the kth order modulus of continuitry of f in the X metric (cf. (2.10), (2.11)). Let E be a Banach ideal space of functions defined on $\overset{+}{\mathbb{R}}$.

The space $\Lambda^{(k)}(X,E)$ is defined as the Banach space of all functions $f \in X$ such that the norm

$$\|f\| \overset{\text{def}}{=} \|f\|_X + \|\omega_X^{(k)}(f;\delta)\|_{E,\delta}$$

is finite.

In suitable assumptions on E, expressed in terms of properties of the Hardy-Littlewood operators

$$(A_k\varphi)(s) = \int_0^s \varphi(t)(s/t)^k dt/t, \quad (B_k\varphi)(s) = \int_s^1 \varphi(t)(s/t)^k dt/t,$$

Calderón, Yu.A. Brudnyĭ and V.K. Shalashov, and A.V. Bukhvalov have developed an interpolation theory for the space $\Lambda^{(k)}(X,E)$, based on a canonical imbedding

$$\Lambda^{(k)}(X,E) \hookrightarrow E(X) \oplus X$$

constructed by Calderón; here $E(X)$ is the Banach space of X-valued vector functions $\varphi : \overset{+}{\mathbb{R}} \to X$ such that $\|\varphi(\cdot)\|_X \in E$, with the norm

$$\|\varphi\| = \| \, \|\varphi(\cdot)\|_X \, \|_E.$$

In Yu.A. Brudnyĭ's paper [1976] and in his paper [1973] coauthored with V.K. Shalashov many theorems on equivalent normings of Lipschitz spaces are established and further imbedding theorems and theorems on extension with preservation of differentiability properties off the limits of a domain with Lipschitz boundary. The proof of these theorems is based on estimates of the moduli of continuity of functions and their derivatives which have an independent interest.

In Brudnyĭ [1976] a general approach is developed for the study of Lipschitz, quasi-Lipschitz and a number of other spaces based on a single technique of local approximations. The *local best approximation of order k of a function $f \in B$*, where B is a Banach space, is the set function

$$E_k(f,U) = \inf_{p \in P_k} \|(f-p)\chi_U\|_B,$$

where χ_U is the characteristic function of the set U:

$$\chi_U(x) = \begin{cases} 1, & x \in U, \\ 0, & x \notin U, \end{cases}$$

and the infimum runs over all polynomials P of degree k. The differential properties of functions are characterized in terms of membership for $E_k(f,U)$ with fixed k (as a function of U) in a suitable ideal space X. The spaces $\mathscr{L}_N^{(k)}(X)$ arising in this way were studied by Yu.A. Brudnyĭ; in particular, he established their relation to the previous Lipschitz spaces and to other spaces such as Morrey spaces, spaces of functions of bounded variation etc.

A special case of the Calderón spaces $\Lambda^{(k)}(X,E)$ are the *generalized Besov spaces $B^{(r)}(X,E)$*. Let us define them.

If u is a positive function on $\overset{+}{\mathbb{R}}$, we denote by uE the ideal Banach space of measurable functions φ on $\overset{+}{\mathbb{R}}$ such that $u\varphi \in E$, with the norm $\|\varphi\|_{uE} \stackrel{\text{def}}{=} \|u\varphi\|_E$.

Let $E = E(\overset{+}{\mathbb{R}})$ be a symmetric space with the measure $d\tau/\tau$, which is an interpolation space between $L_1(\overset{+}{\mathbb{R}}, d\tau/\tau)$ and $L_\infty(\overset{+}{\mathbb{R}})$ (for this it is sufficient tht E is separable or complete).

By a generalized Besov space $B^{(r)}(X,E)$, $r > 0$, $r = \bar{r} + \alpha$, $\bar{r} \in \mathbb{N}_0$, $0 < \alpha \leq 1$, we intend the Banach space of all measurable functions f on $\mathbb{R}^n$ such that the norm

$$\|f\| = \|f\|_X + \sum_{|\nu|=\bar{r}} \|\delta^{-\alpha}\omega_X^{1+\varepsilon}(f^{(\nu)};\delta)\|_{E,\delta},$$

is finite, taking $\varepsilon = 0$ for $0 < \mu < 1$ and $\varepsilon = 1$ for $\mu = 1$.

This appelation for $B^{(r)}(X,E)$ is choosen as one of the first definitions given by O.V. Besov for the spaces $B^{(r)}_{p,q}$ was formulated in terms of moduli of continuity. Moreover, if $X = L_p(\mathbb{R}^n)$, $E = L_q(\overset{+}{\mathbb{R}}, d\tau/\tau)$ then $B^{(r)}(X,E) = B^{(r)}_{p,q}$, that is, the spaces $B^{(r)}(X,E)$ indeed generalize the spaces $B^{(r)}_{p,q}$.

Each space $B^{(r)}(X,E)$ is also a Calderón space (the converse is not true); more precisely

$$B^{(r)}(X,E) = \Lambda^{(k)}(X, \delta^{-r}E), \quad k \geq r + 1 + \varepsilon. \tag{10.10}$$

This identity allows one to study the properties of the spaces $B^{(r)}(X,E)$ (for example, their interpolation properties) on the basis of the study of analogous properties of the spaces $\Lambda^{(k)}(X,E)$ using the imbedding into the space $E(X)$ of vector functions (Calderón, A.V. Bukhvalov). The study of the spaces $B^{(r)}(X,E)$ and the imbedding theorems connected with them was initiated by K.K. Golovkin and continued in the work of A.V. Bukhvalov, M.Z. Berkolaĭko and others.

V.I. Burenkov considered the case when one as X anew takes a space of type $B^{(r)}(X,E)$. This leads to "iterated" spaces of the type $B^{(r_1)}(B^{(r_2)}(X,E),E)$. For $X = L_p = L_p(\mathbb{R}^n)$, $E = L_q^* \equiv L_q(\mathbb{R}^n, d\tau/\tau)$ V.I. Burenkov proved the identity

$$B^{(r_1)}(B^{(r_2)}(L_p, L_q^*), L_q^*) = B^{(r_1+r_2)}(L_p, L_q^*) \equiv B^{(r_1+r_2)}_{p,q}$$

and applied it to the study of the regularity of solutions of partial differential equations.

In the beginning of the 70's there arose the problem of the description of the traces in the Sobolev space $W_p^{(r)}$ with a mixed norm ($p = (p_1, \ldots, p_n)$) on the hyperplane $x_k = 0$ where k is not by necessity equal to n. Ya.S. Bugrov [1971], [1973] obtained results from which follows an exact answer when

$$p_k = p_{k+1} = \ldots = p_n.$$

Thus for $k = 1$ the problem is thereby solved only in the case when the mixed norm reduces to an ordinary one, i.e. when $p_1 = \ldots = p_n$.

M.Z. Berkolaĭko obtained a solution in the general case. Let us mention an interesting consequence of his results. If r and $r - m/2$ are integers, $1 < p < +\infty$, $p \neq 2$, then

$$W^{(r)}_{(\underbrace{2,\ldots,2}_{m},\underbrace{p,\ldots,p}_{n-m})} \overset{\hookrightarrow}{\hookleftarrow} W^{(r-m/2)}_{p}(\mathbb{R}^{n-m}_{\tilde{x}^{(m)}}). \tag{10.11}$$

i.e. an ordinary Sobolev space with $p \neq 2$ arises as the spaces of a Sobolev space with mixed norm.

Let us further remark that the case when the Sobolev space $W^{(r)}_p$ or even the Sobolev-Liouville space $L^{(r)}_p$ coincides exactly with the spaces of traces of some function space was encounters also later in the study of weighted spaces: T.O. Shaposhnikova [1980] proved that the Sobolev-Liouville space $L^{(r)}_p(\mathbb{R}^{n-m})$ is the space of traces of a weighted space with the norm

$$\left\{ \int_{\mathbb{R}^{n-m}} \left(\int_{\mathbb{R}^m} |x^{(m)}|^{2-2r-m}(|\nabla_x u|^2 + |u|^2)dx^{(m)} \right)^{p/2} dx^{(n-m)} \right\}^{1/p}.$$

From M.Z. Berkolaĭko's results it further follows that the space of traces of the Sobolev type space $W^{(r)}_p$, $p = (p_1, \ldots, p_n)$, on the coordinate hyperplanes coincides with a space $B^{(\varrho)}_{p,q}$, generally speaking, only in the case of the hyperplane $x_n = 0$; in the remaining cases one must for the description of the traces introduce new spaces, generalizing the Lizorkin-Triebel spaces $L^{(r)}_{p,q}$ to the case of vectorial $p = (p_1, \ldots, p_n)$.

From the results considered in this section, gotten in the study of various function spaces, it is seen that their application allows one to make a more profound and more refined study of the properties of the functions themselves. Let us illustrate this at the hand of an example.

Let

$$f(x^{(m)}) \in W^{(l)}_4(\mathbb{R}^m).$$

As (cf. (4.17))

$$W^{(l)}_4(\mathbb{R}^m) \hookrightarrow B^{(l)}_{4,4}(\mathbb{R}^m),$$

$$B^{(l+m/4)}_{4,2}(\mathbb{R}^{2m}) \hookrightarrow W^{(l+m/4)}_4(\mathbb{R}^{2m}),$$

then, in view of the imbedding (4.11),

$$B^{(l)}_{4,4}(\mathbb{R}^m) \hookrightarrow B^{(l+m/4)}_{4,2}(\mathbb{R}^{2m}), \tag{10.12}$$

one can extend the function $f(x^{(m)})$ from $\mathbb{R}^m$ to $\mathbb{R}^{2m}$ as a function in $W^{(l+m/4)}_2(\mathbb{R}^{2m})$.

If we now use the imbedding (10.11) it is seen that the same function can be continued to a function of class $W_p^{(l+m/2)}(\mathbb{R}^{2m})$, where $p = (\underbrace{2,\ldots,2}_{m\text{ times}}, \underbrace{4,\ldots,4}_{m\text{ times}})$, in other words, improve the exponent of smoothness to $m/2$, that is, double it. This fact is a reflexion of a general phenomenon concerning the raising of the smoothness exponent of the class (of course, not indefinitely), to which belongs the extension of a given function, at the expense of the corresponding metric, in the case at hand the metric induced by the mixed norm. This fact in turn indicates the expediency of considering mixed norms in practise.

The introduction of general function spaces also widens the possibility to apply methods of the theory of functions to the study of partial differential equations. For example, from Theorem 10.6 it is clear that considering the solution of partial differential equations in $W^{(r)}M_p[L_p]$ and not in the traditional $W_p^{(r)}$ we almost do not lose in the description of the properties of the solutions (for example, for $r - s - n/p > 0$ the boundedness of the partial derivatives of order s) but obtain a possibility to impose boundary conditions in the Nikol'skiĭ spaces $H_p^{(\varrho)}$, which have a much simpler (compared to $B_{p,p}^{(\varrho)}$) constructive description, which is not without importance in the applications. On the other hand, it follows from Theorem 10.8 that one can pose boundary problems also in the case

$$r - \frac{n-m}{p} = 0,$$

but then one has to pose them not in the space $W_p^{(r)}$, but in $W^{(r)}\Lambda_p[L_p]$. Then the boundary conditions are given in $L_p(\mathbb{R}^n)$.

§ 4. Some Problems

In conclusion, we focus on two unsolved problems in the theory of function spaces (Kudryavtsev [1983]). The first of them is internal to the theory of functions, the second pertains to applications of the theory of function spaces to the study of partial differential equations

It would be interesting to know from which properties and in which function spaces one can determine the dimension of the argument appearing in the function spaces in view, i.e. to clarify on how many parameters the functions in it depend. In particular, which spaces of functions in n variables are distinguished, what their internal properties goes, by corresponding spaces of functions of m variables with $m \neq n$ and what are these properties. The problem amounts to an axiomatic description of spaces which are isomorphic to a class of functions depending on n variables (and, for example, defined in the entire space $\mathbb{R}^n$), with n given.

The second problem consists of an exact description of the properties of the traces of functions $u : G \to \mathbb{R}$ such that

$$\int_G \sum_{|k|\le l} |a_k(x)u^{(k)}(x)|^p dx < +\infty,$$

$$a_k(x) > 0, \quad k = (k_1, \dots, k_n) \in \mathbb{N}_0^n, \quad l \in \mathbb{N}, \quad x \in G \subset \mathbb{R}^n$$

in terms of the functions $a_k(x)$ themselves, imposing no restriction on their growth, with the aid of estimates using functions of a lower number of variables (as one usually does nowadays). The space of traces of this class of functions must consist of functions with a smoothness variable whose variation depends on the properties of the functions $a_k(x)$. A solution to this problem would provide a possibility to give necessary and sufficient conditions which one can put on the boundary values in order to be able to solve the first boundary problem for the *linear elliptic equation*

$$\sum_{|k|\le l} (-1)^{|k|}(a_k(x)u^{(k)}(x))^{(k)} = f(x), \quad x \in G \tag{10.13}$$

i.e. an equation such that

$$\sum_{|k|=l} a_k(x)\xi^k > 0, \quad x \in G, \quad \xi = (\xi_1, \dots, \xi_n), \quad \xi^k = \xi_1^{k_1} \dots \xi_n^{k_n}, \quad \xi \ne 0$$

(or even equations such that, alternatively, all terms with lower derivatives vanish at the expence of vanishing of the corresponding coefficients: $a_k(x) \equiv 0$ for $|k| < l$) in the class of all functions u such that

$$A(u) = \int_G \sum_{|k|\le l} a_k(x)(u^{(k)})^2 dx < +\infty.$$

The equation (10.13) is, of course, the Euler equation of the functional

$$K(u) = A(u) - 2(f, u),$$

where

$$(f, u) = \int_G f(x)u(x)dx.$$

Presently this problem is known only in the assumption that the coefficients $a_k(x)$ can be estimated in some sense with the aid of functions depending only on one variable, the distance from the point to the boundary of the domain where the equation is defined.

References*

The following list of references consists of both monographs and papers in periodicals. The books of Kantorovich-Akilov [1984], Kolomogorov-Fomin [1976], Natanson [1974], Dunford-Schwartz [1958] have text book character and are devoted to general questions in the theory of functions or in functional analysis; there one can find a detailed description of the notions employed in this Part. The bulk of the monographs in our list pertains more directly to the questions studied here. We mention in particular Adams [1975], Amanov [1976], Besov-Il'in-Nikol'skiĭ [1975], Burenkov [1966], Kokilashvili [1985], Krasnosel'skiĭ-Rutitskiĭ [1958], Kreĭn-Petunin-Semenov [1978], Kudryavtsev [1959], Kufner [1980], Kufner-John-Fucik [1977], Luxemburg [1955], Maz'ya [1984], Nikol'skiĭ [1977], Stein [1973], Triebel [1977], [1978a], [1978b], [1978c], [1983], Uspenskiĭ-Demidenko-Perepelkin [1984]. We further mention Gol'dshteĭn-Reshetnyak [1983], Samko [1984], Butzer-Behrens [1976], Stein-Weiss [1979], treating questions close to those touched upon in this Part. Finally, there are included some monographs which although mainly devoted to other questions also treat material of interest to us.

The list of original papers does not pretend toward completeness. Moreover, by the number of papers by different authors it is not possible to conclude anything about their contribution to theory under consideration, as many important result by these authors may be contained in the monographs just mentioned. In the list of papers we have only included those paper to which we found it natural to refer in the course of the development of the material of this Part and which, by and large, are not illuminated in sufficient amount of completeness in the monograph literature.

Adams, R. A.
[1975] *Sobolev spaces.* Academic Press, New York - London. Zbl. 314.46030

Amanov, T. I.
[1976] *Spaces of differentiable functions with a dominating mixed derivative.* Nauka, Alma Ata [Russian]

Babich, V. M., Slobodetskiĭ, L. N.
[1956] On the boundedness of the Dirichlet integral. Dokl. Akad. Nauk SSSR *106*, 604-607 [Russian]. Zbl. 72, 50

Bakhvalov, N. S.
[1963] Imbedding theorems for classes of functions with a some bounded derivatives. Vestn. Mosk. Univ., Ser. I Mat. Mekh. *3*, 7-16 [Russian]. Zbl. 122, 113

Benedek, A., Panzone, R.
[1961] The spaces L^p with mixed norm. Duke Math. J. *28*, 301-324. Zbl. 107, 89

Bénilan, Ph., Brezis, H., Crandall, M. G.
[1975] A semilinear elliptic equation in $L^1(\mathbb{R}^n)$. Ann. Sc. Norm. Super., Pisa, Cl. Sci., IV. Ser., *2*, 523-555. Zbl. 314.35077

Bergh, J., Löfström, J.
[1976] *Interpolation spaces. An introduction.* (Grundlehren 223) Springer, Berlin. Zbl. 344.46071

Berkolaĭko, M. Z.
[1983] Imbedding theorems for various metrics and dimensions for generalized Besov spaces. Tr. Mat. Inst. Steklova *161*, 18-28. Zbl. 545.46023. English translation: Proc. Steklov Inst. Math. *161*, 19-31 (1984)

Berkolaĭko, M. Z., Ovchinnikov, V. I.
[1983] Inequalities for entire functions of exponential type in symmetric spaces. Tr. Mat. Inst. Steklov. *161*, 3-17. Zbl. 541.46025. English translation: Proc. Steklov Inst. Math. *161*, 1-17 (1984)

*For the convenience of the reader, references to reviews in Zentralblatt für Mathematik (Zbl.), compiled using the MATH database, have, as far as possible, been included in this bibliography.

Besov, O. V.
[1967] On the conditions for the existence of a classical solution of the wave equation. Sib. Mat. Zh. *8*, 243-256. Zbl. 146, 335. English translation: Sib. Math. J. *8*, 179-188 (1968)
[1969] The behavior of differentiable functions at infinity and the density of finite functions. Tr. Mat. Inst. Steklova *105*, 3-14. Zbl. 202, 396. English translation: Proc. Steklov Inst. Math. *105*, 1-15 (1970)
[1971] Théorèmes de plongement des espaces fonctionnels, in: *Actes Congrès Internat. Math.* 1970, 2, pp. 467-473. Gauthier-Villars, Paris. Zbl. 234.46034
[1974] Estimates for the moduli of continuity of abstract functions defined in a domain. Tr. Mat. Inst. Steklova *131*, 16-24. Zbl. 316.46022. English translation: Proc. Steklov Inst. Math. *131 (1974)*, 15-23 (1975)
[1980] Weighted estimates for mixed derivatives in a domain. Tr. Mat. Inst. Steklova *156*, 16-21. Zbl. 459.46023. English translation: Proc. Steklov Inst. Math. *156*, 17-22 (1983)
[1983] On the density of finite functions in a weighted Sobolev space. Tr. Mat. Inst. Steklova *161*, 29-47. Zbl. 541.46026. English translation: Proc. Steklov Inst. Math. *161*, 33-52 (1984)
[1984] Integral representations of functions and imbedding theorems for domains with the flexible horn conditions. Tr. Mat. Inst. Steklova *170*, 12-30. Zbl. 582.46037. English translation: Proc. Steklov Inst. Math. *170*, 33-38 (1987)
[1985] Estimates for integral moduli of continuity and imbedding theorems for domains with the flexible horn condition. Tr. Mat. Inst. Steklova *172*, 4-15. Zbl. 587.46032. English translation: Proc. Steklov Inst. Math. *172*, 1-13 (1987)
Besov, O. V., Il'in, V. P., Nikol'skiĭ, S. M.
[1975] *Integral representations of functions and imbedding theorems*. Nauka, Moscow. Zbl. 352.46023. English translation: Winston, Washington D.C.; Halsted, New York - Toronto - London 1978
Beurling, A.
[1940] Ensembles exceptionnels. Acta Math. *72*, 1-13. Zbl. 23, 142
Brudnyĭ, Yu. A.
[1971] Spaces defined with the aid of local approximations. Tr. Mosk. Mat. O.-va *24*, 69-132. Zbl. 254.46018. English translation: Trans. Mosc. Math. Soc. *24*, 73-139 (1974)
[1976] Extension theorems for a family of function spaces. Zap. Nauchn. Semin. Leningr. Otd. Mat. Inst. Steklova *56*, 170-172 [Russian]. Zbl. 346.46027
[1977] Approximation spaces, in: *Geometry of linear spaces and operator theory*, pp. 3-30. Yaroslavl' [Russian]. Zbl. 418.46021
Brudnyĭ, Yu. A., Shalashov, V. K.
[1973] Lipschitz function spaces, in: *Metric questions in the theory of functions and transformations 4*, pp. 3-60. Kiev [Russian]. Zbl. 281.46026
Bugrov, Ya. S.
[1957a] Dirichlet's problem for the disk. Dokl. Akad. Nauk SSSR *115*, 639-642 [Russian]. Zbl. 91, 97
[1957b] On imbedding theorems. Dokl. Akad. Nauk SSSR *116*, 531-534 [Russian]. Zbl. 137, 99
[1958] A property of polyharmonic functions. Izv. Akad. Nauk SSSR Ser. Mat. *22*, 491-514 [Russian]. Zbl. 81, 99
[1963a] Imbedding theorems for the H-classes of S. M. Nikol'skiĭ. Sib. Mat. Zh. *5*, 1009-1029 [Russian]. Zbl. 139, 302
[1963b] Differential properties of the solutions of a class of differential equations of higher order. Mat. Sb., Nov. Ser. *63*, 59-121 [Russian]. Zbl. 128, 92
[1966] A theorem on the representation of a class of functions. Sib. Mat. Zh. 7, 242-251. Zbl. 192, 227. English translation: Sib. Math. J. 7, 194-201 (1966)
[1967] Estimates for polyharmonic polynomials in a mixed norm. Izv. Akad. Nauk SSSR Ser. Mat. *31*, 275-288. Zbl. 168, 96. English translation: Math. USSR, Izv. *1*, 259-272 (1968)
[1971] Function spaces with a mixed norm. Izv. Akad. Nauk SSSR Ser. Mat. *35*, 1137-1158.

Zbl. 223.46036. English translation: Math. USSR, Sb. *5*, 1145-1167 (1972)

[1973] Imbedding theorems for function classes with a mixed norm. Mat. Sb., Nov. Ser. *92*, 611-621. Zbl. 287.46042. English translation: Math. USSR, Izv. *21*, 607-618 (1975)

Bukhvalov, A. V.

[1979a] Generalization of the Kolmogorov-Nagumo theorem to tensorial derivatives, in: *Qualitative and approximation methods in the study of operator equations*, pp. 48-65. Yaroslavl' [Russian]. Zbl. 443.46027

[1979b] Continuity of operators in spaces of vector functions with applications to the study of Sobolev spaces and analytic functions in the vectorial case. Dokl. Akad. Nauk SSSR *246*, 524-528. Zbl. 426.47018. English translation: Sov. Math., Dokl. *20*, 480-484 (1979)

[1981] The complex method of interpolation in spaces of vector functions and generalized Besov spaces. Dokl. Akad. Nauk SSSR *260*, 265-269. Zbl. 493.46062. English translation: Sov. Math., Dokl. *24*, 239-243 (1981)

Burenkov, V. I.

[1966] Imbedding and extension theorems for classes of differentiable functions of several variables defined in the whole space, in: Itogi Nauki Tekn. VINITI, Mat. Anal. 1966, pp. 71-155. Zbl. 173, 414. English translation: Progress Math. *2*, 73-161 (1968)

[1969] On the additivity of the spaces W_p^r and B_p^r and on imbedding theorems in domains of general type. Tr. Mat. Inst. Steklova *105*, 30-45. Zbl. 206, 123. English translation: Proc. Steklov Inst. Math. *105*, 35-53 (1971)

[1974a] Sobolev's integral representation and Taylor's formula. Tr. Mat. Inst. Steklov *131*, 33-38. Zbl. 313.46032. English translation: Proc. Steklov Inst. Math. *131*, 33-38 (1975)

[1974b] On the density of infinitely differentiable functions in Sobolev spaces for an arbitrary open set. Tr. Mat. Inst. Steklova *131*, 39-50. Zbl. 313.46033. English translation: Proc. Steklov Inst. Math. *131*, 39-51 (1975)

[1974c] On the approximation of functions in the space $W_p^r(\Omega)$ by finite functions for an arbitrary open set Ω. Tr. Mat. Inst. Steklova *131*, 51-63. Zbl. 322.46041. English translation: Proc. Steklov Inst. Math. *105*, 53-60 (1975)

[1976a] On a way of approximating differentiable functions. Tr. Mat. Inst. Steklova *140*, 27-67. Zbl. 433.46031. English translation: Proc. Steklov Inst. Math. *140*, 27-70 (1979)

[1976b] On extension of a function under preservation of a seminorm. Dokl. Akad. Nauk SSSR *228*, 779-782. Zbl. 355.46014. English translation: Sov. Math., Dokl. *17*, 806-810 (1976)

[1982] Mollifying operators with variable step and their application to approximations by infinitely differentiable functions, in: *Nonlinear Analysis, Function Spaces and Applications* (Pisek 1982), Vol. 2. pp. 5-37 (Teubner-Texte Math. 49). Teubner, Leipzig. Zbl. 536.46021

Burenkov, V. I., Faĭn, B. L.

[1979] On extension of functions in anisotropic classes under preservation of class. Tr. Mat. Inst. Steklova *150*, 52-66. Zbl. 417.46038. English translation: Proc. Steklov Inst. Math. *150*, 55-70 (1981)

Burenkov, V. I., Gold'man, M. L.

[1979] On extension of functions in L_p. Tr. Mat. Inst. Steklova *150*, 31-51. Zbl. 417.46037. English translation: Proc. Steklov Inst. Math. *150*, 33-53 (1981)

[1983] On the connections between norms of operators in periodic and nonperiodic function spaces. Tr. Mat. Inst. Steklova *161*, 47-105. Zbl. 541.46027. English translation: Proc. Steklov Inst. Math. *161*, 53-112 (1984)

Butzer, P. L., Behrens, H.

[1967] *Semi-groups of operators and approximation.* (Grundlehren 145) Springer, Berlin - Heidelberg - New York. Zbl. 164, 437

Calderón, A. P.

[1961] Lebesgue spaces of differentiable functions and distributions, in: Proc. Sympos. Pure Math. 4, pp. 33-49. Am. Math. Soc., Providence. Zbl. 195, 411

Deny, J., Lions, J.-L.

[1953-1954] Les espaces du type Beppo Levi. Ann. Inst. Fourier *5*, 305-370 (1955). Zbl. 65, 99

Douglas, J.
[1931] Solution of the problem of Plateau. Trans. Am. Math. Soc. *33*, 263-321. Zbl. 1, 141
Dubinskiĭ, Yu. A.
[1975] Sobolev spaces of infinite order and the behavior of solutions of some boundary problems under unlimited growth of the order of the equation, Mat. Sb. Nov. Ser. *98*, 163-184. Zbl. 324.46037. English translation: Math. USSR, Sb. *27*, 143-162 (1977)
[1976] Nontrivialness of Sobolev spaces of infinite order in the case of the entire Euclidean space and the torus. Mat. Sb. Nov. Ser. *100*, 436-446. Zbl. 332, 46021. English translation: Math. USSR, Sb. *29*, 393-401 (1978)
[1978] Traces of functions in Sobolev spaces of infinite order and inhomogeneous problems for nonlinear equations. Mat. Sb. Nov. Ser. *106*, 66-84. Zbl. 391.35027. English translation: Math. USSR, Sb. *34*, 627-644 (1978)
[1979] Limits of Banach spaces. Imbedding theorems. Application to Sobolev spaces of infinite order, Mat. Sb. Nov. Ser. *106*, 428-439. Zbl. 421.46024. English translation: Math. USSR, Sb. *38*, 395-409 (1981)
Dunford, N., Schwartz, J.
[1958] *Linear operators. Part 1: General theory.* Interscience: New York - London. Zbl. 84, 104 (Reprint Wiley, New York 1988)
Dzhabrailov, A. D.
[1967a] Properties of functions in some weighted spaces. Tr. Mat. Inst. Steklova *89*, 70-79. Zbl. 172, 165. English translation: Proc. Steklov Inst. Math. *89*, 80-91 (1968)
[1967b] On the theory of imbedding theorems, Tr. Mat. Inst. Steklova *89*, 80-118. Zbl. 166, 397. English translation: Proc. Steklov Inst. Math. *89*, 92-135 (1968)
[1972] Imbedding theorems for some classes of functions whose mixed derivatives satisfy a multiple integral Hölder condition. Tr. Mat. Inst. Steklova *117*, 113-138. Zbl. 253, 46074. English translation: Proc. Steklov Inst. Math. *117*, 135-164 (1974)
Dzhafarov, A. S.
[1963] Imbedding theorems for generalized classes of S. M. Nikolskiĭ. Azerbaidzhan. Gos. Univ. Uch. Zap., Ser. Fiz.-Mat. Nauk, No. 2, 45-49 [Russian]
Faĭn, B. L.
[1976] On the extension of functions from an infinite cylinder with deterioration of the class, Tr. Mat. Inst. Steklova *140*, 277-286. Zbl. 397.46026. English translation: Proc. Steklov Inst. Math. *140*, 303-312 (1979)
Fefferman, Ch., Stein, E. M.
[1972] H^p spaces of several variables. Acta Math. *129*, 137-193. Zbl. 257.46078
Fokht, A.S.
[1982] Weighted imbedding theorems and estimates for the solutions of equations of elliptic type I, II. Diff. Uravn. *18*, 1440-1449; 1927-1938. Zbl. 511.35013. English translation: Differ. Equations *18*, 1024-1032, 1385-1394 (1983)
[1984] On weighted imbedding theorems and their applications to estimates for solutions of equations of elliptic type. Differ. Uravn. *20*, 337-343. Zbl. 553.35024. English translation: Differ. Equations *20*, 270-276 (1984)
Gallouet, Th., Morel, J.-M.
[1983] Une équation semilinéaires elliptiques $L^1(\mathbb{R}^N)$. C. R. Acad. Sci. Paris, Sér. I *296*, 493-496. Zbl. 545.35035
Gil'derman, Yu. I.
[1962] Abstract set functions and S.L. Sobolev's imbedding theorems. Dokl. Akad. Nauk SSSR *144*, 962-964. Zbl. 201, 162. English translation: Sov. Math., Dokl. *4*, 743-747 (1963)
Globenko, I.G.
[1962] Some questions in the theory of imbedding for domains with singularities at the boundary. Mat. Sb. Nov. Ser. *57*, 201-224 [Russian]. Zbl. 103, 82
Gol'dman, M. L.
[1971] Isomorphism of generalized Hölder classes. Differ. Uravn. 7, 1449-1458. Zbl. 258.46032. English translation: Differ. Equations 7, 1100-1107 (1973)

[1979] Description of the traces of some function spaces. Tr. Mat. Inst. Steklova *160*, 99-127. Zbl. 417.46039. English translation: Proc. Steklov Inst. Math. *150*, 105-133 (1981)

[1980] A covering method for the description of general spaces of Besov type. Tr. Mat. Inst. Steklova *156*, 47-81. Zbl. 455.46035. English translation: Proc. Steklov Inst. Math. *156*, 51-87 (1983)

[1984a] On the imbedding of Nikol'skiĭ-Besov spaces with moduli of continuity of general type into Lorentz space. Dokl. Akad. Nauk SSSR *277*, 20-24. Zbl. 598.46023. English translation: Sov. Math., Dokl. *30*, 11-16 (1984)

[1984b] Imbedding theorems for anisotropic Nikol'skiĭ-Besov spaces with moduli of continuity of general type. Tr. Mat. Inst. Steklova *170*, 86-104. Zbl. 578.46025. English translation: Proc. Steklov Inst. Math. *170*, 95-116 (1987)

Gol'dshteĭn, V. M.

[1981] Extension of functions with first generalized derivatives from plane domains. Dokl. Akad. Nauk SSSR *257*, 4512-454. Zbl. 484.46023. English translation: Sov. Math., Dokl. *23*, 255-258 (1981)

Gol'dshteĭn, V. M., Reshetnyak, Yu. G.

[1983] *Introduction to the theory of functions with generalized derivatives and quasiconformal maps*. Nauka, Moscow. Zbl. 591.46021

Golovkin, K. K.

[1962] On equivalent normed fractional spaces. Tr. Mat. Inst. Steklova *66*, 364-383. Zbl. 142, 100. English translation: Transl., II. Ser., Am. Math. Soc. *79*, 53-76 (1969)

[1964a] On the impossibility of certain inequalities between function norms. Tr. Mat. Inst. Steklova *70*, 5-25. Zbl. 119, 106. English translation: Transl., II. Ser., Am. Math. Soc. *67*, 1-24 (1968)

[1964b] Imbedding theorems for fractional functions. Tr. Mat. Inst. Steklova *70*, 38-46. Zbl. 119, 106. English translation: Transl., II. Ser., Am. Math. Soc. *91*, 57-67 (1970)

[1967] On a generalization of the Marcinkiewicz interpolation theorem. Tr. Mat. Inst. Steklova *102*, 5-28. Zbl. 202, 400. English translation: Proc. Steklov Inst. Math. *102*, 1-28 (1970)

[1969] Parametrically normed spaces and normed massives. Tr. Mat. Inst. Steklova *106*, 1-135. Zbl. 223.46015. English translation: Proc. Steklov Inst. Math. *106*, 1-121 (1972)

Golovkin, K. K., Solonnikov, V. A.

[1964/66] Estimates for integral operators with translation invariant norms. I-II, Tr. Mat. Inst. Steklova *70*, 47-58; *72*, 5-30. Zbl. 163, 332, 167, 49. English translation: Transl., II. Ser., Am. Math. Soc. *61*, 97-112 (1968); Proc. Steklov Inst. Math. *92*, 3-32 (1968)

Gossez, J.-P.

[1974] Nonlinear elliptic boundary value problems for equations with rapidly (or slowly) increasing coefficients, Trans. Am. Math. Soc. *190*, 163-205. Zbl. 277.35052

Il'in, V. P.

[1977] Some imbedding theorems for the function spaces $L^{\lambda,\phi,b_s}_{r,p,\theta}$. Zap. Nauchn. Semin. Leningr. Otd. Mat. Inst. Steklova *70*, 49-75. Zbl. 429.46020. English translation: J. Sov. Math. *23*, 1909-1929 (1983)

Il'in, V. P., Shubochkina, T. A.

[1980] Some imbedding theorems for the function spaces $W^l_{p,a,\kappa}(G)$. Zap. Nauchn. Semin. Leningr. Otd. Mat. Inst. Steklova *102*, 27-41. Zbl. 474.46022. English translation: J. Sov. Math. *22*, 1183-1192 (1983)

[1981] On the existence and on the optimal choice of the values of the parameters in the inequalities which guarantee the validity of imbedding theorems, Zap. Nauchn. Semin. Leningr. Otd. Mat. Inst. Steklova *111*, 63-87. Zbl. 479.15016. English translation: J. Sov. Math. *24*, 37-55 (1984)

Jones, P. W.

[1981] Quasiconformal mappings and extendability of functions in Sobolev spaces. Acta Math. *147*, 71-88. Zbl. 489.30017

Kalyabin, G. A.

[1975] Estimates for functions with a finite O. V. Besov seminorm. Differ. Uravn. *11*, 713-717.

Zbl. 309.46022. English translation: Differ. Equations *11*, 538-541 (1976)
[1976] The trace class for a family of periodic weighted classes. Tr. Mat. Inst. Steklova *140*, 169-180. Zbl. 397.46030. English translation: Proc. Steklov Inst. Math. *140*, 183-195 (1979)
[1977a] The trace problem for anisotropic weighted spaces of Liouville type. Izv. Akad. Nauk SSSR, Ser. Mat. *41*, 1138-1160. Zbl. 385.46018. English translation: Math. USSR, Izv. *11*, 1085-1107 (1977)
[1977b] Caracterization of spaces of generalized Liouville differentiation. Mat. Sb. Nov. Ser. *104*, 42-48. Zbl. 373.46044. English translation: Math. USSR, Sb. *33*, 37-42 (1977)
[1978a] The trace class for generalized anisotropic Liouville spaces. Izv. Akad. Nauk SSSR, Ser. Mat. *42*, 305-314. Zbl. 378.46033. English translation: Math. USSR, Izv. *12*, 289-297 (1978)
[1978b] A generalized trace method in the theory of interpolation of Banach spaces. Mat. Sb. Nov. Ser. *106*, 85-93. Zbl. 385.46019. English translation: Math. USSR, Sb. *34*, 645-653 (1978)
[1979] Description of the traces for anisotropic spaces of Triebel-Lizorkin type. Tr. Mat. Inst. Steklova *150*, 160-173. Zbl. 417.46040. English translation: Proc. Steklov Inst. Math. *150*, 169-183 (1981)
[1980] Description of functions in classes of Besov-Lizorkin-Triebel type. Tr. Mat. Inst. Steklova *156*, 82-109. Zbl. 455.46036. English translation: Proc. Steklov Inst. Math. *156*, 89-118 (1983)
[1983] An estimate for the capacity of sets with respect to generalized Lizorkin-Triebel classes and weighted Sobolev classes. Tr. Mat. Inst. Steklova *161*, 111-124. Zbl. 545.46024. English translation: Proc. Steklov Inst. Math. *161*, 119-133 (1984)

Kantorovich, L. V., Akilov, G. P.
[1984] *Functional analysis*. 3rd rev. ed. Nauka, Moscow. English translation: Pergamon, Oxford - London - Edinburgh - New York - Paris - Frankfurt 1964

Kazaryan, G. G.
[1974] On the comparison of differential operators and differential operators of constant strength. Tr. Mat. Inst. Steklova *131*, 94-118. Zbl. 337.47021. English translation: Proc. Steklov Inst. Math. *131*, 99-123 (1975)
[1976] Estimates for differential operators and hypoelliptic operators. Tr. Mat. Inst. Steklova *140*, 130-161. Zbl. 399.35025. English translation: Proc. Steklov Inst. Math. *140*, 141-174 (1979)
[1977] Operators of constant strength with a lower estimate in terms of derivatives and formally hypoelliptic operators. Anal. Math. *3*, 263-289 [Russian]. Zbl. 374.35011
[1979a] A hypoellipticity criterion in terms of the power and strength of operators. Tr. Mat. Inst. Steklova *150*, 128-142. Zbl. 417.35029. English translation: Proc. Steklov Inst. Math. *150*, 135-150 (1981)
[1979b] Comparison of the power of polynomials and their hypoellipticity. Tr. Mat. Inst. Steklova *150*, 143-159. Zbl. 417.35030. English translation: Proc. Steklov Inst. Math. *150*, 151-167 (1981)

Kipriyanov, I. A.
[1967] The Fourier-Bessel transform and imbedding theorems for weighted classes. Tr. Mat. Inst. Steklova *89*, 130-213. Zbl. 157, 219. English translation: Proc. Steklov Inst. Math. *89*, 149-246 (1968)

Klimov, V. S.
[1976] Imbedding theorems and geometric inequalities. Izv. Akad. Nauk SSSR Ser. Mat. *40*, 645-671. Zbl. 332.46022. English translation: Math. USSR, Izv. *10*, 615-638 (1977)

Kocharli, A. F.
[1974] Some weighted imbedding theorems in spaces with nonsmooth boundary. Tr. Mat. Inst. Steklova *131*, 128-146. Zbl. 316.46023. English translation: Proc. Steklov Inst. Math. *131*, 135-162 (1975)

Kokilashvili, V. M.
[1978a] Anisotropic Bessel potentials. Imbedding theorems with an limit exponent. Tr. Mat. Inst. Tbilis. *58*, 134-149 [Russian]. Zbl. 456.46030
[1978b] Maximal inequalities and multipliers in weighted Triebel-Lizorkin spaces. Dokl. Akad.

Nauk SSSR *239*, 42-45. Zbl. 396.46034. English translation: Sov. Math., Dokl. *19*, 272-276 (1978)

[1979] On Hardy inequalities in weighted spaces. Soobshch. Akad. Nauk Gruz. SSR *96*, 37-40 [Russian]. Zbl. 434.26007

[1983] On weighted Lizorkin-Triebel spaces, singular integrals, multipliers, imbedding theorems. Tr. Mat. Inst. Steklova *161*, 125-149. Zbl. 576.46023. English translation: Proc. Steklov Inst. Math. *161*, 135-162 (1984)

[1985a] Maximal functions and integrals of potential type in weighted Lebesgue and Lorentz spaces. Tr. Mat. Inst. Steklova *172*, 192-201. Zbl. 575.42019. English translation: Proc. Steklov Inst. Math. *172*, 213-222 (1987)

[1985b] Weighted inequalities for some integral transformations. Tr. Tbilis. Mat. Inst. Razmadze *76*, 100-106 [Russian]. Zbl. 584.44004

[1985c] Maximal functions and singular integrals in weighted function spaces, Tr. Tbilis. Mat. Inst. Razmadze *80*, 1-116 [Russian]. Zbl. 609.42015

Kolmogorov, A. N., Fomin, S. V.

[1976] *Elements of the theory of functions and functional analysis*. Nauka, Moscow (Zbl. 167, 421 2nd ed. 1968) (5th ed. 1981) English translation of 1st ed.: Graylock, Rochester/Albany 1957/61 (2 vols.). Zbl. 57, 336; Zbl. 90, 87

Kondrashov, V. I.

[1938] On some estimates for families of functions subject to integral inequalities. Dokl. Akad. Nauk SSSR *18*, 235-240 [Russian]. Zbl. 21, 132

[1950] The behavior of functions in L_p^ν on manifolds of different dimensions. Dokl. Akad. Nauk SSSR *72*, 1009-1012 [Russian]. Zbl. 37, 204

Konyushkov, V. I.

[1958] Best approximations of trigonometric polynomials and Fourier coefficients. Mat. Sb. Nov. Ser. *44*, 53-84 [Russian]. Zbl. 81, 284

Korotkov, V. G.

[1962] Abstract set functions and imbedding theorems. Dokl. Akad. Nauk SSSR *146*, 531-534. Zbl. 134, 113. English translation: Sov. Math., Dokl. *3*, 1345-1349 (1963)

[1965] On compactness tests in abstract function spaces and on the complete continuity of the imbedding operator. Dokl. Akad. Nauk SSSR *160*, 530-533. Zbl. 139, 71. English translation: Sov. Math., Dokl. *6*, 132-135 (1965)

Krasnosel'skiĭ, A. F., Rutitskiĭ, Ya. B.

[1958] *Convex functions and Orlicz spaces*. Fizmatgiz, Moscow. Zbl. 84, 101. English translation: Nordhoff, Groningen 1961

Kreĭn, S. G., Petunin, Yu. I., Semenov, E. M.

[1978] *Interpolation of linear operators*. Nauka, Moscow. Zbl. 499.46044. English translation: Transl. Math. Monographs 54, Am. Math. Soc., Providence 1981

Kudrjavcev, L. D. (= Kudryavtsev, L. D.)

[1983] Some topics of the imbedding of function spaces and its applications, in: *Recent trends in Mathematics*, Conf., Rheinhardsbrunn, Oct. 11 - 13, 1982, pp. 183-192. Teubner, Leipzig. Zbl. 521.46025

[1986] On norms of weighted spaces of functions given on infinite intervals, Anal. Math. *12*, 269-282. Zbl. 646.46024

Kudryavtsev, L. D.

[1959] Direct and converse imbedding theorems. Applications and solution of elliptic equation by variational methods. Tr. Mat. Inst. Steklova *55*, 1-182 [Russian]. Zbl. 95, 92

[1966] Imbedding theorems for classes of functions defined in the entire space or a halfspace, I-II. Mat. Sb. Nov. Ser. *69*, 616-639; *70*, 3-35. Zbl. 192, 473. English translation: Transl., II. Ser., Am. Math. Soc. *74* 199-225; 227-260 (1968)

[1967] Solution of the first boundary problem for selfadjoint elliptic equations in the case of unbounded domains. Izv. Akad. Nauk SSSR, Ser. Math. *31*, 1179-1199. Zbl. 153, 426. English translation: Math. USSR, Izv. *1*, 1131-1151 (1969)

[1972] On polynomial traces and on moduli of smoothness of functions of several variables.

Tr. Mat. Inst. Steklova *117*, 180-211. Zbl. 244.46039. English translation: Proc. Steklov Inst. Math. *117*, 215-250 (1974)

[1975a] On the stablization of functions at infinity and on the solutions of differentail equations. Differ. Uravn. *11*, 332-357. Zbl. 299.35013. English translation: Differ. Equations *11*, 253-272 (1976)

[1975b] On the stabilization of functions at infinity and at a hyperplane. Tr. Mat. Inst. Steklova *134*, 124-141. Zbl. 337.46033. English translation: Proc. Steklov Inst. Math. *134*, 141-160 (1977)

[1980] On the construction of functions of finite functions which approximate functions in weighted classes. Tr. Mat. Inst. Steklova *156*, 121-129. Zbl. 457.46027. English translation: Proc. Steklov Inst. Math. *156*, 131-139 (1983)

[1983a] On the question of polynomial traces. Tr. Mat. Inst. Steklova *161*, 149-157. Zbl. 597.46029. English translation: Proc. Steklov Inst. Math. *161*, 163-170 (1984)

[1983b] On a variational method for the search of generalized solutions of differential equations in function spaces with a power weight. Differ. Uravn. *19*, 1723-1740. Zbl. 553.35016. English translation: Differ. Equations *19*, 1282-1296 (1983)

Kufner, A.

[1980] *Weighted Sobolev spaces.* (Teubner-Texte Math. 31.) Teubner, Leipzig (licens. ed. Wiley 1985). Zbl. 455.46034

Kufner, A., John, O., Fucik, S.

[1977] *Function spaces.* Academia, Prague. Zbl. 364.46022

Kuznetsov, Yu. V.

[1974] Some inequalities for fractional seminorms. Tr. Mat. Inst. Steklova *131*, 147-157. Zbl. 329.46034. English translation: Proc. Steklov Inst. Math. *131*, 153-164 (1975)

[1976] On the pasting of functions in the space $W^r_{p,\theta}$. Tr. Mat. Inst. Steklova *140*, 191-200. Zbl. 397.46028. English translation: Proc. Steklov Inst. Math. *140*, 209-220 (1979)

Lacroix, M.-Th.

[1975] Espaces de traces des espaces de Sobolev-Orlicz, J. Math. Pures Appl., IX. Sér. *53*, 439-458. Zbl. 275.46027

Lions, J.-L., Magenes, E.

[1968] *Problèmes aux limites non homogènes et applications, 1-2.* Dunod, Paris. Zbl. 165, 108

Lizorkin, P. I.

[1960] Boundary behavior of functions in "weighted" classes, Dokl. Akad. Nauk SSSR *132*, 514-517. Zbl. 106, 308. English translation: Sov. Math., Dokl. *1*, 589-593 (1960)

[1967a] A theorem of Littlewood-Paley type for multiple Fourier integrals. Tr. Mat. Inst. Steklova *89*, 214-230. Zbl. 159, 174. English translation: Proc. Steklov Inst. Math. *89*, 247-267 (1968)

[1967b] On multipliers of the Fourier integral in the spaces $L_{p,\theta}$, Tr. Mat. Inst. Steklova *89*, 231-248. Zbl. 159, 174. English translation: Proc. Steklov Inst. Math. *89*, 269-290 (1968)

[1969] Generalized Liouville differentiation and the method of multipliers in the theory of imbedding of classes of differentiable functions. Tr. Mat. Inst. Steklova *105*, 89-167. Zbl. 204, 437. English translation: Proc. Steklov Inst. Math. *105*, 105-202 (1971)

[1970] Multipliers of Fourier integrals and estimate for convolutions in spaces with mixed norm. Applications, Izv. Akad. Nauk SSSR, Ser. Math. *34*, 218-247. Zbl. 193, 214. English translation: Math. USSR, Izv. *4*, 225-255 (1971)

[1972] Operators connected with fractional differentiation and classes of differentiable functions. Tr. Mat. Inst. Steklova *117*, 212-243. Zbl. 253.46071. English translation: Proc. Steklov Inst. Math. *117*, 251-286 (1974)

[1974] Properties of functions in the spaces $\Lambda^r_{p,\theta}$. Tr. Mat. Inst. Steklova *131*, 158-181. Zbl. 317.46029. English translation: Proc. Steklov Inst. Math. *131*, 165-188 (1975)

[1975] On functional characteristics of the interpolation spaces $(L_p(\Omega), W^1_p(\Omega))_{\theta,p}$. Tr. Mat. Inst. Steklova *134*, 180-203. Zbl. 337.46033. English translation: Proc. Steklov Inst. Math. *134*, 203-227 (1977)

[1976] Interpolation of weighted L_p spaces. Tr. Mat. Inst. Steklova *140*, 201-211. Zbl. 397.46029. English translation: Proc. Steklov Inst. Math. *140*, 221-232 (1979)

[1977] On bases and multipliers in the spaces $B^r_{p,\theta}(M)$. Tr. Mat. Inst. Steklova *143*, 88-104. Zbl.468.46020. English translation: Proc. Steklov Inst. Math. *143*, 93-110 (1980)

Lizorkin, P. I., Nikol'skiĭ, S. M.

[1969] Compactness of sets of differentiable functions. Tr. Mat. Inst. Steklova *105*, 168-177. Zbl. 204, 438. English translation: Proc. Steklov Inst. Math. *105*, 203-215 (1971)

[1981] Coercivity properties of elliptic equations with degeneracy. A variational method. Tr. Mat. Inst. Steklova *157*, 90-118. Zbl. 475.35050. English translation: Proc. Steklov Inst. Math. *157*, 95-125 (1983)

[1983] Coercivity properties of elliptic equations with degeneracy and generalized right hand side. Tr. Mat. Inst. Steklova *161*, 157-183. Zbl. 541.35035. English translation: Proc. Steklov Inst. Math. *161*, 171-198 (1984)

Lizorkin, P. I., Otelbaev, M.

[1979] Imbedding and compactness theorems for Sobolev type spaces with weights, I-II. Mat. Sb. Nov. Ser. *108*, 358-377; *112*, 56-85. Zbl. 405.46025; Zbl. 447.46027. English translation: Math. USSR, Sb. *36*, 331-349 (1981); *40*, 51-77 (1981)

Luxemburg, W. A.

[1955] *Banach functional spaces.* Proefschr. doct. techn. wet., Techn. Hochschule te Delft. Zbl. 68, 92

Magenes, E.

[1964] Spazi di interpolazioni ed equazioni a derivate parziali, in: *Atti del VII. Congresso dell'Unione Matematica Italiana*, pp. 134-197. Edizioni Cremonese, Roma. Zbl. 173, 158

Marcinkiewicz, J.

[1939] Sur l'interpolation d'opérateurs, C.R. Acad. Sci. Paris *208*, 1272-1273. Zbl. 21, 16

Maz'ya, V. G.

[1979] On the summability with respect to an arbitrary measure of functions in the spaces of S.L. Sobolev - L.N. Slobodetskiĭ, Zap. Nauchn. Semin. Leningr. Otd. Mat. Inst. Steklova *92*, 192-202 [Russian]. Zbl. 431.46023

[1980] Integral representations for functions satisfying homogeneous boundary condition and its applications, Izv. Vyssh. Uchebn. Zaved., Mat. 1980 No. 2, 34-44. Zbl. 433.46033. English translation: Sov. Math. *24*, No. 2, 35-44 (1980)

[1981] On extension of functions in S.L. Sobolev spaces, Zap. Nauchn. Semin. Leningr. Otd. Mat. Inst. Steklova *113*, 231-236. Zbl. 474.46020. English translation: J. Sov. Math. *22*, 1851-1855 (1983)

[1983] On functions with finite Dirichlet integral in domains with a cusp point at the boundary, Zap. Nauchn. Semin. Leningr. Otd. Mat. Inst. Steklova *126*, 117-137. Zbl. 527.46025. English translation: J. Sov. Math. *27*, 2500-2514 (1984)

[1984] *S.L. Sobolev spaces.* LGU, Leningrad. English translation: Springer-Verlag, Berlin - Heidelberg - New York 1985

Mazya, W. (= Maz'ya, V.G.), Nagel, J.

[1978] Über äquivalente Normierung der anisotropen Funktionalräume $H^p(\mathbb{R}^n)$, Beitr. Anal. *12*, 7-17. Zbl. 422.46029

Meyers, N. G., Serrin, J.

[1964] H=W, Proc. Natl. Acad. Sci. U.S.A. *51*, 1055-1056. Zbl. 123, 305

Mikhaĭlov, V. P.

[1967] On the behavior at infinity of a class of polynomials, Tr. Mat. Inst. Steklova *91*, 59-81. Zbl. 185, 339. English translation: Proc. Steklov Inst. Math. *91*, 61-82 (1969)

Morrey, Ch.B.

[1966] *Multiple integrals in the calculus of variations.* (Grundlehren 130) Springer, Berlin - Heidelberg - New York. Zbl. 142, 387

Natanson, I. P.

[1974] *Theory of functions of a real variable.* 1st ed. Nauka, Moscow. Zbl. 39, 28. German translation: Akademie-Verlag, Berlin 1954 (5th ed. 1981)

Nečas, J.

[1967] *Les méthodes directes en théorie des equations elliptiques.* NCAV, Prague

Nikol'skaya, N. S.
[1973] Approximation of differentiable functions of several variables by Fourier sums in the metric L_p. Dokl. Akad. Nauk SSSR *208*, 1282-1285. Zbl. 295.41012. English translation: Sov. Math., Dokl. *14*, 262-266 (1973)
Nikol'skiĭ, S. M.
[1951] Inequalities for entire functions of finite degree and their application to the theory of differentiable functions of several variables, Tr. Mat. Inst. Steklova *38*, 244-278. Zbl. 49, 323. English translation: Transl., II. Ser., Am. Math. Soc. *80*, 1-38 (1969)
[1953] On the question on solving the polyharmonic equation by variational methods, Dokl. Akad. Nauk SSSR *88*, 409-411 [Russian]. Zbl. 53, 74
[1956] On the extension of functions of several variables with preservation of differentiability properties, Mat. Sb. Nov. Ser. *40*, 243-268. Zbl. 73, 26. English translation: Transl., II. Ser., Am. Math. Soc. *83*, 159-188 (1969)
[1956/58] Boundary properties of functions defined in domains with corner points, I-III, Mat. Sb. Nov. Ser. *40*, 303-318; *44*, 127-144; *45*, 181-194. Zbl. 73, 277; Zbl. 80, 85; Zbl. 133, 362. English translation: Transl., II. Ser., Am. Math. Soc. *83*, 101-120, 121-141, 143-157 (1969)
[1958] The variational problem of Hilbert, Izv. Akad. Nauk SSSR Ser. Mat. *22*, 599-630 [Russian]. Zbl. 86, 307
[1961] On the question of inequalities between partial derivatives, Tr. Mat. Inst. Steklova *64*, 147-164 [Russian]. Zbl. 121, 293
[1960/61] On a property of the classes H_p^r, Ann. Univ. Budapest No. 3-4, 205-216 [Russian]. Zbl. 111, 305
[1970] Variational problems of equations of elliptic type with degeneracy at the boundary, Tr. Mat. Inst. Steklova *150*, 212-238. Zbl. 416.35032. English translation: Proc. Steklov Inst. Math. *150*, 227-254 (1981)
[1977] *Approximation of function of several variables and imbedding theorems*. 2nd edition, Nauka, Moscow. Zbl. 496.46020. English translation: (Grundlehren 205) Springer, Berlin - Heidelberg - New York 1975)
Nikolsky, S. M. (= Nikol'skiĭ, S. M.), Lions, J.-L., Lisorkin, P. I. (= Lizorkin, P. I.)
[1965] Integral representations and isomorphism properties of some classes of functions, Ann. Sc. Norm. Sup. Pisa Sci. Fis. Mat., III. Ser., *19*, 127-178. Zbl. 151, 180
Nikol'skiĭ, Yu. S.
[1969] On imbedding theorems in weighted classes, Tr. Mat. Inst. Steklova *105*, 178-189. Zbl. 207, 434. English translation: Proc. Steklov Inst. Math. *105*, 217-231 (1971)
[1971] On extensions of weighted spaces of differentiable functions, Sibirsk. Mat. Zh. *12*, 158-70. Zbl. 226.46037. English translation: Sib. Math. J. *12*, 113-122 (1971)
[1974] On the behavior at infinity of functions with given differential-difference properties in L_p, Tr. Mat. Inst. Steklova *131*, 182-198. Zbl. 317.46028. English translation: Proc. Steklov Inst. Math. *131*, 189-205 (1975)
[1979] Inequalities between various seminorms of differentiable functions in several variables, Tr. Mat. Inst. Steklova *150*, 239-264. Zbl. 416.46023. English translation: *150*, 255-280 (1981)
Ovchinnikov, V. I.
[1976] Interpolation theorems resulting from Grothendieck's inequality, Funktsional. Anal. i Prilozhenya *10*, 45-54. Zbl. 342.46024. English translation: Funct. Anal. Appl. *10*, 287-294 (1977)
Peetre, J.
[1966] Espaces d'interpolation et théorème de Soboleff. Ann. Inst. Fourier, *16*, No. 1, 279-317. Zbl. 151, 79
[1975a] On the spaces of Triebel-Lizorkin type. Ark. Mat. *13*, 123-130. Zbl. 302.46021.
[1975b] The trace of Besov spaces – a limiting case. Technical report, Lund
[1976] *New thoughts on Besov spaces*. (Duke Univ. Math. Ser. 1) Duke University, Durham. Zbl. 356.46038
Perepelkin, V. G.
[1976] Integral representations for functions belonging to weighted S. L. Sobolev classes in

domains and some applications, I-II. Sib. Mat. Zh. *17*, 119-140; 318-330. Zbl. 322.46027; Zbl. 344.46075. English translation: Sib. Math. J. *17*, 96-112; 248-257 (1976)

Pigolkina, T. S.

[1969] On the theory of weighted classes. Tr. Mat. Inst. Steklova *105*, 201-212. Zbl. 207, 433. English translation: Proc. Steklov Inst. Math. *105*, 246-260 (1971)

Pokhozhaev, S. I.

[1965] On S. L. Sobolev's imbedding theorem in the case $pl = n$, in: *Proc. Sci. Tech. Conf. on Adv. Scienc. 1964-1965*, Ser. Mat. Min. Vyssh. i Sred. Spets. Obraz. RSFSR, MEI, Moscow [Russian]

Pólya, G.

[1931] Bemerkungen zur Interpolation und zur Näherungstheorie der Balkenbiegung. Z. Angew. Math. Mech. *11*, 445-449. Zbl. 3, 273

Portnov, V. R.

[1964] Two imbedding theorems for the spaces $L^{(l)}_{p,f}(\Omega \times \mathbb{R}_+)$. Dokl. Akad. Nauk SSSR *155*, 761-764. Zbl. 139, 71. English translation: Sov. Math., Dokl. *5*, 514-517 (1964)

[1965] A density theorem for finite functions in weighted classes. Dokl. Akad. Nauk SSSR *160*, 545-548. Zbl. 139, 71. English translation: Sov. Math., Dokl. *6*, 149-152 (1965)

Potapov, M. K.

[1968] On some conditions for mixed derivatives to be in L_p. Mathematica (Cluj) *10*, 355-367 [Russian]. Zbl. 184, 95

[1969] On the imbedding and coincidence of some classes of functions. Izv. Akad. Nauk SSSR, Ser. Mat. *33*, 840-860. Zbl. 187, 57. English translation: Math. USSR, Izv. *3*, 795-813 (1971)

[1980] Imbedding theorems in a mixed metric. Tr. Mat. Inst. Steklova *156*, 143-156. Zbl. 454.46030. Proc. Steklov Inst. Math. *156*, 155-171 (1983)

Poulsen, E. T.

[1962] Boundary values in functional spaces. Math. Scand. *10*, 45-52. Zbl. 104, 332

Radyno, Ya. V.

[1985] Differential equations in a scale of Banach spaces. Differ. Uravn. *21*, 1412-1422. Zbl. 592.34038 Differ. Equations *21*, 971-979 (1985)

Reshetnyak, Yu. G.

[1971] Some integral representations of differentiable functions, Sib. Mat. Zh. *12*, 420-432. Zbl. 221.46031. English translation: Sib. Math. J. *12*, 299-307 (1971)

[1980] Integral representations of functions in nonsmooth domains, Sib. Mat. Zh. *21*, 108-116. Zbl. 455.35025. English translation: Sib. Math. J. *21*, 833-839 (1981)

Samko, S. G.

[1984] *Hypersingular integrals and their applications*. Izd. Univ.: Rostov n/D [Russian]. Zbl. 577.42016

Sedov, V. N.

[1972a] Weighted spaces. Imbedding theorems, Differ. Uravn. *8*, 1452-1462. Zbl. 253.46077. English translation: Differ. Equations *8*, 1118-1126 (1974)

[1972b] Weighted spaces. Boundary conditions, Differ. Uravn. *8*, 1835-1847. Zbl. 264.46031 Differ. Equations *8*, 1415-1424 (1974)

Shaposhnikova, T. O.

[1980] Equivalent normings in functions spaces with fractional or functional smoothness, Sib. Mat. Zh. *21*, 184-196. Zbl. 438.46024. English translation: Sib. Math. J. *21*, 450-460 (1981)

Slobodetskiĭ, L. N.

[1958a] Sobolev spaces of fractional order and their applications to boundary value problems for partial differential equations, Dokl. Akad. Nauk SSSR *118*, 243-246 [Russian]. Zbl. 88, 303

[1958b] Generalized S.L. Sobolev spaces and their applications to boundary problems with partial derivatives, Uch. Zap. Leningr. Gos. Ped. Inst. No. 197, 54-112. Zbl. 192, 228. English translation: Transl., II. Ser., Am. Math. Soc. *57*, 207-275 (1966)

Sobolev, S. L.

[1935] Cauchy's problem in a function space, Dokl. Akad. Nauk SSSR *3*, 291-294 [Russian]

[1938] On a theorem of functional analysis, Mat. Sb. Nov. Ser. *4*, 471-497 [Russian]. Zbl. 12, 406

[1959] Some generalizations of the imbedding theorem, Fundam. Math. *47*, 277-324. Zbl. 127, 68. English translation: Transl., II. Ser., Am. Math. Soc. *30*, 295-344

[1962] *Some applications of functional analysis to mathematical physics.* 2nd edition, SO Akad. Nauk SSSR, Novosibirsk. Zbl. 123, 90. English translation: Transl. Math. Monographs Vol.7, Am. Math. Soc. 1963 German translation: Akademie-Verlag, Berlin 1964

[1963] On the density of finite functions in the space $L_p^m(E_n)$, Sib. Mat. Zh. *4*, 673-682 [Russian]. Zbl. 204, 438

[1974] *Introduction to the theory of cubature formulae.* Nauka, Moscow [Russian]. Zbl. 294,65013

Solonnikov, V. A.

[1960] On some properties of the space M_p^l of fractional order, Dokl. Akad. Nauk SSSR *134*, 282-285. Zbl. 133, 70. English translation: Sov. Math., Dokl. *1*, 1071-1074 (1960)

[1972] On some inequalities for functions in the classes $\vec{W}_p(\mathbb{R}^n)$, Zap. Nauchn. Sem. Leningr. Otd. Mat. Inst. Steklova *27*, 194-210. Zbl. 339.46025. English translation: J. Sov. Math. *3*, 549-564 (1975)

Stein, E. M.

[1970] *Singular integrals and differentiability properties of functions.* Princeton University Press, Princeton. Zbl. 207, 135 (Russian translation: Mir, Moscow 1973)

Stein, E. M., Weiss, G.

[1971] *Introduction to Fourier analysis on Euclidean spaces.* Princeton University Press, Princeton. Zbl. 232,42007

Tandit, B. V.

[1980a] On traces of functions on weighted spaces, Funktsional. Anal. i Prilozhen. *14*, 83-85. Zbl. 462.42012. English translation: Func. Anal. Appl. *14*, 317-319 (1981)

[1980b] Boundedness properties of functions in the space $W_{p,\phi}^r$, Tr. Mat. Inst. Steklova *156*, 223-232. Zbl. 457.46032. English translation: Proc. Steklov Inst. Math. *156*, 245-254 (1983)

[1980c] On the traces of functions in the class $W_{p,\phi}^r$, Differ. Uravn. *16*, 2062-2074. Zbl. 462.46022. English translation: *16*, 1321-1329 (1981)

Terekhin, A. P.

[1979] Mixed q-integral p-variation and theorems on equivalence and imbedding of classes of functions with mixed moduli of smoothness, Tr. Mat. Inst. Steklova *150*, 306-319. Zbl. 416.46018. English translation: Proc. Steklov Inst. Math. *150*, 323-336 (1981)

Thorin, G. O.

[1948] *Convexity theorems generalizing those of M. Riesz and Hadamard with some applications.* Thesis, Lund (also: Commun. Semin. Math. Univ. Lund *9*, 1-58). Zbl. 34, 204

Triebel, H.

[1973] Interpolation theory for function spaces of Besov type defined in domains, I, II, Math. Nachr. *57*, 51-85; *58*, 63-86. Zbl. 282.46023; Zbl. 282.46024

[1975] Interpolation properties of ε-capacity and of diameters. Geometric characteristics of the imbedding of function spaces of Sobolev-Besov type, Mat. Sb. Nov. Ser. *98*, 27-41. Zbl. 312.46043. English translation: Math. USSR, Sb. *27*, 23-37 (1977)

[1976] Spaces of Kudrjavcev type, I, II, Math. Anal. Appl. *56*, 253-277; 278-287. Zbl. 345.46031; Zbl. 345.46032

[1977] *Fourier analysis and function spaces.* (Teubner-Texte Math. 7) Teubner, Leipzig. Zbl. 345.42003

[1978a] *Spaces of Besov-Hardy-Sobolev-type.* (Teubner-Texte Math. 15) Teubner, Leipzig. Zbl. 408.46024

[1978b] On Besov-Hardy-Sobolev spaces in domains and regular elliptic boundary value problems. The case $0 < p \leq \infty$, Commun. Partial Diff. Equations *3*, 1083-1164. Zbl. 403.35034

[1978c] *Interpolation theory. Function spaces. Differential operators.* Deutsch. Verl. Wiss., Berlin. Zbl. 387.46033

[1983] *Theory of function spaces.* Birkhäuser, Basel - Boston - Stuttgart; Akademische Verl.-Ges., Leipzig. Zbl. 546.46027

Troisi, M.

[1969] Problemi elliptici con dati singolari. Ann. Mat. Pura Appl. *83*, 363-407. Zbl. 194, 419

Ul'yanov, P. L.

[1967] On the imbedding of some function spaces, Mat. Zametki *1*, 405-414. Zbl. 165, 389. English translation: Math. Notes *1*, 270-276 (1968)

[1968] Imbedding of some function classes H_p^{ω}, Izv. Akad Nauk SSSR Ser. Mat. *32*, 649-686. Zbl. 176, 433. English translation: Math. USSR, Izv. *2*, 601-637 (1969)

Uspenskiĭ, S. V.

[1961] On imbedding theorems for weighted classes, Tr. Mat. Inst. Steklova *60*, 282-303. Zbl. 119, 101. English translation: Transl., II. Ser., Am. Math. Soc. *87*, 121-145 (1970)

[1966] Imbedding and extension theorems for a class of spaces, I-II, Sib. Mat. Zh. *7*, 192-199; 409-418. Zbl. 152, 128. English translation: Sib. Math. J. *7*, 154-161; 333-342 (1966)

[1967] On mixed derivatives of functions summable with respect to a weight, Differ. Uravn. *3*, 139-154. Zbl. 152, 129. English translation: Differ. Equations *3*, 70-77 (1971)

Uspenskiĭ, S. V., Deminenko, G. B., Perepelkin, V. G.

[1984] *Imbedding theorems and applications to differential equations.* Nauka, Novosibirsk [Russian]. Zbl. 575.46030

Vasharin, A. A.

[1959] Boundary properties of function in the class $W_{2,\alpha}^1$ and their applications to the solution of a boundary problem in mathematical physics, Izv. Akad. Nauk SSSR, Ser. Mat. *23*, 421-454 [Russian]. Zbl. 92, 102

Yakovlev, G. N.

[1961] Boundary properties of a class of functions, Tr. Mat. Inst. Steklova *60*, 325-349 [Russian]. Zbl. 116, 316

[1967] On traces of functions in the space W_p^l on piecewise smooth surfaces, Mat. Sb. Nov. Ser. *74*, 526-543. Zbl. 162, 448. English translation: Math. USSR, Sb. *3*, 481-497 (1969)

II. Classes of Domains, Measures and Capacities in the Theory of Differentiable Functions

V.G. Maz'ya

Translated from the Russian
by J. Peetre

Contents

Introduction 143

Chapter 1. The Influence of the Geometry of the Domain on the Properties of Sobolev Spaces 143

§ 1. The Connection Between Imbedding Theorems and Isoperimetric Inequalities 143

1.1. The Equivalence of the Brunn Inequality to an Isoperimetric Inequality 143

1.2. The Best Constant in Sobolev's Inequality 145

1.3. Estimating the Integral with Respect to an Arbitrary Measure (the Case $p = 1$) 147

1.4. Estimating the Integral with Respect to an Arbitrary Measure (the Case $p > 1$) 149

1.5. Other Inequalities for the L_p-Norm of the Gradient 152

§ 2. Sobolev Spaces in "Bad" Domains 156

2.1. Introduction 156

2.2. The Classes J_α 159

2.3. The Case $p > 1$ 163

§ 3. Integral Inequalities for Functions on Riemannian Manifolds . . 172

§ 4. On the Space of Functions of Bounded Variation 176

§ 5. Extension of a Function in Sobolev Space on a Domain to $\mathbb{R}^n$. 181

§ 6. Sobolev Spaces in Domains Which Depend in a Singular Manner on Parameters 184

Chapter 2. Inequalities for Potentials and Their Applications to the Theory of Spaces of Differentiable Functions 189

§ 1. Estimates for Riesz and Bessel Potentials 189

1.1. On Summability in Measure of Functions in the Spaces $h_p^{(l)}, b_p^{(l)}$ etc. 189
1.2. Weighted Estimates for Potentials 196
§ 2. Multipliers in Spaces of Differentiable Functions 197

Chapter 3. Imbedding Theorems for Spaces of Functions Satisfying Homogeneous Boundary Conditions 199
§ 1. Inequalities for Functions with Noncomplete Homogeneous Cauchy Data 199
§ 2. "Massive" Sets of Zeros of Functions in Sobolev Spaces 202
§ 3. Imbedding of the Space $\overset{\circ}{L}{}_p^{(l)}(\Omega)$ into Other Function Spaces . . . 204
§ 4. On the Space $\overset{\circ}{H}{}_p^{(l)}(\Omega, \nu)$ 207
References 209

Introduction

In the past thirty years there has accumulated a large amount of information about conditions which are necessary and sufficient for various properties of spaces of Sobolev type to hold true. It is question of boundedness and compactness criteria for imbedding operators characterizing the domain or the weight functions, of tests for the possibility of extending functions from the domain to $\mathbb{R}^n$, of conditions asserting the density of one space of differentiable functions in another etc. An adequate description of the properties of function spaces has made it necessary to introduce new classes of domains of definition for the functions, or classes of measures entering in the norms. In this connection the universal importance of the notion of capacity of a set became manifest.

The present Part is mainly devoted to this circle of ideas. It is intended as a sequel of the survey by L.D. Kudryavtsev and S.M. Nikol'skiĭ but can be read independently.

Chapter 1

The Influence of the Geometry of the Domain on the Properties of Sobolev Spaces

§ 1. The Connection Between Imbedding Theorems and Isoperimetric Inequalities

1.1. The Equivalence of the Brunn Inequality to an Isoperimetric Inequality. Already the ancient Greeks knew that among all plane figures of given area the circle has the least perimeter. Formalized and generalized in different directions in the course of the past hundred years, this classical fact has found numerous applications in geometry, in functional analysis and in mathematical physics. In particular, more recently it was realized that it had a tight connection with the theory of Sobolev spaces.

Let g be an open set in $\mathbb{R}^n$ with compact smooth boundary (an *admissible set*). We state the *multidimensional version of the isoperimetric property of the circle* in the form of an inequality

$$[\text{mes}_n(g)]^{(n-1)/n} \leq n^{-1} v_n^{-1/n} s(\partial g), \tag{1.1}$$

where mes_n and s are the volume and the $(n-1)$-dimensional area, $v_n = \pi^{n/2}/\Gamma(1+n/2)$ being the total volume of the n-dimensional unit ball.

In 1960 Federer, Fleming and the author discovered that the geometric inequality (1.1) is equivalent to the integral inequality

$$\|u\|_{L_{n/(n-1)}(\Omega)} \leq n^{-1} v_n^{-1/n} \|\text{grad}\, u\|_{L_1(\Omega)}, \tag{1.2}$$

where Ω is an open subset of $\mathbb{R}^n$ and u an arbitrary element of the space $C_0^\infty(\mathbb{R}^n)$ of infinitely differentiable functions with compact supports in Ω. This inequality with different constant was obtained in 1887 by Brunn (Über Ovale und Eiflächen, Inaug. Diss., München) and once more in 1958 by Gagliardo.

Let us show that (1.2) follows from (1.1). To this end we rewrite inequality (1.2) in terms of integrals with respect to the level sets of the function $|u|$. Put $M_t = \{x : |u(x)| \geq t\}$ and $E_t = \{x : |u(x)| = t\}$. It is clear that

$$\int_\Omega |u(x)|^{n/(n-1)} dx = \int_\Omega \int_0^{|u(x)|} d(t^{\frac{n}{n-1}}) dx.$$

Changing the order of integration we bring the right hand side in the form

$$\int_0^\infty \int_{M_t} dx d(t^{\frac{n}{n-1}}).$$

Thus

$$\|u\|_{L_{n/(n-1)}(\Omega)} = \left[\frac{n}{n-1} \int_0^\infty \text{mes}_n(M_t) t^{\frac{1}{n-1}} dt\right]^{(n-1)/n}. \tag{1.3}$$

In order to transform the right hand side of (1.2) we use the well-known formula

$$\int_\Omega \Phi(x)|\text{grad } u(x)| dx = \int_0^\infty dt \int_{E_t} \Phi(x) ds(x), \tag{1.4}$$

whose validity most convincingly can be verified using the following heuristic reasoning. The volume element dx equals the product $dlds$, where dl is the element of the force line and ds the area element of the equipotential surface passing through the point x. It is clear that $|\text{grad}\, u(x)| = |\partial u/\partial l| = dt/dl$. Integrating first over the surface E_t and then with respect to t we obtain (1.4). A rigorous foundation of formula (1.4) for Lipschitz functions can be found in Burago-Zalgaller [1980].

In our case $\Phi = 1$ and (1.4) reduces to the following *representation* of *the variation of a function of n variables*

$$\|\text{grad}\, u\|_{L_1(\Omega)} = \int_0^\infty s(E_t) dt. \tag{1.5}$$

We should bear in mind that the set of those levels t for which $\{x \in E_t : \text{grad}\, u(x) = 0\} \neq \emptyset$, has measure zero, so that E_t is a smooth surface bounding M_t for almost all $t > 0$. Upon applying to M_t the classical isoperimetric inequality (1.1) we find

$$n^{-1}v_n^{-1/n}\|\text{grad}\, u\|_{L_1(\Omega)} \geq \int_0^\infty [\text{mes}_n(M_t)]^{(n-1)/n}dt. \tag{1.6}$$

It remains to compare the right hand parts of (1.3) and (1.6). As the function $t \mapsto \text{mes}_n(M_t)$ is nonincreasing, then

$$t[\text{mes}_n(M_t)]^{(n-1)/n} \leq \int_0^t [\text{mes}_n(M_t)]^{(n-1)/n}dt$$

and therefore the right hand side of the identity does not exceed

$$\left\{\frac{n}{n-1}\int_0^\infty \left[\int_0^t [\text{mes}_n(M_\tau)]^{(n-1)/n}d\tau\right]^{1/(n-1)} [\text{mes}_n(M_t)]^{(n-1)/n}dt\right\}^{(n-1)/n}$$

$$= \int_0^\infty [\text{mes}_n(M_t)]^{(n-1)/n}dt.$$

Now inequality (1.2) follows from (1.3) and (1.6).

The implication (1.2) $\Rightarrow$ (1.3) is almost selfevident. Indeed, let us plug into (1.2) the function $x \mapsto u_\varepsilon(x) = \alpha[\text{dist}(x, g)]$ where $t \mapsto \alpha(t)$ is a nonincreasing differentiable function in $[0, 1]$ which equals 1 on $t = 0$ and 0 on $t > \varepsilon$ (ε being a sufficiently small number). In view of (1.4) we have the inequality

$$\int_\Omega |\text{grad}\ u_\varepsilon|dx = \int_0^\varepsilon \alpha'(t)s(\partial g_t)dt,$$

where $g_t = \{x : \text{dist}(x, g) < t\}$. Since $s(\partial g_t) \to s(\partial g)$ as $t \to 0$ then

$$\int_\Omega |\text{grad}\ u_\varepsilon|dx \to s(\partial g) \quad \text{as} \quad \varepsilon \to 0.$$

On the other hand

$$\|u_\varepsilon\|_{L_{n/(n-1)}(\Omega)} \geq [\text{mes}_n(g)]^{(n-1)/n}.$$

Thus (1.1) is a consequence of (1.2).

It follows from the preceding argument that the constant $n^{-1}v_n^{-1/n}$ is exact.

1.2. The Best Constant in Sobolev's Inequality. Using the isoperimetric inequality (1.1) one can find the best constant in *Sobolev's inequality*

$$\|u\|_{L_{pn/(n-p)}(\Omega)} \leq K(n, p)\|\text{grad}\ u\|_{L_p(\Omega)}. \tag{1.7}$$

Here $n > p > 1$ and as before $u \in C_0^\infty(\Omega)$.

Let us give a sketch of the proof omitting the technical details.

We write $|u|$ in the form of a superposition $\lambda(v)$, where $v(x)$ is the volume of the set bounded by the level surface of $|u|$ passing through the point x. Then

$$\|u\|_{L_{pn/(n-p)}(\Omega)} = \left\{\int_0^{\text{mes}_n(\Omega)} [\lambda(v)]^{pn/(n-p)}dv\right\}^{(n-p)/pn}. \tag{1.8}$$

On the other hand, by virtue of (1.4),

$$\|\text{grad } u\|_{L_p(\Omega)} = \left\{ \int_0^{\text{mes}_n(\Omega)} |\lambda'(v)|^p \int_{v(x)=v} |\text{grad } v(x)|^{p-1} ds(x) dv \right\}^{1/p}. \tag{1.9}$$

Let us now remark that by Hölder's inequality we have for the area $s(v)$ of the surface $\{x : v(x) = v\}$ the estimate

$$[s(v)]^p = \left(\int_{v(x)=v} \left(\frac{dv}{dl}\right)^{(p-1)/p} \left(\frac{dl}{dv}\right)^{(p-1)/p} ds \right)^p$$

$$\leq \int_{v(x)=v} \left(\frac{dv}{dl}\right)^{p-1} ds \left(\int_{v(x)=v} \frac{dlds}{dv} \right)^{p-1}$$

so that, as $\int_{v(x)=v} dlds = dv$, we find

$$\int_{v(x)=v} |\text{grad } v(x)|^{p-1} ds(x) \geq [s(v)]^p. \tag{1.10}$$

This estimate together with (1.1), (1.8) and (1.9) shows that Sobolev's inequality follows from the following inequality for functions of one variable vanishing at infinity:

$$\left\{ \int_0^\infty [\lambda(v)]^{pn/(n-p)} dv \right\}^{(p-n)/pn} \leq K(n,p) n v_n^{1/n} \left\{ \int_0^\infty |\lambda'(v)|^p v^{p(n-1)/n} dv \right\}^{1/p}. \tag{1.11}$$

Moreover, inserting in (1.7) a function u which depends only on the distance from the origin gives the equivalence of (1.7) and (1.11). Thus

$$\sup_{u \in C_0^\infty(\Omega)} \|u\|_{L_{pn/(n-p)}(\Omega)} / \|\text{grad } u\|_{L_p(\Omega)} =$$

$$= \sup_\lambda n^{-1} v_n^{-1} \frac{\{\int_0^\infty [\lambda(v)]^{pn/(n-p)} dv\}^{(n-p)/pn}}{\{\int_0^\infty |\lambda'(v)|^p v^{p(n-1)/n} dv\}^{1/p}}.$$

So, the question of the exact constant in (1.7) is reduced to a one dimensional variational problem. This problem was solved exactly by Bliss in 1930 using the classical methods of the calculus of variations. The exact upper bound is reached by any function of the form

$$\lambda(v) = (a + b v^{p/n(p-1)})^{1-n/p}, \quad a, b = \text{const} > 0,$$

and equals

$$\pi^{-1/2} n^{-1/p} \left(\frac{p-1}{n-p}\right)^{(p-1)/p} \left\{ \frac{\Gamma(1+n/2)\Gamma(n)}{\Gamma(n/p)\Gamma(1+n-n/p)} \right\}^{1/n}. \tag{1.12}$$

1.3. Estimating the Integral with Respect to an Arbitrary Measure (the Case $p = 1$). The proof given in Sect. 1.1 of the equivalence of inequality (1.2) and the isoperimetric inequality (1.1) carries over, with trivial modifications, to the general situation when we estimate the norm of a function $u \in C_0^\infty(\Omega)$ in L_q with respect to an arbitrary nonnegative measure μ. As a result we obtain the following exhaustive statement.

Theorem 1.1. *Let μ be a measure on Ω and consider for $q \geq 1$ the space $L_q(\Omega, \mu)$ with norm*

$$\|u\|_{L_q(\Omega,\mu)} = \left\{ \int_\Omega |u(x)|^q d\mu(x) \right\}^{1/q}.$$

The inequality

$$\|u\|_{L_q(\Omega,\mu)} \leq C \|\operatorname{grad} u\|_{L_1(\Omega)} \tag{1.13}$$

holds for all $u \in C_0^\infty(\Omega)$ iff for each admissible set $g, \bar{g} \in \Omega$, holds

$$[\mu(g)]^{1/q} \leq Cs(\partial g). \tag{1.14}$$

Example 1.1. Let $\alpha \in [0,1]$ and set $\mu(g) = \int_g |x|^{-\alpha} dx$. Using (1.1) it is easy to verify that

$$[\mu(g)]^{\frac{n-1}{n-\alpha}} \leq (n-\alpha)^{\frac{1-n}{n-\alpha}} (nv_n)^{\frac{\alpha-1}{n-\alpha}} s(\partial g),$$

where equality holds if g is a ball. Therefore for all $u \in C_0^\infty(\mathbb{R}^n)$ holds the inequality

$$\left(\int_{\mathbb{R}^n} |u(x)|^{\frac{n-\alpha}{n-1}} \frac{dx}{|x|^\alpha} \right)^{\frac{n-1}{n-\alpha}} \leq (n-\alpha)^{\frac{1-n}{n-\alpha}} (nv_n)^{\frac{\alpha-1}{n-\alpha}} \|\operatorname{grad} u\|_{L_1(\mathbb{R}^n)}.$$

It does not require any new ideas to establish that the multiplicative inequality

$$\|u\|_{L_q(\Omega,\mu)} \leq C \|\operatorname{grad} u\|_{L_1(\Omega)}^\delta \|u\|_{L_r(\Omega,\nu)}^{1-\delta}, u \in C_0^\infty(\Omega),$$

containing two measure μ and ν, is for $\delta + (1-\delta)r^{-1} \geq q^{-1}$, $\delta \in [0,1]$, equivalent to the inequality

$$[\mu(g)]^{1/q} \leq \text{const} \cdot [s(\partial g)]^\delta [\nu(g)]^{(1-\delta)/r}.$$

Analogously, the inequality

$$\|u\|_{L_q(\Omega,\mu)} \leq C(\|\operatorname{grad} u\|_{L_1(\Omega)} + \|u\|_{L_r(\Omega,\nu)}),$$

where $u \in C_0^\infty(\Omega)$ and $q \geq 1 \geq r > 0$, holds iff for all admissible sets $g, \bar{g} \subset \Omega$, holds

$$[\mu(g)]^{1/q} \leq \text{const} \cdot (s(\partial g) + [\nu(g)]^{1/r}).$$

In the case $\Omega = \mathbb{R}^n$ it suffices to verify inequality (1.14) for balls only. Indeed, one has

Theorem 1.2. *Inequality (1.13), where $\Omega = \mathbb{R}^n$ and $q \geq 1$, holds iff for each ball $\mathscr{B}_\varrho(x) = \{y : |y - x| < \rho\}$ holds the estimate*

$$[\mu(\mathscr{B}_\varrho(x))]^{1/q} \leq \text{const} \cdot \varrho^{n-1}. \tag{1.15}$$

The exact constant C in (1.13) satisfies the relation $C \sim K$, where

$$K = \sup_{x \in \mathbb{R}^n, \varrho > 0} \varrho^{1-n} [\mu(\mathscr{B}_\varrho(x))]^{1/q}. \tag{1.16}$$

Here and in what follows the equivalence sign $\sim$ means that the ratio between the left hand and the right hand sides is separated from zero and bounded by constants depending only on n, q, p, l etc. Such constants will always be written $c, c_1, c_2, \ldots$.

In view of Theorem 1.1 it suffices to derive (1.14) from (1.15). Let g be an admissible set. Each point $x \in g$ is the center of the ball $\mathscr{B}_r(x)$ such that

$$\frac{\operatorname{mes}_n(\mathscr{B}_r(x) \cap g)}{\operatorname{mes}_n(\mathscr{B}_r(x))} = \frac{1}{2}$$

(this ratio is a continuous function of r which equals 1 for small r and tends to 0 as $r \to \infty$). The reader will of course have no doubts about the validity of the obvious inequality

$$s(\mathscr{B}_r(x) \cap \partial g) \geq \alpha_n r^{n-1},$$

where α_n is a positive constant depending only on n. By the classical *theorem of Besicovich* there exists a sequence of nonintersecting balls $\{\mathscr{B}_{r_j}(x)\}_{j \geq 1}$ such that

$$g \subset \bigcup_{j=1}^{\infty} \mathscr{B}_{3r_j}(x_j).$$

Therefore

$$[\mu(g)]^{1/q} \leq \Big[\sum_j \mu(\mathscr{B}_{3r_j}(x_j))\Big]^{1/q} \leq \sum_j [\mu(\mathscr{B}_{3r_j}(x_j))]^{1/q} \leq$$

$$\leq K \sum_j (3r_j)^{n-1} \leq 3^{n-1}\alpha_n^{-1} K \sum_j s(\mathscr{B}_{3r_j}(x_j) \cap \partial g)$$

and finally

$$[\mu(g)]^{1/q} \leq 3^{n-1}\alpha_n^{-1} K \cdot s(\partial g).$$

1.4. Estimating the Integral with Respect to an Arbitrary Measure (the Case $p > 1$). In order to establish the necessary and sufficient condition for the validity of the inequality

$$\|u\|_{L_q(\Omega,\mu)} \leq D\|\text{grad } u\|_{L_p(\Omega)}, \tag{1.17}$$

where $q \geq p > 1$, we have to introduce the notion of p-capacity.

Let F be a compact subset of Ω. The *p-capacity of F with respect to Ω* is defined to be the number

$$p\text{-cap}_\Omega(F) = \inf\left\{\int_\Omega |\text{grad } u|^p dx : u \in C_0^\infty(\Omega), u \geq 1 \text{ on } F\right\}. \tag{1.18}$$

We will write $\text{cap}_\Omega(F)$ instead of $2\text{-cap}_\Omega(F)$. The set function cap_Ω differs only by a factor $(n-2)v_n/n$ from the *harmonic capacity*, defined as the least upper bound of $v(F)$ over the set of all measures v supported by F and satisfying the condition

$$\int G(P,Q)dv(Q) \leq 1,$$

where G is Green's function of the domain Ω. If $\Omega = \mathbb{R}^3$ then it is just question of the *electrostatic capacity* of F.

It follows from the definition (1.18) that p-capacity is a nondecreasing function in F and a nonincreasing one in Ω. We have *Choquet's inequality*

$$p\text{-cap}_\Omega(F_1 \cap F_2) + p\text{-cap}_\Omega(F_1 \cup F_2) \leq p\text{-cap}_\Omega(F_1) + p\text{-cap}_\Omega(F_2)$$

for arbitrary compact sets F_1 and F_2 in Ω. It is easy to check that p-capacity is continuous from the right. This means that for each $\varepsilon > 0$ there exists a neighborhood G, $F \subset G \subset \overline{G} \subset \Omega$ such that for each compact set F_1, $F \subset F_1 \subset G$ holds the inequality

$$p\text{-cap}_\Omega(F_1) \leq p\text{-cap}_\Omega(F) + \varepsilon.$$

Let E be an arbitrary subset of Ω. The *inner* and the *outer p-capacities* are defined as the numbers

$$p\text{-}\underline{\text{cap}}_\Omega(E) = \sup_{F\subset E} p\text{-cap}_\Omega(F), \quad F \text{ compact in } \Omega,$$

$$p\text{-}\overline{\text{cap}}_\Omega(E) = \inf_{G\supset E} p\text{-}\underline{\text{cap}}_\Omega(G), \quad G \text{ open in } \Omega.$$

It follows from general Choquet theory that for each Borel set both capacities coincide. Their common value is called the p-capacity and will be written $p\text{-cap}_\Omega(E)$.

Minimizing the right hand side of formula (1.9) over all functions $\lambda \in C_0^\infty[0, \text{mes}_n(\Omega)]$ subject to the inequality $\lambda(v) \geq 1$ for $v \leq \text{mes}_n(F)$, we obtain yet another expression for the p-capacity

$$p\text{-cap}_\Omega(F) = \inf\left\{\int_{\text{mes}_n(F)}^{\text{mes}_n(\Omega)} \left[\int_{v(x)=v} |\text{grad } v(x)|^{p-1} ds(x)\right]^{1/(1-p)} dv\right\}^{1-p}. \tag{1.19}$$

Here the infimum is taken over all functions u in the definition (1.18). This useful identity is known as *Dirichlet's principle with prescribed level surfaces*, Pólya-Szegö [1951].

Estimating the integral over the level surface $v(x) = v$ with the aid of (1.10), we derive from (1.19) the following useful inequality

$$p\text{-cap}_\Omega(F) \geq \inf\left\{\int_{\text{mes}_n(F)}^{\text{mes}_n(\Omega)} \frac{dv}{[s(v)]^{p/(p-1)}}\right\}^{1-p}. \tag{1.20}$$

From this and (1.1) we obtain *isoperimetric inequalities for the p-capacity*

$$p\text{-cap}_\Omega(F) \geq n v_n^{p/n} \left|\frac{p-n}{p-1}\right|^{p-1} \cdot$$

$$\cdot \left|[\text{mes}_n(\Omega)]^{(p-n)/n(p-1)} - [\text{mes}_n(F)]^{(p-n)/n(p-1)}\right|^{1-p}, \quad \text{if} \quad p \neq n,$$

$$p\text{-cap}_\Omega(F) \geq n^n v_n \left[\log \frac{\text{mes}_n(\Omega)}{\text{mes}_n(F)}\right]^{1-n}, \quad \text{if} \quad p = n.$$

In particular, if $n > p$ then

$$p\text{-cap}_{\mathbb{R}^n}(F) \geq n v_n^{p/n} \left(\frac{n-p}{p-1}\right)^{p-1} [\text{mes}_n(F)]^{(n-p)/n}.$$

If Ω and F are concentric balls, then the three preceding estimates come as identities.

The application of p-capacity to imbedding theorems is based on the inequality

$$\int_0^\infty p\text{-cap}_\Omega(M_t) t^{p-1} dt \leq \left(\frac{p}{p-1}\right)^{p-1} \int_\Omega |\text{grad}\, u|^p dx, \tag{1.21}$$

which plays the same role as formula (1.5).

For $p > 1$ the estimate (1.21) is obtained in the following manner. In view of (1.9)

$$\|\text{grad}\ u\|^p_{L_p(\Omega)} = \int_0^{\psi(\text{mes}_n \Omega)} |\frac{d}{d\psi} \lambda(v(\psi))|^p d\psi, \tag{1.22}$$

where ψ is a new independent variable defined by the formula

$$\psi(v) = \int_v^{\text{mes}_n(\Omega)} \left[\int_{v(x)=v} |\text{grad}\, v(x)|^{p-1} ds(x)\right]^{1/(1-p)} dv.$$

In view of Hardy's inequality the right hand side of (1.22) majorizes the number

$$\left(\frac{p-1}{p}\right)^p \int_0^{\psi(\text{mes}_n(\Omega))} \left[\frac{\lambda(v(\psi))}{\psi}\right]^p d\psi = \frac{(p-1)^{p-1}}{p^p} \int_0^{\psi(\text{mes}_n(\Omega))} \psi^{1-p} d[\lambda(v(\psi))]^p.$$

It remains to apply the identity (1.19) according to which

$$p\text{-cap}_\Omega(M_{\lambda(v(\psi))}) \leq \psi^{1-p}.$$

As a simple consequence of formula (1.5) we have the representation

$$1\text{-cap}_\Omega(F) = \inf_g s(\partial g),$$

where g is any admissible set with $\Omega \supset \bar{g} \supset g \supset F$. From this and from (1.5) it follows that inequality (1.21) holds also for $p = 1$.

Now we can obtain a twosided estimate for the constant D in inequality (1.17).

Theorem 1.2. (i) *Assume that there exists a constant $\mathscr{A}$ such that for each compact set $F \subset \Omega$ holds*

$$[\mu(F)]^{\alpha p} \leq \mathscr{A} \cdot p\text{-cap}_\Omega(F), \tag{1.23}$$

where $p \geq 1, \alpha > 0, \alpha p \leq 1$. Then inequality (1.17) holds for all $u \in C_0^\infty(\Omega)$, with $q = \alpha^{-1}$ and $D \leq p(p-1)^{(1-p)/p}\mathscr{A}^{1/p}$.

(ii) *If (1.17) holds for each $u \in C_0^\infty(\Omega)$, with $q > 0$, $p \geq 1$ and a constant D independent of u, then inequality (1.23) holds for all compact sets F, $F \subset \Omega$, with $\alpha = q^{-1}$ and $\mathscr{A}^{1/p} \leq D$.*

Reasoning as in Sect. 1.1 we conclude that

$$\|u\|_{L_q(\Omega,\mu)} \leq \left(\int_0^\infty [\mu(M_t)]^{p/q} d(t^p) \right)^{1/p}.$$

Applying subsequently (1.23) and (1.21) we arrive at statement (i). In order to get (ii) we plug into (1.17) an arbitrary function u in $C_0^\infty(\Omega)$ such that $u \geq 1$ on a compact set F. Then $[\mu(F)]^{1/q} \leq D\|\text{grad}\, u\|_{L_p(\Omega)}$. Minimizing the right hand side concludes the proof.

For $p = 2$ Theorem 1.2 has applications to the quadratic form

$$Q[u,u] = \int_\Omega |\text{grad}\, u|^2 dx - \int_\Omega |u|^2 d\mu(x), \quad u \in C_0^\infty(\Omega),$$

appearing in the theory of the Schrödinger operator. With the aid of it one can derive conditions for positivity and semiboundedness and further conditions for the finiteness and discreteness of the spectrum of this operator. For instance, the inequality

$$\lim_{\varepsilon \to +0} \sup \left\{ \frac{\mu(F)}{\text{cap}_\Omega(F)} : F \subset \Omega, \text{diam}(F) < \varepsilon \right\} < \frac{1}{4}$$

is sufficient for the semiboundedness of the form Q in $L_2(\Omega)$. On the other hand, if Q is semibounded in $L_2(\Omega)$ then the left hand side of the last inequality is not larger than one.

1.5. Other Inequalities for the L_p-Norm of the Gradient. Let us remark that in the case $\Omega = \mathbb{R}^n, q > p > 1$, inequality (1.23), which is necessary and sufficient for (1.17), is fulfilled for all compact sets F iff it holds for any ball (cf. Theorem 1.2). This follows from a theorem by D.R. Adams to which we return in § 1 of Chap. 2.

Next we give a condition, formulated in terms of p-capacity, for the validity of some integral inequalities involving the L_p-norm of the gradient of a function in $C_0^\infty(\Omega)$, Maz'ya [1985].

(i) Condition (1.23) is necessary and sufficient for the inequality

$$\|u\|_{L_q(\Omega,\mu)} \le C \|\text{grad } u\|^{\delta}_{L_p(\Omega)} \|u\|^{1-\delta}_{L_r(\Omega,\mu)},$$

where $0 < q \le q^* = \alpha^{-1}$ provided $\alpha p \le 1$ or $0 < q < q^* = \alpha^{-1}$ provided $\alpha p > 1; r \in (0, q), \delta = q^*(q-r)/q(q^*-r)$.

(ii) Inequality (1.17) with $p > q \ge 1$ holds true iff

$$\sup_{\{S\}} \sum_{j=-\infty}^{+\infty} \left(\frac{\mu_j^{p/q}}{\gamma_j} \right)^{q/(p-q)} < \infty.$$

Here $S = \{g_j\}_{j=-\infty}^{+\infty}$ is an arbitrary sequence of admissible subsets of Ω such that $\bar{g}_j \subset g_{j+1}; \mu_j = \mu(g_j); \gamma_j = p\text{-cap}_{g_{j+1}}(g_j)$.

A sufficient condition for (1.17) with $p > q \ge 1$ is the inequality

$$\int_0^{\mu(\Omega)} \left[\frac{\tau}{\nu(\tau)} \right]^{q/(p-q)} d\tau < \infty,$$

where $\nu(t) = \inf p\text{-cap}_\Omega(g)$, where the infimum is taken over all admissible sets g, $\bar{g} \subset \Omega$ subject to the condition $\mu(g) \ge t$.

(iii) The inequality

$$\|u\|_{L_q(\Omega,\mu)} \le C(\|\text{grad } u\|_{L_p(\Omega)} + \|u\|_{L_r(\Omega,\nu)}),$$

where $q \ge p \ge r, p > 1$, holds true iff for all admissible sets g and G with $\bar{g} \subset G \subset \bar{G} \subset \Omega$ we have the estimate

$$[\mu(g)]^{1/q} \le \text{const} \left\{ [p\text{-cap}_G(g)]^{1/p} + [\nu(G)]^{1/r} \right\}.$$

Clearly, the following simpler condition is sufficient for the validity of the last inequality

$$[\mu(g)]^{1/q} \le \text{const} \left\{ [p\text{-cap}_\Omega(g)]^{1/p} + [\nu(g)]^{1/r} \right\},$$

where only sets g intervene. However, one can show that this condition is not necessary.

(iv) The multiplicative inequality

$$\|u\|_{L_q(\Omega,\mu)} \leq C \|\text{grad } u\|^{\delta}_{L_p(\Omega)} \|u\|^{1-\delta}_{L_r(\Omega,\nu)},$$

where $p > 1, 1/q \leq (1-\delta)/r + \delta/r, r, q > 0, 0 \leq \delta \leq 1$, holds true iff for any two admissible sets g and G, $\bar{g} \subset G \subset \bar{G} \subset \Omega$, holds the estimate

$$[\mu(g)]^{1/q} \leq \text{const} \cdot [p\text{-cap}_G(g)]^{\delta/p} [\nu(G)]^{(1-\delta)/r}$$

(v) Let us state a sufficient and a similar, though different necessary condition for the validity of the inequlity

$$\int_{\Omega} |u|^p d\mu \leq c \int_{\Omega} |\text{grad}\, u|^p dx, \quad u \in C_0^{\infty}(\Omega), \tag{1.24}$$

where μ is a *signed* measure (and not a positive measure as before).

Let μ^+ and μ^- be the positive and the negative parts of μ and let g and G be admissible sets with $\bar{g} \subset G \subset \bar{G} \subset \Omega$. If for some $\varepsilon \in (0, 1)$ holds

$$\mu^+(g) \leq C_\varepsilon\, p\text{-cap}_G(g) + (1-\varepsilon)\mu^-(G),$$

where $C_\varepsilon = \text{const}$, then inequality (1.24) is valid. A necessary condition for (1.24) is the estimate

$$\mu^+(g) \leq C\, p\text{-cap}_G(g) + \mu^-(G).$$

We remark that in all the inequalities in (i)-(iv) one can replace the expression $\|\text{grad } u\|^p_{L_p(\Omega)}$ by the integral

$$\int_{\Omega} [\Phi(x, \text{grad}\, u(x))]^p dx,$$

where $(x, y) \mapsto \Phi(x, y)$ is a function on $\mathbb{R}^n \times \mathbb{R}^n$, positive homogeneous of degree one, continuous in y for a.a. x and measurable in x for all y. The role of p-capacity is then taken over by its generalization

$$(p, \Phi)\text{-cap}(F) = \inf \left\{ \int_{\Omega} [\Phi(x, \text{grad}\, u(x))]^p dx \, : u \in C_0^{\infty}(\Omega), u \geq 1 \text{ on } F \right\}.$$

Although criteria formulated in terms of the (p, Φ)-capacity have a considerable interest, their verification in concrete situations often turns out to be troublesome. Also even for rather simple functions Φ one does not know estimates of the (p, Φ)-capacity in terms of measures. Therefore the necessary and sufficient conditions mentioned do not lessen the value of other methods for the study of integral inequalities, not based on capacity. Let us turn to a recent result by Fabes, Kenig and Serapioni, obtained by such methods and susceptible of a simple formulation.

Theorem 1.3. *Let* $1 < p < \infty$ *and let* w *be a measurable nonnegative function on* $\mathbb{R}^n$ *satisfying* the Muckenhoupt condition A_p

$$\sup \left(\frac{1}{|\mathscr{B}_\rho|} \int_{\mathscr{B}_\rho(x)} w(y)dy \right) \left(\frac{1}{|\mathscr{B}_\rho|} \int_{\mathscr{B}_\rho(x)} \frac{dy}{[w(y)]^{1/(p-1)}} \right)^{p-1} < \infty, \tag{1.25}$$

where the supremum runs over all ball $\mathscr{B}_\rho(x)$ *and* $|\mathscr{B}_\rho|$ *stands for the volume of a ball of radius* ϱ. *Let further* Ω *be an open subset of* $\mathbb{R}^n$.

There exist positive constants C *and* δ *such that for all* $u \in C_0^\infty(\Omega)$ *and* $q \in [p, \frac{pn}{n-1} + \delta]$ *we have the estimate*

$$\left(\int_\Omega |u|^q w dx \right)^{1/q} \leq C \left(\int_\Omega |\text{grad} u|^p w dx \right)^{1/p}. \tag{1.26}$$

The same authors also obtained inequality (1.26) with the weight

$$w(x) = |\det f'(x)|^{1-2/n},$$

where $f : \mathbb{R}^n \to \mathbb{R}^n$ is a quasiconformal mapping. This weight, generally speaking, does not satisfy condition A_2, so that A_p is not a necessary condition for (1.26).

So far there are no results of this nature regarding inequalities of the type (1.26) with two different weights in the left hand and the right hand members. An exception is the case $n = 1$, to which case we now pass. It is question of the inequality

$$\left[\int_0^\infty |u(x)|^q d\mu(x) \right]^{1/q} \leq C \left[\int_0^\infty |u'(x)|^p d\nu(x) \right]^{1/p}, \tag{1.27}$$

where u is an absolutely differentiable function on $[0, \infty)$, whose derivative u' is a Borel function belonging to $L_p((0, \infty), \nu)$, $u(0) = 0$, μ and ν being measures on $(0, \infty)$; the absolutely continuous part of ν will be denoted by ν^*.

Theorem 1.4. (Cf. [14].) *(i) If* $1 \leq p \leq q \leq \infty$ *then inequality (1.27) holds iff*

$$B = \sup_{r>0} [\mu([r, \infty])]^{1/q} \left[\int_0^r \left(\frac{d\nu^*}{dx} \right)^{-1/(p-1)} dx \right]^{(p-1)/p} < \infty. \tag{1.28}$$

Moreover, if C *denotes the best constant in (1.27) then*

$$B \leq C \leq B \left(\frac{q}{q-1} \right)^{(p-1)/p} q^{1/q}.$$

If $p = 1$ *or* $q = \infty$ *then* $B = C$.

If $q = \infty$ *then condition (1.28) means that* $B = \sup\{r > 0 : \mu([r,\infty)) > 0\} < \infty$ *and* $dv^*/dx > 0$ *for almost all* $x \in (0,\infty)$.

(ii) If $1 \le q < p \le \infty$ *then inequality (1.27) holds iff*

$$B = \left\{ \int_0^\infty \left[\mu([x,\infty)) \left(\int_0^x \left(\frac{dv^*}{dy}\right)^{-p/(p-1)} dy \right)^{q-1} \right]^{p/(p-q)} \left(\frac{dv^*}{dx}\right)^{-p/(p-1)} dx \right\}^{\frac{p-q}{pq}} < \infty. \tag{1.29}$$

The exact constant C *in (1.27) is connected with* B *via the inequalities*

$$\left(\frac{p-q}{p-1}\right)^{(q-1)/q} q^{1/q} \cdot B \le C \le \left(\frac{p}{p-1}\right)^{(q-1)/q} q^{1/q} \cdot B.$$

The change of variables $(0,+\infty) \ni x \mapsto y = x - x^{-1} \in (-\infty,\infty)$ allows us to derive immediately from (1.28) and (1.29) a necessary and sufficient condition for the validity of the inequality

$$\left[\int_{\mathbb{R}^1} |u(x)|^q d\mu(x) \right]^{1/q} \le C \left[\int_{\mathbb{R}^1} |u'(x)|^p dv(x) \right]^{1/p}.$$

We conclude this section by three not very well-known concrete weighted inequalities, each of which indicates a possible path for generalizations.

(i) For all $u \in C_0^\infty(\mathbb{R}^n)$ holds the estimate

$$\int_{\mathbb{R}^n} \frac{|u(x)|^p dx}{(x_{n-1}^2 + x_n^2)^{1/2}} \le c \int_{\mathbb{R}^n} |x_n|^{p-1} |\operatorname{grad} u(x)|^p dx.$$

It is almost evident that one can not here omit the term x_{n-1}^2 in the left hand side even if we impose the supplementary condition $u = 0$ on the hyperplane $x_n = 0$.

(ii) Let $n > 2, m \le n$. For all $u \in C_0^\infty(\mathbb{R}^n)$, if $m = 1$ also requiring that $u(x) = 0$ for $x_n = 0$, we have the estimate

$$\|u\|_{L_{2n/(n-2)}}^2 \le c \left(\int_{\mathbb{R}^n} |\operatorname{grad} u|^2 dx - \frac{(m-2)^2}{4} \int_{\mathbb{R}^n} \frac{u^2}{|y|^2} dx \right), \tag{1.30}$$

where $y = (x_1, \ldots, x_m)$. This is an improvement of the well-known Hardy's inequality, which guarantees only the positiveness of the right hand side of (1.30). Analogous estimates, involving the norm of the gradient in $L_p(\mathbb{R}^n)$, apparently are not known.

(iii) Let $x' = (x_1, \ldots, x_{n-1})$. For all $u \in C_0^\infty(\mathbb{R}^n)$ holds the inequality

$$\int_{\mathbb{R}^{n-1}} [u(x',0)]^2 dx' \le c \int_{\mathbb{R}^n} \left[(|x_n| + |x'|^2) \left(\frac{\partial u}{\partial x_n}\right)^2 + |\operatorname{grad}_{x'} u|^2 \right] dx. \tag{1.31}$$

It is easy to see that the presence of the term $|x'|^2$ is essential for the validity of (1.31). The estimate (1.31) is exact in the sense that the space of restrictions to $\mathbb{R}^{n-1} = \{x \in \mathbb{R}^n : x_n = 0\}$ of functions with the right hand side of (1.31) uniformly bounded is not relatively compact in $L_2(\mathscr{B}_1^{(n-1)})$, where $\mathscr{B}_\varrho^{(n-1)} = \{x' \in \mathbb{R}^{n-1} : |x'| < \varrho\}$.

§ 2. Sobolev Spaces in "Bad" Domains

2.1. Introduction. In this section, in contradistinction to § 1 of this Chapter, we consider functions in Sobolev spaces imposing no boundary assumptions whatsoever. The idea of the equivalence of imbedding theorems and isoperimetric inequalities turns out to be useful also here. We have in view the necessary and sufficient conditions on the domains, or the weight functions, under which the imbedding operators are bounded or compact.

Let us define three function spaces to be studied in this section.

Let $\mathrm{grad}_l u$ be the gradient of order l of the function u taken in the sense of the theory of distributions, i.e. $\mathrm{grad}_l u = \{\partial^l u/\partial x_1^{\alpha_1} \dots \partial x_n^{\alpha_n}\}$.

We denote by $L_p^{(l)}(\Omega)$ the space of functions in the open set $\Omega \subset \mathbb{R}^n$ such that $\mathrm{grad}_l u \in L_p(\Omega)$. If Ω is a domain then one can introduce in $L_p^{(l)}(\Omega)$ the norm

$$\|\mathrm{grad}_l u\|_{L_p(\Omega)} + \|u\|_{L_p(\omega)},$$

where ω is any nonempty bounded set, contained in Ω together with its boundary.

Let us further introduce the *space* $W_p^{(l)}(\Omega) = L_p^{(l)}(\Omega) \cap L_p(\Omega)$ equipped with the norm

$$\|u\|_{W_p^{(l)}(\Omega)} = \|\mathrm{grad}_l u\|_{L_p(\Omega)} + \|u\|_{L_p(\Omega)}.$$

Put also $\tilde{W}_p^{(l)}(\Omega) = \bigcap_{k=0}^{l} L_p^{(k)}(\Omega)$ as well as

$$\|u\|_{\tilde{W}_p^{(l)}(\Omega)} = \sum_{k=0}^{l} \|\mathrm{grad}_k u\|_{L_p(\Omega)}.$$

As functions in $L_p^{(l)}(\Omega)$, $W_p^{(l)}(\Omega)$ and $\tilde{W}_p^{(l)}(\Omega)$ may behave rather badly near $\partial\Omega$, these spaces turn out to be more sensitive to singularities of the boundary than spaces of functions satisfying homogeneous Cauchy conditions on $\partial\Omega$.

In S.L. Sobolev's theorems the domain of definition of functions satisfies the so-called *cone condition* (each inner point is the vertex of a cone of given height and opening and situated inside the domain). The simple example of the triangle $\Omega = \{x : 0 < x_1 < 1, 0 < x_2 < x_1^\nu\}$ and the function $u(x) = x_1^{-\mu}, \mu > 0$, defined on Ω, shows that the cone condition is essential for the validity of this theorem. On the other hand, a glance at Fig. 1 reveals that it is not necessary for the imbedding of $W_p^{(1)}(\Omega)$ into $L_{2p/(2-p)}(\Omega), 2 > p$. Indeed, joining Ω with

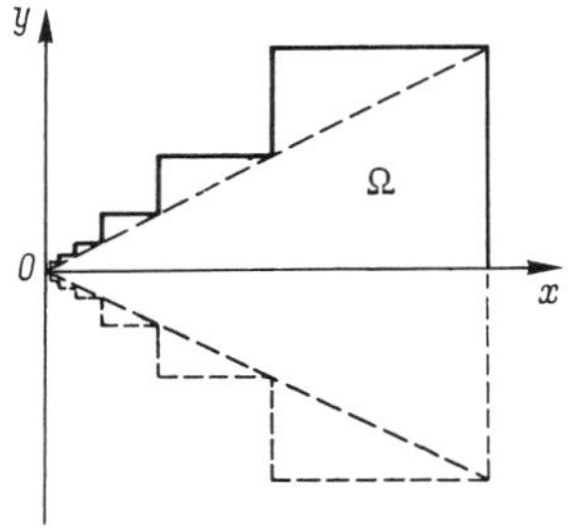

Fig. 1

its mirror image we obtain a new domain subject to the cone condition to which by Sobolev's theorem this imbedding holds true. Therefore, the same holds also for the original domain, although the cone condition is not satisfied there.

Let us now remark that even before S.L. Sobolev's results one has known that special integral inequalities hold under rather weak assumptions on the domain. For example, the *inequality of Friedrichs*

$$\int_\Omega u^2 dx \le C\left(\int_\Omega (\operatorname{grad} u)^2 dx + \int_{\partial\Omega} u^2 ds\right)$$

was proved under the sole assumption that Ω is a bounded domain for which the Gauss-Ostrogradskiĭ formula holds true. In the classical monograph by Courant and Hilbert it is shown that for domains, which locally can be given in Cartesian coordinates with the aid of continuous functions, one has *Poincaré's inequality*

$$\int_\Omega u^2 dx \le C \int_\Omega (\operatorname{grad} u)^2 dx + \frac{1}{\operatorname{mes}_n(\Omega)}\left(\int_\Omega u dx\right)^2$$

and, further, *Rellich's lemma* concerning the compactness in $L_2(\Omega)$ of the unit ball on $W_2^{(1)}(\Omega)$.

That some restrictions are necessary in order to guarantee the last inequality was shown by Nikodym (1933) in a paper devoted to the study of functions with a finite Dirichlet integral. Nikodym gave an example of a domain for which not every such function is square integrable, in other words $W_2^{(1)}(\Omega) \neq L_2^{(1)}(\Omega)$. This domain is the union of the rectangles (cf. Fig. 2)

$$A_k = \{(x, y) : 2^{1-k} - 2^{-1-k} < x < 2^{1-k}, 2/3 < y < 1\},$$

$$B_k = \{(x, y) : 2^{1-k} - \varepsilon_k < x < 2^{1-k}, 1/3 \le y \le 2/3\},$$

$$C = \{(x, y) : 0 < x < 1, 0 < y < 1/3\},$$

where $\varepsilon_k \in (0, 2^{-k-1})$ and $k = 1, 2, \ldots$.

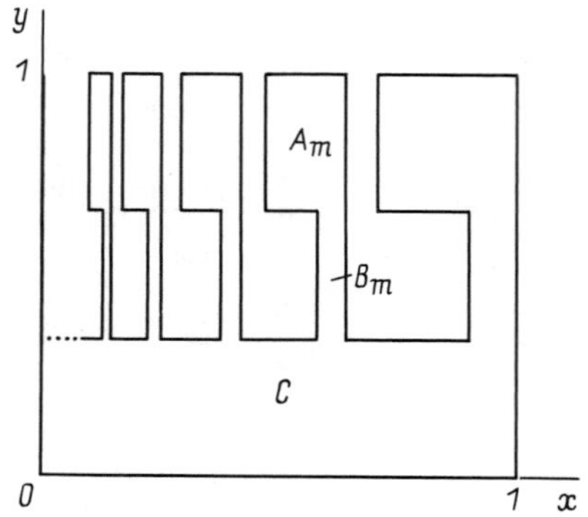

Fig. 2

Pick positive numbers α_k such that the series

$$\sum_{k=1}^{\infty} \alpha_k^2 \operatorname{mes}_2(A_k) \tag{1.32}$$

diverges. Let u be a continuous function which equals α_k on A_k, vanishes on C and is linear on B_k. In view of the divergence of the series (1.32) u does not belong to $L_2(\Omega)$ but one can take the numbers ε_k so small that its Dirichlet integral, equal to

$$\sum_{k=1}^{\infty} \int\!\!\int_{B_k} \left(\frac{\partial u}{\partial y}\right)^2 dxdy,$$

is convergent.

Another example, of historical interest, of a domain for which neither Poincaré's inequality nor the compactness of the imbedding of $W_2^{(1)}(\Omega)$ into $L_2(\Omega)$ hold, is given in the aforementioned monograph by Courant and Hilbert and will be treated in what follows (see Sect. 2.3 of Chap. 1).

But now let us turn to the domain, depicted in Fig. 3, for which, quite surprisingly, it does not follow from the quadratic summability of the function and all its second derivatives that the gradient is square summable, that is, $W_2^{(2)}(\Omega) \neq \tilde{W}_2^{(2)}(\Omega)$. It is easy to see that the function

$$u(x,y) = \begin{cases} 0 & \text{in } P, \\ 4^m (y-1)^2 & \text{in } S_m (m = 1, 2, \ldots), \\ 2^{m+1}(y-1) - 1 & \text{in } P_m (m = 1, 2, \ldots). \end{cases}$$

belongs to the space $W_2^{(2)}(\Omega)$ but $\|\operatorname{grad} u\|_{L_2(\Omega)} = \infty$.

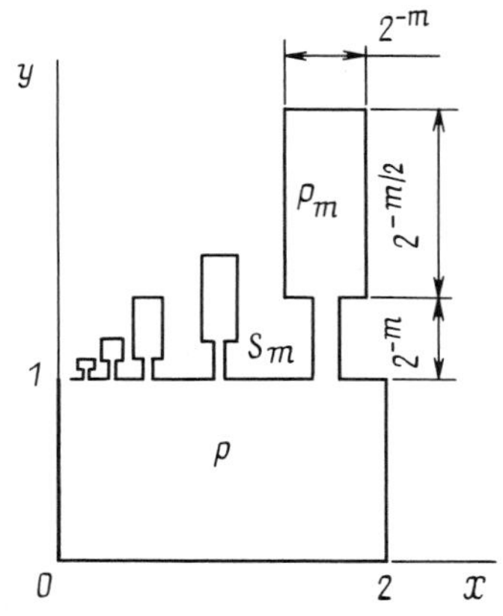

Fig. 3

2.2. The Classes J_α. In connection with what has been said there arises naturally the question of the description of the classes of domains for which various properties of the imbedding operator are equivalent. Reasonings analogous to those used in Sect. 1.1, § 1 of this Chapter allow us to completely solve this problem for $l = 1$.

In the case $p = 1$ the role of inequality (1.1) is taken by a relative isoperimetric inequality connecting the volume of admissible sets $g, g \subset \Omega$ with the area of the surfaces $\partial_i g = \Omega \cap \partial g$. Contrary to § 1 of this Chapter we mean now by *admissible* subsets bounded subsets g of Ω such that $\partial_i g$ is a smooth surface. Let us introduce the corresponding classes of domains Ω.

We say that an open set Ω belongs to the class J_α $(\alpha \geq (n-1)/n)$ if there exists a constant $M \in (0, \operatorname{mes}_n(\Omega))$ such that

$$\mathscr{A}_\alpha(M) \stackrel{\text{def}}{=} \sup_{\{g\}} \frac{[\operatorname{mes}_n(g)]^\alpha}{s(\partial_i g)} < \infty, \tag{1.33}$$

where $\{g\}$ is the family of all admissible subsets of Ω with $\operatorname{mes}_n(g) \leq M$.

We denote by $\lambda_M(\mu)$ the least upper bound of the numbers $s(\partial_i g)$ taken over all admissible sets $g \subset \Omega$ satisfying the condition $\mu \leq \operatorname{mes}_n(g) \leq M$, where $M \in (0, \operatorname{mes}_n(\Omega))$. If $M = \operatorname{mes}_n(\Omega)/2$ we will use the notation λ in place of λ_M.

It is easy to characterize the *class* J_α in terms of the function λ_M. Namely, $\Omega \in J_\alpha$ iff

$$\varliminf_{\mu \to +0} \mu^{-\alpha} \lambda_M(\mu) > 0. \tag{1.34}$$

Condition (1.33) or, equivalently, (1.34) characterizes the local properties of the boundary of Ω. Let us make this explicit. If Ω is a domain with a sufficiently smooth boundary then, as is readily seen, the area of the surface $\partial g \cap \Omega$ for any arbitrary set g of sufficiently small volume is (up to a constant factor) majorized by $s(\partial_i g)$. Therefore, by virtue of the classical isoperimetric inequality (1.1), holds

$$[\mathrm{mes}_n(g)]^{(n-1)/n} \leq \mathrm{const} \cdot s(\partial_i g),$$

that is, Ω belongs to the class $J_{(n-1)/n}$. If there are "cups" on $\partial\Omega$, which are directed inside Ω, then it is intuitively clear, and it can be rigorously established that the last inequality remains in force. If the cusps are directed out from Ω then there exists a sequence of sets $\{g_m\}_{m=1}^{\infty}$ such that

$$\lim_{m\to\infty} \frac{[\mathrm{mes}_n(g_m)]^{(n-1)/n}}{s(\partial_i g_m)} = \infty.$$

Moreover, inequality (1.32) can now be fulfilled with some $\alpha > (n-1)/n$. The exponent α measures in this case the sharpness of the cusp.

It is easy to see that the union of finitely many bounded domains of class J_α belongs to the same class.

With the aid of spherical symmetrization with respect to a ray (Pólya-Szegö [1951]) one can show that for the unit ball Ω holds

$$\mathscr{A}_{(n-1)/n}(v_n/2) = \left(\frac{v_n}{2}\right)^{(n-1)/n} v_{n-1}^{-1},$$

where the extremal sets g are semi-balls. As every bounded set starshaped with respect to a ball is a bilipschitzian image of a ball, it follows that such a set is in the class $J_{(n-1)/n}$. The same is true for each bounded domain satisfying the cone condition, because every such domain is the union of finitely many domains starshaped with respect to a ball.

It turns out that the class J_α as well as the function λ are useful in the study of the conditions for the validity of the *generalized Poincaré inequality*

$$\inf_{c\in\mathbb{R}^1} \|u-c\|_{L_q(\Omega)} \leq C\|\mathrm{grad}\, u\|_{L_p(\Omega)}, \tag{1.35}$$

where Ω is a domain of finite volume. It follows from Banach's isomorphism theorem that this inequality holds at the same time for the imbedding of $L_p^1(\Omega)$ into $L_q(\Omega,\mu)$.

In the case $p = 1$ an argument analogous to the one in Subsection 1.1 of § 1, Chapter 1 gives the following definitive result.

Theorem 2.1. *Let Ω be a domain with* $\mathrm{mes}_n(\Omega) < \infty$.

1) If $\Omega \in J_\alpha$, where $\alpha \in [(n-1)/n, 1]$, then inequality (1.35) holds for all functions $u \in L_1^1(\Omega)$, with $p = 1, q = \alpha^{-1}$ and $C \leq \mathscr{A}_\alpha(\frac{1}{2}\mathrm{mes}_n(\Omega))$.

2) If (1.35), with $p = 1$, holds for all $u \in L_1^1(\Omega)$ then $\alpha = q^{-1}$ and $C2^{(q-1)/q} \geq \mathscr{A}_\alpha(\frac{1}{2}\mathrm{mes}_n(\Omega))$.

We say that a set Ω belongs to the *class* $\overset{\circ}{J}_{\alpha}, \alpha > (n-1)/n$ if

$$\lim_{M\to 0} \sup \left([\mathrm{mes}_n(g)]^{\alpha} / s(\partial_i g) \right) = 0, \tag{1.36}$$

where the supremum is taken over all admissible subsets g of Ω satisfying the condition $\mathrm{mes}_n(g) \leq M$.

It is clear that condition (1.36) is equivalent to the identity

$$\lim_{\mu\to 0} \mu^{-\alpha} \lambda_M(\mu) = \infty,$$

where M is any fixed number in the interval $(0, \mathrm{mes}_n(\Omega))$.

In terms of the class $\overset{\circ}{J}_{\alpha}$ one can formulate the following compactness criterion.

Theorem 2.2. *Let Ω be a domain and assume that* $\mathrm{mes}_n(\Omega) < \infty$. *Then the imbedding of $L_1^{(1)}(\Omega)$ into $L_q(\Omega)$, where $n/(n-1) > q \geq 1$, is compact iff Ω is in the class $\overset{\circ}{J}_{1/q}$.*

We give some examples of domains in the classes J_{α} and $\overset{\circ}{J}_{\alpha}$.

Example 2.1. Let us consider a domain Ω, which equals the sum of the squares $Q_m = \{2^{-m-1} \leq x \leq 3 \cdot 2^{-m-2}, 0 < y < 2^{-m-2}\}$ and the rectangles $R_m = \{3 \cdot 2^{-m-2} < x < 2^{-m}, 0 < y < 1\}$ $(m = 0, 1, \ldots)$ (Fig. 4). One can show that for such an Ω one has the relation $\lambda(\mu) \sim \mu$. Therefore it belongs to J_1 and not to $\overset{\circ}{J}_1$. From this and Theorems 2.1 and 2.2 follows that one has a noncompact imbedding $L_1^{(1)}(\Omega) \subset L_1(\Omega)$.

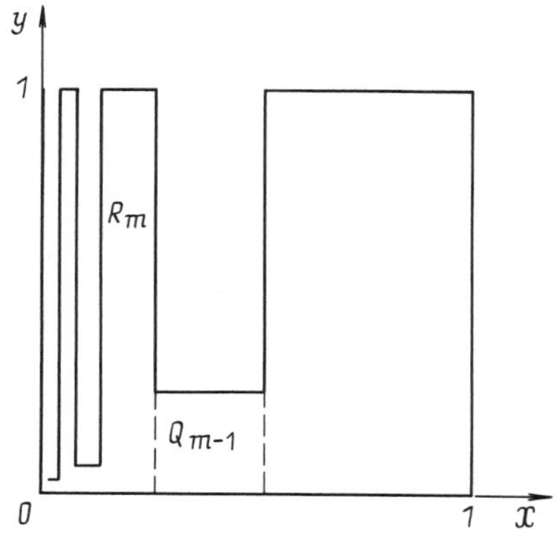

Fig. 4

Example 2.2. Let Ω be an n-dimensional "whirlpool" $\{x = (x', x_n), |x'| < f(x_n), 0 < x_n < 1\}$, where f is a continuously differentiable nonnegative convex function on $[0, 1]$, $f(0) = 0$ (Fig. 5). For this domain holds

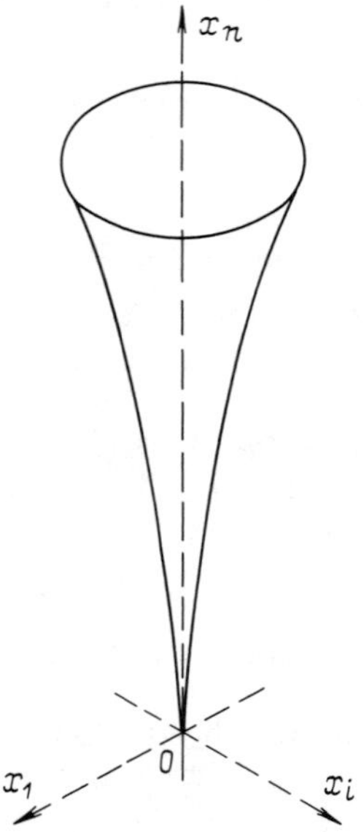

Fig. 5

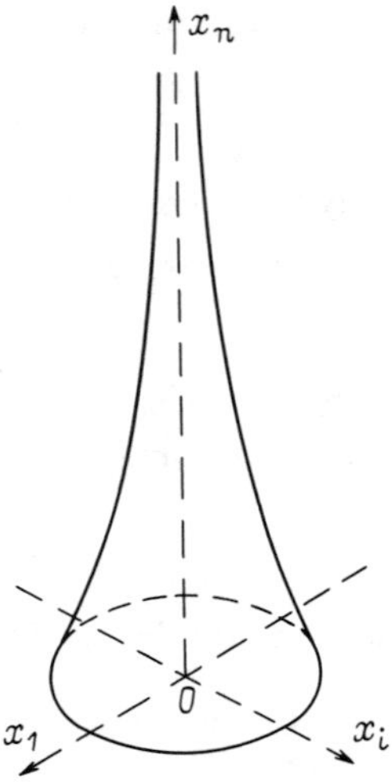

Fig. 6

$$c[f(t)]^{n-1} \le \lambda\left(v_{n-1}\int_0^t [f(\tau)]^{n-1}d\tau\right) \le [f(t)]^{n-1} \tag{1.37}$$

for sufficiently small t. It readily follows from this that the domain $\{x : 0 < x_n < 1, |x'| < x_n^{\kappa}\}$, $\kappa \ge 1$, belongs to J_α with $\alpha \ge \kappa(n-1)/[\kappa(n-1)+1]$. The same domain is in $\overset{\circ}{J}_\alpha$ iff $\alpha > \kappa(n-1)/[\kappa(n-1)+1]$. Therefore $L_1^{(1)}(\Omega) \subset L_q(\Omega)$ if $q \le [\kappa(n-1)+1]/\kappa(n-1)$ and this imbedding is compact if $q < [\kappa(n-1)+1]/\kappa(n-1)$.

Example 2.3. Let us consider a tube narrowing at infinity, $\Omega = \{x = (x', x_n), |x'| < f(x_n)\}$, where f is a positive convex continuously differentiable function on $[0, \infty]$ (Fig. 6). One can show that for sufficiently large t holds the inequality

$$c[f(t)]^{n-1} < \lambda\left(v_{n-1}\int_t^\infty [f(\tau)]^{n-1}d\tau\right) \le [f(t)]^{n-1}. \tag{1.38}$$

Therefore the domain $\{x : |x'| < (1 + x_n)^{-\kappa}, 0 < x_n < \infty\}$, $\kappa(n-1) > 1$, belongs to J_α if $\alpha = \kappa(n-1)/[\kappa(n-1)-1]$. The same domain is in $\overset{\circ}{J}_\alpha$ iff $\alpha > \kappa(n-1)/[\kappa(n-1)-1]$.

Example 2.4. As an example of a bounded domain in J_α with $\alpha > 1$ we may quote the spiral depicted in Fig. 7: $\Omega = \{\varrho e^{i\theta} : 1 - (8+\theta)^{1-\beta} > \varrho > 1 - (8+\theta)^{1-\beta} - c(8+\theta)^{-\beta}, 0 < \theta < \infty\}$, $\beta > 1$, where $\lambda(\mu) \sim \mu^\alpha$, with $\alpha = \beta/(\beta-1)$. Thus, $\Omega \in J_{\beta/(\beta-1)}$ and $\Omega \in \overset{\circ}{J}_\alpha$ if $\alpha > \beta/(\beta-1)$.

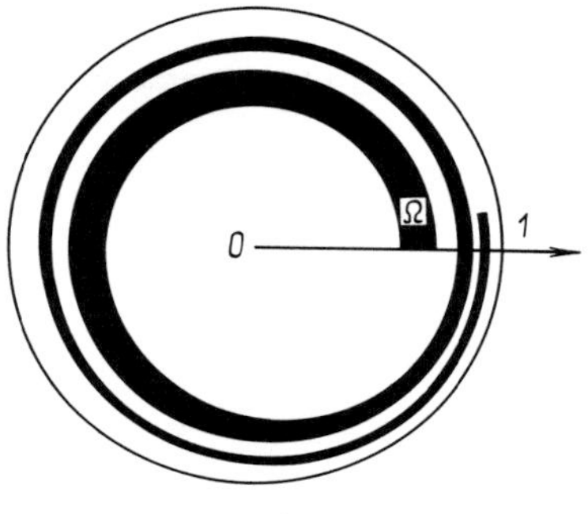

Fig. 7

2.3. The Case $p > 1$. In order to get results analogous to Theorems 2.1 and 2.2 for the space $L_p^{(1)}$ we must introduce other classes of sets defined with the aid of the so called *p*-conductance, which plays the same role as *p*-capacity in Sect. 1 of Chap. 1.

We denote by F a relatively closed (in Ω) bounded subset of the open set Ω and by G a bounded open subset of Ω. We assume that $F \subset G$. The set $K = G \backslash F$ will be referred to as a *conductor*.

By the *p-conductance* of the conductor K we mean the number

$$c_p(K) = \inf \int_\Omega |\operatorname{grad} f|^p dx,$$

where the infimum is taken over all functions $f \in C^{0,1}(\Omega)$ such that $f(x) \geq 1$ for $x \in F$ and $f(x) \leq 0$ for $x \in \Omega \backslash G$.

It is clear that in the case $\bar{G} \subset \Omega$ one has $c_p(K) = p\text{-cap}_G(F)$. The properties of *p*-conductance are to some extent parallel to those of *p*-capacity. For instance, one has the following generalization of inequality (1.21):

$$\int_0^\infty c_p(K_t) t^{p-1} dt \leq \left(\frac{p}{p-1}\right)^{p-1} \int_\Omega |\operatorname{grad} u|^p dx, \tag{1.39}$$

where u is an arbitrary function in $C^1(\Omega)$, equal to zero outside some bounded subset $G \subset \Omega$, and K_t the conductor $G \backslash \{x \in \Omega : |u(x)| \geq t\}$.

Let us denote by $I_{p,\alpha}, p \geq 1, \alpha > 0$, the class of domains which satisfy the following condition. There exists a constant $M \in (0, \mathrm{mes}_n(\Omega))$ such that

$$\mathscr{A}_{p,\alpha}(M) \overset{\text{def}}{=} \sup_{\{K\}} \frac{[\mathrm{mes}_n(F)]^\alpha}{[c_p(K)]^{1/p}} < \infty,$$

where $\{K\}$ is the family of conductors $K = G \backslash F$ in Ω with positive p-conductance, subject to the condition $\mathrm{mes}_n(G) \leq M$.

If $\alpha < (n-p)/np, n > p$ then the class $I_{p,\alpha}$ is empty. If Ω is a union of domains of class $I_{p,\alpha}$ then it belongs to the same class. The classes $I_{1,\alpha}$ and J_α coincide.

Let $\nu_{M,p}(t)$ be the infimum of $c_p(K)$ taken over the set of all conductors $K = G \backslash F$ satisfying the two conditions $\mathrm{mes}_n(F) \geq t, \mathrm{mes}_n(G) \leq M$.

It is clear that if $p \geq 1$ then

$$\mathscr{A}_{p,\alpha}(M) = \sup_{0<t\leq M} t^\alpha [\nu_{M,p}(t)]^{-1/p}.$$

It follows that Ω belongs to $I_{p,\alpha}$ iff

$$\varliminf_{t\to+0} t^{-\alpha p} \nu_{M,p}(t) > 0.$$

From Dirichlet's principle with prescribed level surfaces for p-conductance (cf. (1.19)) we obtain the inequality

$$\nu_{M,p}(t) \geq \left(\int_t^M [\lambda_m(\sigma)]^{p/(1-p)} d\sigma \right)^{1-p}, \tag{1.40}$$

which in turn implies the inclusion $J_{\alpha+(p-1)/p} \subset I_{p,\alpha}$ as well as the estimate

$$\mathscr{A}_{p,\alpha}(M) \leq \left(\frac{p-1}{p\alpha} \right)^{(p-1)/p} \mathscr{A}_{1,\alpha+(p-1)/p}(M).$$

The expediency of the introduction of the class $I_{p,\alpha}$ is illustrated by the following theorem. In the proof essential use is made of inequality (1.39).

Theorem 2.3. *1) Assume that $\Omega \in I_{p,\alpha}$ and let q be a positive number satisfying one of the conditions: (i) $q \leq q^* = \alpha^{-1}$ if $p \leq q^*$ or (ii) $q < q^*$ if $p > q^*$. Then holds for any function $u \in L_p^1(\Omega) \cap L_s(\Omega)$*

$$\|u\|_{L_q(\Omega)} \leq (C_1 \|\mathrm{grad}\, u\|_{L_p(\Omega)} + C_2 \|u\|_{L_s(\Omega)})^{1-\kappa} \|u\|^\kappa_{L_r(\Omega)}, \tag{1.41}$$

where $s < q^*, r < q$, $\kappa = r(q^*-q)/q(q^*-r)$, $C_2 = cM^{(s-q^*)/sq^*}, C_1 \leq c\mathscr{A}_{p,\alpha}(M)$.

2) Let $q^* > 0$ and assume that inequality (1.41) holds for all $u \in L_p^{(1)}(\Omega) \cap L_s(\Omega)$ and some $q \in (0, q^*]$, with $0 < s < q^*, 0 < r < q^*, \kappa = r(q^*-q)/q(q^*-r)$. Then $\Omega \in I_{p,\alpha}$ with $\alpha = 1/q^*$ and if in the definition of the class $I_{p,\alpha}$ we take the constant M to be $M = cC_2^{sq^*/(s-q^*)}$, where c is a sufficiently small positive number depending only on p, q^*, s, then $C_1 \geq c\mathscr{A}_{p,\alpha}(M)$.

If $p > q^* = \alpha^{-1}$, then membership for Ω in $I_{p,\alpha}$ is necessary but not sufficient for the validity of (1.41) with the limit exponent $q = q^*$. A necessary and sufficent condition in this case is that Ω belongs to a new *class* $H_{p,\alpha}$, which we define as follows.

Let $\alpha p > 1$. Put

$$\mathscr{B}_{p,\alpha}(M) = \sup_{\{S\}} \left[\sum_{j=-\infty}^{\infty} \frac{[\mathrm{mes}_n(G_j)]^{\alpha p/(p-1)}}{[c_p(G_{j+1} \backslash \mathrm{clos}_\Omega G_j)]^{1/(\alpha p-1)}} \right]^{\alpha - 1/p},$$

where M is a constant in the interval $(0, \mathrm{mes}_n(\Omega))$ and $\{S\}$ is the family of all nondecreasing sequences S of open bounded sets G_j, $-\infty < j < \infty$, contained in Ω such that $\mathrm{mes}_n(\cup_j G_j) \le M$; by clos_Ω we intend the relative closure with respect to Ω. We say that Ω belongs to the class $H_{p,\alpha}$ if $\mathscr{B}_{p,\alpha}(M) < \infty$ for some $M \in (0, \mathrm{mes}_n(\Omega))$.

It is easy to see that the condition

$$\int_0^M \left[\frac{\tau}{\nu_{M,p}(\tau)} \right]^{1/(\alpha p-1)} d\tau < \infty \tag{1.42}$$

is sufficient for Ω to be in $H_{p,\alpha}$. It follows from this and from (1.40) that $\Omega \in H_{p,\alpha}$ provided

$$\int_0^M \left[\frac{\tau}{\lambda_M(\tau)} \right]^{p/(\alpha p-1)} d\tau < \infty.$$

Let us remark that if $p_1 > p \ge 1$ and $\alpha_1 - p_1^{-1} = \alpha - p^{-1}$ then $I_{p,\alpha} \subset I_{p_1,\alpha_1}$ and $H_{p,\alpha} \subset H_{p_1,\alpha_1}$.

If Ω is a domain of finite volume then the space $L_p^{(1)}(\Omega)$ is imbedded in $L_q(\Omega)$ if, and if $q \ge 1$, iff $\Omega \in I_{p,1/q}, p \le q$, or $\Omega \in H_{p,1/q}, p > q$.

More exactly, one has the following result concerning the generalized Poincaré inequality (1.35), analogous to Theorem 2.1.

Theorem 2.4. *Let Ω be a domain with* $\mathrm{mes}_n(\Omega) < \infty$.

1) If $\mathscr{A}_{p,\alpha}(\frac{1}{2}\mathrm{mes}_n(\Omega)) < \infty$ if $\alpha p \le 1$ or $\mathscr{B}_{p,\alpha}(\frac{1}{2}\mathrm{mes}_n(\Omega)) < \infty$ if $\alpha p > 1$, then inequality (1.35) holds for functions $u \in L_p^{(1)}(\Omega)$, with $q = \alpha^{-1}$, $C \le \mathscr{A}_{p,\alpha}(\frac{1}{2}\mathrm{mes}_n(\Omega))$ if $\alpha p \le 1$ and $C \le c\mathscr{B}_{p,\alpha}(\frac{1}{2}\mathrm{mes}_n(\Omega))$ if $\alpha p > 1$.

2) If there exists a constant C such that inequality (1.35), with $q \ge 1$, holds for all functions $u \in L_p^{(1)}(\Omega)$ then $C \ge c\mathscr{A}_{p,\alpha}(\frac{1}{2}\mathrm{mes}_n(\Omega))$ with $\alpha^{-1} = q \ge p$.

Let us turn to some examples of domains of class $I_{p,\alpha}$ and $H_{p,\alpha}$.

Example 2.5. For the domain Ω in Fig. 5, already treated in Example 2.2, one has the relation

$$\nu_{M,p}\left(v_{n-1} \int_0^x [f(\tau)]^{n-1} d\tau \right) \sim \left(\int_x^1 [f(\tau)]^{(1-n)/(p-1)} d\tau \right)^{1-p}$$

(cf. inequality (1.37)). It follows that $\Omega \in I_{p,\alpha}$ iff

$$\overline{\lim_{x\to+0}} \left(\int_0^x [f(\tau)]^{n-1} d\tau \right)^{\alpha p/(p-1)} \int_x^1 [f(\tau)]^{(1-n)/(p-1)} d\tau < \infty.$$

In particular, the domain $\Omega^{(\kappa)} = \{x : |x'| < ax_n^\kappa, 0 < x_n < 1\}$, where $\kappa/(p-1) > (n-1)$, one has for small t the inequality $c_1 t^{\alpha p} \le \nu_{M,p}(t) \le c_2 t^{\alpha p}$, where $\alpha = [\kappa(n-1)+1-p]/p[\kappa(n-1)+1]$. Thus, $\Omega^{(\kappa)}$ is in $I_{p,\alpha}$ and it satisfies $L_p^{(1)}(\Omega^{(\kappa)}) \subset L_{1/\alpha}(\Omega^{(\kappa)})$.

At the hand of the domain $\Omega^{(\kappa)}$ it is convenient to illustrate the plausible idea that the summability exponent $pn/(n-p)$ in Sobolev's theorem might be maintained also for "bad" domains, provided one instead of the Lebesgue measure mes_n considers Lebesgue measure with a weight degenerating at "bad" points of the boundary. One can show that the space $L_p^{(1)}(\Omega^{(\kappa)})$ imbeds into the space of functions with the finite norm

$$\left(\int_{\Omega^{(\kappa)}} |u(x)|^{pn/(n-p)} x_n^\sigma dx \right)^{(n-p)/pn},$$

where $n > p \ge 1$ and $\sigma = p(n-1)(\kappa-1)/(n-p)$. Here the exponent in the weight is exact.

Example 2.6. Let Ω be the tube in Example 2.3 (Fig. 6). With the aid of the estimate (1.38) we convince ourselves that

$$\nu_{M,p}\left(v_{n-1} \int_x^\infty [f(\tau)]^{n-1} d\tau \right) \sim \left(\int_0^x [f(\tau)]^{(1-n)/(p-1)} d\tau \right)^{1-p}. \qquad (1.43)$$

Consequently $\Omega \in I_{p,\alpha}$ iff

$$\overline{\lim_{x\to+\infty}} \left(\int_x^\infty [f(\tau)]^{n-1} d\tau \right)^{\alpha p/(p-1)} \cdot \int_0^x [f(\tau)]^{(1-n)/(p-1)} d\tau < \infty. \qquad (1.44)$$

From (1.42) and (1.43) we deduce at once that $\Omega \in H_{p,\alpha}, p\alpha > 1$ if

$$\int_0^\infty \left([f(x)]^{(n-1)(\alpha-1-1/p)} \int_x^\infty [f(t)]^{n-1} dt \right)^{p/(\alpha p-1)} dx < \infty.$$

In particular, for the domain $\Omega = \{x : |x'| < (1+x_n)^{-\kappa}, 0 < x_n < \infty\}$, $\kappa(n-1) > 1$, holds for small t the inequality $c_1 t^{\alpha p} \le \nu_{M,p}(t) \le c_2 t^{\alpha p}$, where $\alpha = [\kappa(n-1)-1+p]/p[\kappa(n-1)-1]$. Therefore it is question of a domain of class $I_{p,\alpha}, p\alpha > 1$. It is easy to see that $\Omega \notin H_{p,\alpha}$, and as $H_{p,\alpha+\varepsilon} \supset I_{p,\alpha}$ for any $\varepsilon > 0$, therefore by Theorem 2.4 the space $L_p^{(1)}(\Omega)$ is imbedded in $L_{1/(\alpha+\varepsilon)}(\Omega)$, although it is not contained in $L_{1/\alpha}(\Omega)$. Thus, in contrast to Sobolev's theorem,

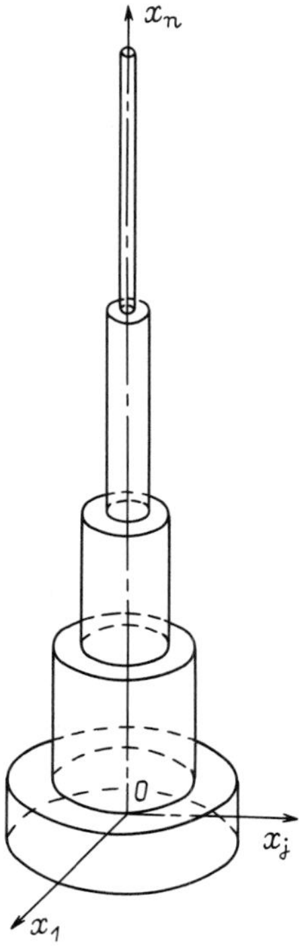

Fig. 8

here the imbedding operator from $L_p^{(1)}(\Omega)$ into $L_q(\Omega)$ with the limit exponent q is not bounded.

It follows from (1.44) that the domain $\Omega = \{x : |x'| < \exp(-cx_n), 0 < x_n < \infty\}$ is in $I_{p,1/p}$, that is, we have $L_p^{(1)}(\Omega) \subset L_p(\Omega)$. On the other hand, there exist domains narrowing at infinity as fast as we wish which are not in $I_{p,1/p}$. Among them there is, as depicted in Fig. 8, the union of the cylinders $G_j = \{x : |x'| < b_j, |x_n - \alpha_j| < a_j\}$, abutting each other, where $\sum_j a_j b_j^{n-1} < \infty$ and $a_j \to \infty$.

Example 2.7. Consider in particular the domain Ω in Nikodym's counterexample (Fig. 2). Let us assume in addition that $\varepsilon_m = \delta(2^{-m-1})$, where δ is a Lipschitz function on $[0, 1]$ such that $\delta(2t) \sim \delta(t)$. One can verify that for small M one has the relations $\lambda_M(t) \sim \delta(\mu)$ and $\nu_{M,p}(\mu) \sim \delta(\mu)$. Hence $\Omega \in J_\alpha, \alpha \geq 1$ iff

$$\varliminf_{t\to+0} t^{-\alpha}\delta(t) > 0.$$

Analogously, membership for Ω in $I_{p,\alpha}$ is equivalent to the inequality

$$\varliminf_{t\to+0} t^{-\alpha p}\delta(t) > 0$$

(as $\delta(t) \le ct$ holds $\alpha p \ge 1$). One can show that the Nikodym domain is in $H_{p,\alpha}$ with $\alpha p > 1$ iff

$$\int_0^1 \left[\frac{t}{\delta(t)}\right]^{1/(\alpha p-1)} dt < \infty.$$

In view of Theorem 2.4 the last inequality is equivalent to inequality (1.35) with $q = 1/\alpha$.

The question of the compactness of the imbedding of $L_p^{(1)}(\Omega)$ into $L_q(\Omega)$ also admits a complete solution. We consider only the case $p \le q$.

Theorem 2.5. *Assume that* $\mathrm{mes}_n(\Omega) < \infty$ *and* $p \le q$. *A necessary and sufficient condition for the imbedding* $L_p^{(1)}(\Omega) \subset L_q(\Omega)$ *to be compact is that*

$$\lim_{M\to 0} \sup \frac{[\mathrm{mes}_n(F)]^{p/q}}{c_p(K)} = 0, \tag{1.45}$$

where the supremum is taken over of the set of conductors $K = G \backslash F$ *in* Ω *which satisfy the condition* $\mathrm{mes}_n(G) \le M$.

If $p = 1$ then (1.45) is equivalent to the inclusion $\Omega \in \overset{\circ}{J}_{1/q}$.

For example, for the domain shown in Fig. 9 the imbedding operator from $L_p^{(1)}(\Omega)$ into $L_p(\Omega)$ is compact for $\alpha < p+1$ but only bounded for $\alpha = p+1$ and unbounded for $\alpha > p+1$.

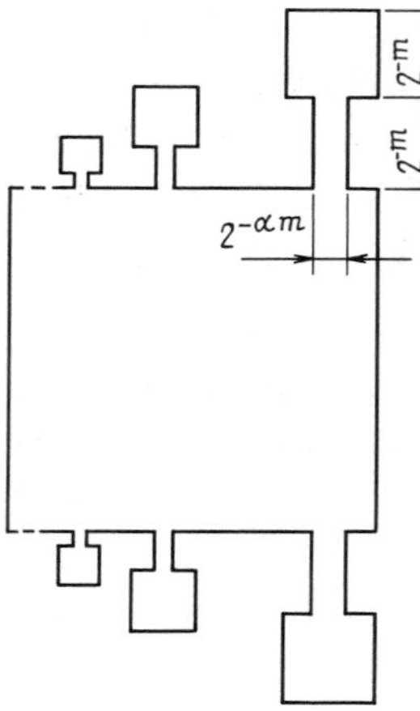

Fig. 9

For the domain in Example 2.3 and 2.6 compactness of the imbedding $L_p^{(1)}(\Omega) \subset L_{1/\alpha}(\Omega)$ is equivalent to the requirement that the limit in (1.44) equals zero.

Theorem 2.4, 2.5 and other similar results lead in the case $p = 2$ to necessary and sufficient conditions for the solvability and the discreteness of the spectrum of Neumann's problem in domains with nonregular boundaries.

Let Ω be a domain of finite volume and let a_{ij} be measurable functions in Ω, $1 \leq i, j \leq n$. Assume that there exists a constant $c > 1$ such that for almost all $x \in \Omega$ and all vectors $\xi = (\xi_1, \dots, \xi_n)$ holds the inequality

$$c^{-1}|\xi|^2 \leq \sum_{i,j=1}^{n} a_{ij}\xi_i\xi_j \leq c|\xi|^2.$$

We define *the operator N_q of Neumann's problem* for the differential operator

$$u \mapsto -\sum_{i,j=1}^{n} \frac{\partial}{\partial x_i}\left(a_{ij}\frac{\partial u}{\partial x_j}\right)$$

by the following two requirements: 1) $u \in L_2^{(1)}(\Omega) \cap L_q(\Omega), N_q u \in L_{q'}(\Omega), q+q' = qq'$; 2) for all $v \in L_2^{(1)}(\Omega) \cap L_q(\Omega)$ holds the identity

$$\int_\Omega v N_q u dx = \int_\Omega \sum_{i,j=1}^{N} a_{ij} \frac{\partial u}{\partial x_i}\frac{\partial v}{\partial x_j} dx.$$

The map $u \mapsto N_q u$ is closed. It is clear that the range $R(N_q)$ of the operator N_q is contained in the set $L_q(\Omega) \ominus 1$ of functions in $L_q(\Omega)$ orthogonal to the unit function on Ω. It is not hard to see that $R(N_q) = L_{q'}(\Omega) \ominus 1$ iff inequality (1.35) with $p = 2$ holds for all $v \in L_2^{(1)}(\Omega) \cap L_q(\Omega)$. It follows readily from this and from Theorem 2.4 that *Neumann's problem $N_q u = f$* is uniquely solvable (up to a constant term) in $L_2^{(1)}(\Omega) \cap L_q(\Omega)$ for all $f \in L_{q'}(\Omega) \ominus 1$ iff $\Omega \in I_{2,1/q}$ if $q \geq 2$ and iff $\Omega \in H_{2,1/q}$ if $q < 2$.

In view of Theorem 2.5 the spectrum of N_2 is discrete iff Ω satisfies condition (1.45) with $p = 2$.

As for the time being no generalization of inequality (1.39) is known where the gradient of u is replaced with the lth order gradient, we cannot generalize Theorem 2.3–2.5 to the case of the space $L_p^{(l)}(\Omega)$. However, one can derive sufficient conditions for the compactness of the imbedding operators from $L_p^{(l)}(\Omega)$ into $L_q(\Omega)$ from the previous theorems using induction over the number of derivatives. We mention one simple result of this kind.

Theorem 2.6. *If Ω is a domain of finite volume in $J_\alpha, \alpha \leq 1, lp(1-\alpha) < 1$ and $q = p/[1 - pl(1-\alpha)]$, then holds the inequality*

$$\inf \|u - P\|_{L_q(\Omega)} \leq C\|\text{grad}_l u\|_{L_p(\Omega)}, \tag{1.46}$$

where the infimum is taken over all polynomials P of degree at most $l-1$. The imbedding of $L_p^{(l)}(\Omega)$ into $L_q(\Omega)$ is compact if $\Omega \in \overset{\circ}{J}_\alpha$.

Inequality (1.46) holds true and the imbedding $L_p^{(l)}(\Omega) \subset L_q(\Omega)$ is compact for any $q > 0$, provided $pl(1-\alpha) = 1$, and if $q < p/[1 - pl(1-\alpha)]$, provided $\alpha > 1$.

Let us turn to the question of boundedness and compactness of the imbedding operator from the space $L_p^{(1)}$ into $C(\Omega) \cap L_\infty(\Omega)$, where Ω is a domain of bounded volume.

Introduce the function γ_p by the identity

$$\gamma_p(\rho) = \inf_{y \in \Omega} c_p(\Omega_\rho(y) \backslash y), \tag{1.47}$$

where $\Omega_\rho(y) = \{x \in \Omega : | x - y | < \rho\}$. In the case $p \leq n$ the right hand side of (1.47) vanishes and therefore it is only meaningful to consider γ_p for $p > n$.

Theorem 2.7. *(i) The imbedding operator $L_p^{(1)}(\Omega) \subset C(\Omega) \cap L_\infty(\Omega)$ is bounded iff the function γ_p does not vanish identically.*

(ii) For the compactness of this operator it is necessary and sufficient that γ_p satisfies the condition $\gamma_p(\rho) \to \infty$ as $\rho \to +0$.

For the domain in Example 2.2 both criteria in Theorem 2.7 are equivalent to the condition

$$\int_0^\infty \frac{d\tau}{[f(\tau)]^{(n-1)/(p-1)}} < \infty.$$

Thus, here, as in Sobolev's theorem, the imbedding operator from $L_p^{(1)}(\Omega)$ into $C(\Omega) \cap L_\infty(\Omega)$ is compact iff it is bounded. Besides, one should not expect that for domains with bad boundaries the simultaneous fulfillment of both properties always takes place. For example, for the domain in Fig. 10, the imbedding operator in question is continuous but not compact.

A necessary condition for the imbedding operator $L_p^{(1)}(\Omega) \subset C(\Omega) \cap L_\infty(\Omega)$ to be continuous is that

$$\int_0^\infty \frac{d\mu}{[\lambda(\mu)]^{p/(p-1)}} < \infty. \tag{1.48}$$

This, likewise, secures the compactness of the imbedding.

Condition (1.48) is in particular fulfilled for domains in J_α with $p(1-\alpha) > 1$.

It might be natural to expect that a weakening of the assumptions on the domain in imbedding theorems of Sobolev type can be compensated by supplementary assumptions on the boundary behavior of the function. The question of the influence of such assumptions on the inequalities between various function classes has so far been studied only uncompletely. We give some results obtained in this direction.

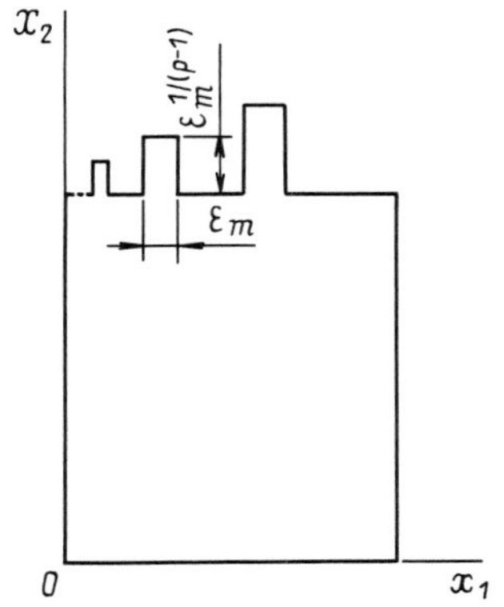

Fig. 10

For a function u in $C(\partial\Omega)$ set

$$\|u\|_{L_r(\partial\Omega)} = \left(\int_{\partial\Omega} |u|^r ds \right)^{1/r},$$

where ds is $(n-1)$-dimensional Hausdorff measure. Denote by $W_{p,r}^{(1)}(\Omega, \partial\Omega)$ the completion of the set of functions defined on Ω with bounded supports which belong to the intersection $L_p^{(1)}(\Omega) \cap C^\infty(\Omega) \cap C(\bar{\Omega})$ in the norm

$$\|\text{grad } u\|_{L_p(\Omega)} + \|u\|_{L_r(\partial\Omega)}.$$

Reasoning as in Sect. 1.1, § 1, Chap. 1 one can show that for any set Ω whatsoever the functions in $W_{1,1}^{(1)}(\Omega, \partial\Omega)$ are summable in Ω with exponent $n/(n-1)$. Moreover, we have the inequality

$$\|u\|_{L_{n/(n-1)}(\Omega)} \leq \frac{\Gamma(1+n/2)}{n\sqrt{\pi}} \|u\|_{W_{1,1}^{(1)}(\Omega,\partial\Omega)} \tag{1.49}$$

with an exact constant. Inserting in (1.49) instead of u the function $|u|^{p(n-1)/(n-p)}$, $1 < p < n$, and applying Hölder's inequality we establish the continuity of the imbedding operator $W_{p,p(n-1)/(n-p)}^{(1)}(\Omega, \partial\Omega) \subset L_{pn/(n-p)}(\Omega)$. Analogously, for an arbitrary open set Ω of finite volume we obtain from (1.49) the following strengthening of Friedrichs's inequality

$$\|u\|_{L_q(\Omega)} \leq C\|u\|_{W_{p,r}^{(1)}(\Omega,\partial\Omega)}, \tag{1.50}$$

where $1 \leq r < p(n-1)/(n-p)$ and $q = rn/(n-1)$.

One can also show that for any open set Ω the imbedding operator $W_{p,r}^{(1)}(\Omega, \partial\Omega) \subset L_q(\Omega)$ is completely continuous if $1 < r < p(n-1)/(n-p)$ and $q < rn/(n-1)$.

Let us check that the exponent q in (1.50) cannot be increased if we do not impose the supplementary conditions on Ω.

Example 2.8. Consider a domain Ω which consists of the union of the halfball $B^- = \{x : x_n < 0, |x| < 1\}$, a sequence of balls B_m $(m = 1, 2, \dots)$ and narrow tubes C_m connecting B_m and B^- (Fig. 11). Let ρ_m be the radius of B_m and h_m the height of C_m. Let u_m be a piecewise linear function equal to one in B_m and to zero outside $B_m \cup C_m$. Assume that there exists a constant C such that for all functions u_m inequality (1.50) is fulfilled. It is clear that then

$$\operatorname{mes}_n(B_m)^{1/q} \le C\left(h_m^{-1}\operatorname{mes}_n(C_m)^{1/p} + s(\partial B_m \cup \partial C_m)^{1/r}\right).$$

As we may take the first term in the right hand side as small as we wish provided we decrease the thickness of the tube C_m then $\rho_m^{n/q} = O(\rho_m^{(n-1)/p})$. Consequently, we have $q \le rn/(n-1)$.

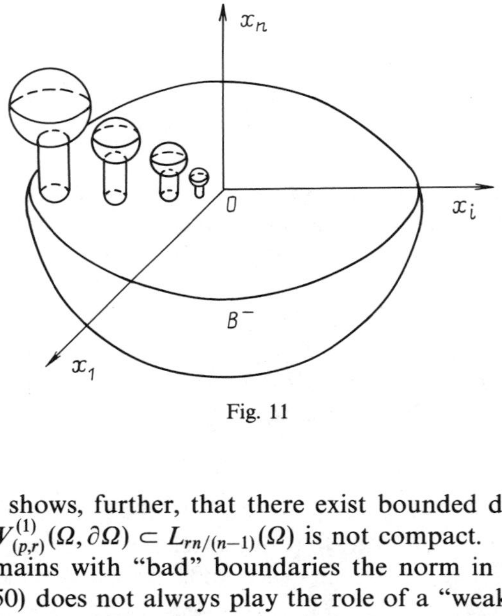

Fig. 11

This example shows, further, that there exist bounded domains such that the imbedding $W^{(1)}_{(p,r)}(\Omega, \partial\Omega) \subset L_{rn/(n-1)}(\Omega)$ is not compact.

Thus, for domains with "bad" boundaries the norm in the space $L_r(\partial\Omega)$ in inequality (1.50) does not always play the role of a "weak complement" to the norm of the gradient in $L_p(\Omega)$: the exponent p may depend on the degree of summability of the function on the boundary of Ω.

§ 3. Integral Inequalities for Functions on Riemannian Manifolds

Until now we have restricted our attention to Sobolev spaces in subdomains of $\mathbb{R}^n$. However, the methods of proof of many of the previous results do not use in a specific way the properties of the Euclidean space. Let us turn, for instance, to Theorem 1.2. The assumption $\Omega \subset \mathbb{R}^n$ is not essential for its proof: formula (1.4) is also valid for functions on a Riemannian manifold, while the rest of the argument has a general character. Let Ω be an open subset of a Riemannian manifold M and let $\mu = \text{mes}_n$ be the n-dimensional measure on subsets of M. Then by virtue of Theorem 1.2 we have the following twosided estimate:

$$\inf_{\{F\}} \frac{[p\text{-cap}_\Omega(F)]^{1/p}}{[\text{mes}_n(F)]^{1/q}} \ge \inf_{u \in C_0^\infty(\Omega)} \frac{\|\text{grad } u\|_{L_p(\Omega)}}{\|u\|_{L_q(\Omega)}} \ge \frac{(p-1)^{(p-1)/p}}{p} \inf_{\{F\}} \frac{[p\text{-cap}_\Omega(F)]^{1/p}}{[\text{mes}_n(F)]^{1/q}},$$

where $q \ge p$ and $\{F\}$ is the family of all compact subsets of Ω.

Let Λ be a function on $(0, \text{mes}_n(\Omega))$ such that for any open set g with compact closure $\bar{g} \subset \Omega$ and smooth boundary one has the *isoperimetric inequality*

$$s(\partial g) \ge \Lambda(\text{mes}_n(g)). \tag{1.51}$$

Then, in view of (1.19)

$$p\text{-cap}_\Omega(F) \ge \left(\int_{\text{mes}_n(F)}^{\text{mes}_n(\Omega)} \frac{dv}{[\Lambda(v)]^{p/(p-1)}} \right)^{1-p}$$

and, consequently,

$$\inf_{u \in C_0^\infty(\Omega)} \frac{\|\text{grad } u\|^p_{L_p(\Omega)}}{\|u\|^p_{L_q(\Omega)}} \ge \frac{(p-1)^{p-1} p^{-p}}{\sup_{0<t<\text{mes}_n(\Omega)} \left(t^{p/q} \left(\int_t^{\text{mes}_n(\Omega)} \frac{dv}{[\Lambda(v)]^{p/(p-1)}} \right)^{p-1} \right)}. \tag{1.52}$$

Choosing for Λ a power function we derive the estimate

$$\inf_{u \in C_0^\infty(\Omega)} \frac{\|\text{grad } u\|^p_{L_p(\Omega)}}{\|u\|^p_{L_q(\Omega)}} \ge \frac{1}{pq^{p-1}} \left(\inf_{\{g\}} \frac{s(\partial g)}{[\text{mes}_n(g)]^{1+(p-q)/pq}} \right)^p. \tag{1.53}$$

In the case $p = q = 2$ the left hand side of the last inequality coincides with the first eigenvalue $\lambda_1(\Omega)$ of Dirichlet's problem for Laplace operator for M. Estimates of eigenvalues for the Dirichlet and Neumann problem by geometric characteristics of the manifold have been the subject of many investigations (cf. [8]). A significant role here was played by Cheeger's inequality from 1970

$$\lambda_1(\Omega) \ge \frac{1}{4} \left(\inf_{\{g\}} \frac{s(\partial g)}{\text{mes}_n(g)} \right)^2, \tag{1.54}$$

which coincides with (1.53) if $p = q = 2$. The constant $1/4$ in (1.54) cannot be improved (Buser). The above proof of (1.53) shows that *Cheeger's inequality* is a simple consequence of the estimate

$$\lambda_1(\Omega) \ge \frac{1}{4} \inf_{F \subset \Omega} \frac{\text{cap}_\Omega(F)}{\text{mes}_n(F)}, \tag{1.55}$$

which was established by the author in 1962. Inequality (1.52) for $p = q = 2$ can be rewritten as

$$\frac{1}{\lambda_1(\Omega)} \leq 4 \sup_{0<t<\text{mes}_n(\Omega)} \left(t \int_t^{\text{mes}_n(\Omega)} \frac{dv}{[\lambda(v)]^2} \right). \tag{1.56}$$

Specializing the function Λ in (1.51) one can derive from it strengthenings of known lower bounds for the eigenvalue $\lambda_1(\Omega)$.

Let us now turn to an example. Let Ω be a subdomain of a two dimensional Riemannian manifold M, whose Gauss curvature K does not exceed $-\alpha^2$ on Ω, $\alpha = \text{const} > 0$. Let us show that

$$\frac{1}{\lambda_1(\Omega)} \leq \frac{4}{\alpha^2} \left(1 - (1 + \alpha^2 \text{mes}_2(\Omega)/4)^{-1/2} \right). \tag{1.57}$$

This is a sharpening of the *inequality of Mc Kean* (1970), $\lambda_1(\Omega) \geq \alpha^2/4$, which follows from (1.54). Indeed, as is known, in the hypothesis $K \leq -\alpha^2$ inequality (1.51) is fulfilled with $\Lambda(v) = (4\pi v + \alpha^2 v^2)^{1/2}$ (Burago-Zalgaller [1980]). Plugging this function into (1.56) we find

$$\frac{1}{\lambda_1(\Omega)} \leq \frac{4}{\alpha^2} \max_{0<t<A} (t \log \frac{A(t+1)}{(A+1)t}) = \frac{4}{\alpha^2} \Psi^{-1}(\frac{A}{A+1}),$$

where $A = \alpha^2 \text{mes}_2(\Omega)/4\pi, \Psi(y) = ye^{1-y}, 0 \leq y \leq 1$, Ψ^{-1} being the inverse of Ψ. Now (1.57) follows from the obvious inequality $\Psi^{-1}(s) \leq 1 - (1-s)^{1/2}$.

Analogous considerations can be done also in connection with other integral inequalities, in particular, with the generalized Poincaré inequality (1.35) for functions on a domain on a Riemannian manifold. Then the powers of summability do not depend only on the amount of irregularity of the boundary but also on the geometry of the manifold. As before, the properties of the imbedding operator are fixed by the constants in the corresponding isoperimetric inequalities between measures and capacities, so the main difficulty consists of the verification of these inequalities.

Recently this theme has given much attention, to a considerable extent, thanks to the applications to physics and geometry. Let us turn to some such results.

As the isoperimetric inequality (1.1) also is true for subsets of the Lobachevskiĭ space H^n of constant negative curvature, the estimate (1.2) with an exact constant remains in force for functions on H^n.

For subsets g of a smooth k-dimensional submanifold of $\mathbb{R}^n$ without boundary one has established the inequality

$$v(g) \leq c_k (s(\partial g) + Q(g))^{k/(k-1)}, \tag{1.58}$$

where v and s are the k-dimensional volume and the $(k-1)$-dimensional area and Q the absolute integral mean curvature (Michael, Simon 1973). Repeating almost word by word the argument in Sect. 1.1, we find that the isoperimetric inequality is equivalent to the integral inequality

$$\left(\int_M |u|^{k/(k-1)}dv\right)^{(k-1)/k} \leq c_k\left(\int_M |\mathrm{grad}\, u|dv + \int_M |uH|dv\right), \tag{1.59}$$

where H is the mean curvature. If M is a manifold with boundary ∂M, then one must add to the right hand side of (1.59) the norm of the trace of u in $L_1(\partial M)$. The constant c_k depends only on k. Inequality (1.58) with an exact constant has so far not been obtained even for minimal surfaces ($H = 0$). Probably, in this case the exact value of c_k coincides with the constant in the classical isoperimetric inequality for subsets of $\mathbb{R}^k$. The best value for c_2 in (1.58) found so far for domains g on a minimal surface in $\mathbb{R}^3$ is $1/2\pi$ but its exact value is attained only for simply or doubly connected g.

Let us remark that an isoperimetric inequality of the type (1.58) and the corresponding imbedding theorem have likewise been proved for objects generalizing k-dimensional surfaces in $\mathbb{R}^n$ such as currents and varifolds (cf. [1]).

Inequality (1.58) (and therefore also its consequence (1.59)) can be extended to manifolds M smoothly imbedded in an n-dimensional Riemannian manifold K (Hoffman and Spruk 1975). Then some natural geometrical restrictions arise, which are connected with the volume of M, the injectivity radius of K and its sectional curvature.

The question of exact constants in inequalities of Sobolev type for functions on a Riemannian manifold have been studied by Aubin. He proved, in particular, that for any compact two dimensional Riemannian manifold M there exists a constant $A(p)$ such that for all $u \in W_p^1(M), 1 \leq p \leq 2$, one has

$$\|u\|_{L_{2p/(2-p)}(M)} \leq K(2,p)\|\mathrm{grad}\ u\|_{L_p(M)} + A(p)\|u\|_{L_p(M)}, \tag{1.60}$$

where $K(n,p)$ is the best constant in (1.7), as given in (1.12). For manifolds of arbitrary dimension one does not have very complete results. There the necessity follows naturally from attempts to solve special nonlinear elliptic equations on Riemannian manifolds, for example, in connection with the well-known *Yamabe problem* concerning the conformal equivalence of an arbitrary Riemannian metric with a metric of constant scalar curvature. The issue is to find a number λ such that the equation

$$4\frac{n-1}{n-2}\Delta\Phi + R\Phi = \lambda\Phi^{(n+2)/(n-2)} \tag{1.61}$$

has a positive solution on a Riemannian manifold (M,g) of dimension $n \geq 3$. Here $R = R(x)$ is the scalar curvature defined by the metric g and λ the constant scalar curvature of the conformal metric.

Yamabe's paper (1960) devoted to this problem contains an error and until quite recently the problem was still open. The best results were obtained by

Aubin. He stated them in terms of the best constant of a certain inequality of Sobolev type. The nonlinear spectral problem (1.61) can be associated with a variational problem where one looks for the infimum J of the functional

$$\int_M (|\text{grad}\, u|^2 + \frac{n-2}{4(n-1)} R u^2) ds / \|u\|^2_{L_{2n/(n-2)}(M)},$$

defined in the space $W_2^{(1)}(M)$. Aubin proved that $J \leq [K(n,2)]^{-2}$ and that Yamabe's problem is positively solved if this inequality is sharp.

Let us further remark that for the n-dimensional ball S of unit measure Aubin proved the inequality

$$\|u\|^2_{L_{2n/(n-2)}(S)} \leq [K(n,2)]^2 \|\text{grad}\ u\|^2_{L_2(S)} + \|u\|^2_{L_2(S)}$$

with an exact constant for both terms in the right hand side. Only in 1986 Gil-Medrano gave a solution to Yamabe's problem, proving the inequality $J < [K(n,2)]^{-2}$ for a manifold M not coinciding with the unit ball (J. Func. Anal. *66*, 42-53 (1986)).

§4. On the Space of Functions of Bounded Variation

In the 60's the family of spaces of differentiable functions was joined by the space $BV(\Omega)$, which turned out to be most useful in geometric measure theory, the calculus of variations and in the theory of quasilinear partial differential equations.

A function u defined in a domain $\Omega \subset \mathbb{R}^n$ belongs to the space $BV(\Omega)$ if its gradient taken in distribution sense is a *vector measure*. Otherwise put, the functional $C_0^\infty(\Omega) \ni g \mapsto (\text{grad}\, u, g)$ admits the estimate

$$|(\text{grad}\, u, g)| \leq C \max |g|,$$

where C is a constant not depending on g. The variation of the measure $\text{grad}\, u$ will be denoted by $\langle u \rangle_{BV(\Omega)}$.

Not all functions in $BV(\Omega)$ can be approximated by smooth functions in the seminorm $\langle u \rangle_{BV(\Omega)}$, because the completion of $BV(\Omega) \cap C^\infty(\Omega)$ in this seminorm leads to the Sobolev space $L_1^{(1)}(\Omega)$. However, functions in $BV(\Omega)$ can be approximated by functions in $C^\infty(\Omega)$ in the following weaker sense. If $u \in BV(\Omega)$ then there exists a sequence $\{u_m\}_{m\geq 1}$ of functions in $C^\infty(\Omega)$ such that $u_m \to u$ in $L_{loc}(\Omega)$ and

$$\lim_{m\to\infty} \int_\Omega |\text{grad}\, u_m| dx = \langle u \rangle_{BV(\Omega)}.$$

Thus, as in the case of the space $L_p^{(l)}(\Omega)$, also $BV(\Omega)$ can be equipped with the norm

$$\|u\|_{BV(\Omega)} = \langle u \rangle_{BV(\Omega)} + \|u\|_{L_1(\omega)},$$

where ω is a nonempty open set such that $\bar{\omega}$ is contained in Ω.

The space $BV(\Omega)$ is intimately connected with the notion of perimeter, which also in a significant way explains its role in Analysis.

The *perimeter of a measurable set* $E \subset \mathbb{R}^n$ (in the sense of Cacciopoli and De Giorgi) is defined to be the number

$$P(E) = \inf \varliminf_{m\to\infty} s(\partial\Pi_m),$$

where the infimum is taken over all sequences of polyhedra Π_m converging to E in volume. The last thing means that the volume of the symmetric difference $(\Pi_m \backslash E) \cup (E \backslash \Pi_m)$ goes to zero.

One can show that the characteristic function χ_E of E belongs to $BV(\mathbb{R}^n)$ iff $P(E) < \infty$. Moreover $P(E) = \langle \chi_E \rangle_{BV(\mathbb{R}^n)}$.

The perimeter $P(E)$ does not exceed the Hausdorff measure $s(\partial E)$ but it may happen that $P(E) < s(\partial E)$. In order to make explicit a deeper connection between the perimeter and the measure s the folowing generalization of the normal to a smooth surface will be expedient.

The unit vector v will be called an (exterior) *normal in the sense of Federer* to the set E at the point x if

$$\lim_{\rho\to 0} \rho^{-n} \mathrm{mes}_n(\{y : y \in E \cap \mathscr{B}_\rho(x), (y-x)v > 0\}) = 0,$$

$$\lim_{\rho\to 0} \rho^{-n} \mathrm{mes}_n(\{y : y \in \mathscr{B}_\rho(x) \backslash E, (y-x)v < 0\}) = 0.$$

The set of points $x \in E$ at which a normal to E exists is called *reduced boundary* of E and will be denoted by $\partial^* E$.

A central result in the theory of the perimeter is the following

Theorem 4.1. (De Giorgi, Federer) *If* $P(E) < \infty$ *then the set* $\partial^* E$ *is measurable with respect to* s *and* $\mathrm{var\,grad}\,\chi_E$. *Further* $\mathrm{var\,grad}\,\chi_E(\partial E \backslash \partial^* E) = 0$ *and for any set* $B \subset \partial^* E$ *holds*

$$(\mathrm{grad}\,\chi_E)(B) = -\int_B v(x) s(dx), \quad \mathrm{var\,grad}\,\chi_E(B) = s(B).$$

From this follows immediately that $P(E) = s(\partial^* E)$.

A simple consequence of Theorem 4.1 is the *Gauss-Ostrogradskiĭ formula* for a set E of finite perimeter:

$$\int_E \mathrm{grad}\, u(x) dx = \int_{\partial^* E} u(x) v(x) s(dx),$$

where u is any function satisfying a Lipschitz condition.

The inner part of the perimeter of a measurable set E with respect to a domain Ω is defined by the formula

$$P_\Omega(E) = \langle \chi_{E\cap\Omega} \rangle_{BV(\Omega)}.$$

If $\chi_{E\cap\Omega} \notin BV(\Omega)$ we set $P_\Omega(E) = \infty$.

For an arbitrary measurable set $E \subset \Omega$ of finite volume there exists a sequence of sets $E_i \subset \Omega$ such that each $\partial E_i \backslash \partial\Omega$ is a C^∞-smooth submanifold of $\mathbb{R}^n$ (in general non compact) and $\chi_{E_i} \to \chi_E$ in $L_1(\Omega)$ and further $P_\Omega(E_i) \to P_\Omega(E)$.

We require also the *set function* $P_{C\Omega}$, that is the outer part (with respect to Ω) of the perimeter of a set, defined as follows

$$P_{C\Omega}(E) = \inf_{G\supset C\Omega} P_G(E).$$

It is easy to see that for a set E with finite perimeter

$$P_\Omega(E) + P_{C\Omega}(E) = P(E).$$

The following result comprises a formula analogous to (1.5).

Theorem 4.2. (Fleming, Richel) *For every locally summable function u on Ω holds*

$$\langle u\rangle_{BV(\Omega)} = \int_{-\infty}^{\infty} P_\Omega(N_t)dt, \tag{1.62}$$

where $N_t = \{x : u(x) > t\}$.

Formula (1.62) allows us to carry over to the case of the seminorm $\langle u\rangle_{BV(\Omega)}$ all that has been said in Subsection 2.2 of § 2 of this Chapter about estimates of the seminorm $\|\operatorname{grad} u\|_{L_1(\Omega)}$.

For functions of class $BV(\Omega)$ one can introduce a notion of trace on the boundary of the domain. By the *rough trace* of a function $u \in BV(\Omega)$ we mean the function u^* defined on the reduced boundary $\partial^*\Omega$ by the formula

$$u^*(x) = \sup\{t : x \in \partial^* N_t, P(N_t) < \infty\}.$$

It is clear that if a function u has a limit at the point $x \in \partial^* E$ then $u^*(x) = \lim_{y\to x} u(y)$. One can show that if $s(\partial\Omega) < \infty$ then the rough trace is an s-measurable function on $\partial^* E$. If $s(\partial\Omega) = P(\Omega) < \infty$, then for each function $u \in BV(\Omega)$ one has the following generalization of inequality (1.49):

$$\|u\|_{L_{n/(n-1)}(\Omega)} \leq n v_n^{-1/n}(\langle u\rangle_{BV(\Omega)} + \|u^*\|_{L_1(\partial\Omega)})$$

with the exact constant $n v_n^{-1/n}$.

In the same assumptions on the domain Ω one has for any function $u \in BV(\Omega)$ with summable rough trace the Gauss-Ostrogradskiĭ formula

$$\operatorname{grad} u(\Omega) = \int_{\partial^*\Omega} u^*(x)\nu(x)s(dx), \tag{1.63}$$

where ν is the outer normal to Ω at x.

We say that a function u has the *trace* $\tilde{u}(x)$ at the point $x \in \partial\Omega$ if it is summable in a neighborhood of this point and if the limit

$$\tilde{u}(x) = \lim_{\rho \to 0} \frac{1}{\operatorname{mes}_n(\Omega_\rho(x))} \int_{\Omega_\rho(x)} u(y)dy,$$

exists, where $\Omega_\rho(x) = \{y \in \Omega : |x-y| < \rho\}$.

In the assumption that $s(\partial\Omega) = P(\Omega) < \infty$, each function $u \in BV(\Omega)$ with summable rough trace has s-almost everywhere on $\partial\Omega$ a trace $\tilde{u}$ which coincides with the rough trace u^*. This statement leads to the following theorem on the existence and summability of $\tilde{u}$.

Theorem 4.3. *Assume that $s(\partial\Omega) = P(\Omega) < \infty$. Then*
(i) *if for each measurable set $E, E \subset \Omega$, holds*

$$\min\{P_{C\Omega}(E), P_{C\Omega}(\Omega \backslash E)\} \leq Q\, P_\Omega(E), \tag{1.64}$$

then each function $u \in BV(\Omega)$ admits s-almost everywhere on $\partial\Omega$ a trace $\tilde{u}$, and

$$\inf_{c \in \mathbb{R}^1} \int_{\partial\Omega} |\tilde{u} - c| s(dx) \leq Q\, \langle u \rangle_{BV(\Omega)}; \tag{1.65}$$

(ii) *if for each function $u \in BV(\Omega)$ which almost everywhere on $\partial\Omega$ admits a trace $\tilde{u}$ subject to inequality (1.65) with a constant Q independent of u, then (1.64) holds for all measurable sets $E \subset \Omega$.*

The hypothesis $s(\partial\Omega) = P(\Omega) < \infty$ is essential for the validity of this theorem.

If $s(\partial\Omega) < \infty$ then each function u in the intersection $BV(\Omega) \cap L_\infty(\Omega)$ admits a trace $\tilde{u}$, which by what has been said coincides with u^*, and the Gauss-Ostrogradskiĭ formula (1.63) holds true (Yu.D. Burago, V.G. Maz'ya 1967).

The requirement (1.64) has a global character: it is not satisfied, for example, by any nonconnected set Ω. This is essential, because it is question of inequality (1.65). If we however put (1.65) in the following (equivalent, provided Ω is connected) form

$$\int_{\partial\Omega} |\tilde{u}| s(dx) \leq \text{const}\, \|u\|_{BV(\Omega)},$$

then the role of inequality (1.64) is taken by the local requirement

$$\sup_{x \in \Omega} \limsup_{\rho \to +0} \left\{ \frac{P_{C\Omega}(E)}{P_\Omega(E)} : E \subset \Omega_\rho(x) \right\} < \infty. \tag{1.66}$$

The following result on continuation to the boundary of the domain was established by Anzellotti and Giaquinta in 1978.

Theorem 4.4. (i) *For each function* $\varphi \in L_\infty(\partial^*\Omega)$ *there exists a function* $u \in L_1^{(1)}(\Omega) \cap L_\infty(\Omega)$ *whose trace on* $\partial^*\Omega$ *equals* φ.

(ii) *If condition (1.66) is fulfilled, then for any function* $\varphi \in L_1(\partial^*\Omega)$ *there exists a function* $u \in W_1^{(1)}(\Omega)$ *whose trace on* $\partial^*\Omega$ *equals* φ *and we have the inequality*

$$\|u\|_{W_1^{(1)}(\Omega)} \le C\|\varphi\|_{L_1(\partial^*\Omega)},$$

where C *depends on* Ω *but not on* φ.

Thus, for domains Ω subject to condition (1.66) the space of traces on $\partial^*\Omega$ of functions in $BV(\Omega)$ and $W_1^{(1)}(\Omega)$ coincides with $L_1(\partial^*\Omega)$.

For sets $E \subset \Omega$ we introduce the characteristic

$$\tau_\Omega(E) = \inf_{\{B : B\cap\Omega = E\}} P_{C\Omega}(B).$$

In terms of the set function τ_Ω we can formulate the following criterion for extension from $BV(\Omega)$ to $BV(\mathbb{R}^n)$.

Theorem 4.5. (Yu.D. Burago, V.G. Maz'ya 1967) (i) *If for every function* $u \in BV(\Omega)$ *there exists an extension* $\tilde{u} \in BV(\mathbb{R}^n)$ *such that*

$$\langle \tilde{u} \rangle_{BV(\mathbb{R}^n)} \le C\langle u \rangle_{BV(\Omega)}, \tag{1.67}$$

then for any set $E \subset \Omega$ *holds*

$$\tau_\Omega(E) \le (C-1)P_\Omega(E). \tag{1.68}$$

(ii) *If inequality (1.68) holds for all sets* $E \subset \Omega$, *then for each function* $u \in BV(\Omega)$ *there exists an extension* $\tilde{u} \in BV(\mathbb{R}^n)$ *such that (1.67) holds true.*

Roughly speaking, $\tau_\Omega(E)$ may be viewed as the area of a soap film placed in the exterior of Ω and suspended on that part of the boundary of E which belongs to $\partial\Omega$. Therefore (1.68) has a simple geometric meaning. Let l be an arbitrary closed contour on the boundary of a three dimensional body Ω and let s_e and s_i be the areas of the films, suspended on l and positioned respectively outside and inside Ω. Condition (1.68) means that the ratio s_e/s_i is bounded irrespective of l. If $n = 2$ then it is question of the ratio between the distance between two arbitrary points of $\partial\Omega$ measured outside and inside Ω.

In some special cases it is possible to find the constant C in the "isoperimetric" inequality (1.68). For example, for a plane convex domain $C = 1 + s(\partial\Omega)/2h$, where h is the minimal length of those segments whose endpoints divide $\partial\Omega$ into arcs of equal length. For the n-dimensional unit ball holds $C = 1 + nv_n/2v_{n-1}$. If Ω is an n-dimensional convex domain, then $C - 1 = Q$, where Q is the constant in (1.64).

§ 5. Extension of a Function in Sobolev Space on a Domain to $\mathbb{R}^n$

Let us pass to the question of extending functions in the space $W_p^{(l)}(\Omega)$ to $\mathbb{R}^n$.

We say that Ω belongs to *the class* $EW_p^{(l)}$, $1 \leq p < \infty$, $l = 1, 2, \ldots$ if there exists a continuous linear operator $\mathscr{E} : W_p^{(l)}(\Omega) \to W_p^{(l)}(\mathbb{R}^n)$ such that $\mathscr{E}u|_\Omega = u$ for all $u \in W_p^{(l)}(\Omega)$.

It was early realized that domains with sufficiently smooth boundary are in $EW_p^{(l)}$. Moreover, each domain with a Lipschitzian boundary belongs to this class.

The most complete results concerning domains in $EW_p^{(l)}$ classes have been obtained in the two dimensional case. S.K. Vodop'yanov, V.M. Gol'dshteĭn and T.G. Latfullin proved (cf. [2]) that a simply connected plane domain belongs to the class $EW_2^{(1)}$ iff its boundary is a *quasicircle* (that is, it is the image of a circle under a quasiconformal map of the plane onto itself). By a theorem of Ahlfors this last condition is equivalent to the following inequality

$$|x - z| \leq c|x - y|, \quad c = \text{const}, \tag{1.69}$$

where x, y are arbitrary points on $\partial\Omega$ and z an arbitrary point on the arc of $\partial\Omega$ of minimal length connecting x and y.

The *Ahlfors condition* is sufficient for a bounded plane domain to be in $EW_p^{(l)}$ for all $p \in (1, \infty), l = 1, 2, \ldots$. Moreover, if Ω and $\mathbb{R}^2 \backslash \bar{\Omega}$ are both in $EW_p^{(l)}$ then it follows that $\partial\Omega$ is quasicircular. However, the class of Jordan curves bounding domains in $EW_p^{(l)}$ (with the exception of $EW_2^{(1)}$) is not exhausted by quasicircles. Consider for instance the domain in Fig. 12. The corresponding upper and lower "teeth" may approach each other so fast that Ahlfors's condition (1.69) is not fulfilled. Then Ω is not a quasicircle. The "teeth" almost leave the form unchanged and may be chosen so as to decrease in a geometric progression. Then the curve $\partial\Omega$ has finite length and is Lipschitz except near the origin. One can show that Ω is in $EW_p^{(1)}$ for $p \in [1, 2)$ and that its complement $\mathbb{R}^2 \backslash \bar{\Omega}$ belongs to $EW_p^{(1)}$ for $p > 2$.

A necessary and sufficient condition for extension of functions in $W_p^{(1)}(\Omega)$, $p \neq 2$, to $\mathbb{R}^2$ without changing the class is not known. It is easy to guess the following not very visual condition, which is a direct analogue of the criterion (1.68) for extension of functions in $BV(\Omega)$. For each conductor $K^{(i)}$ in Ω there exists a conductor $K^{(e)}$ in $\mathbb{R}^2 \backslash \bar{\Omega}$ abutting to it (cf. Fig. 13) such that

$$c_p(K^{(e)}) \leq c_p(K^{(i)}).$$

It cannot be excluded that this requirement is indeed sufficient, but this has not been proved.

Jones (1980) introduced a certain class of n-dimensional domains in $EW_p^{(l)}$ which is wider than the class of Lipschitz domains and for $n = 2$ coincides

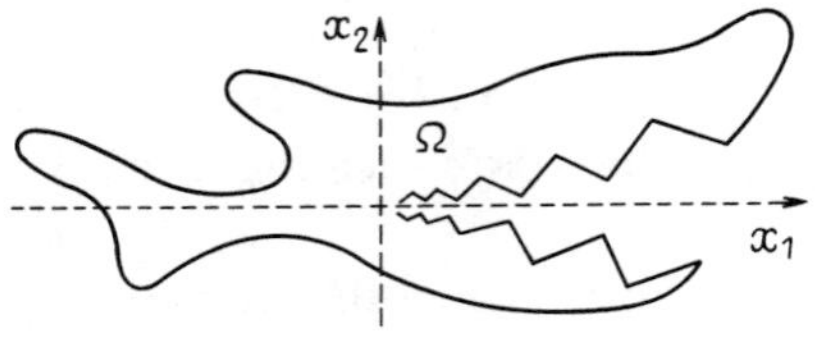

Fig. 12

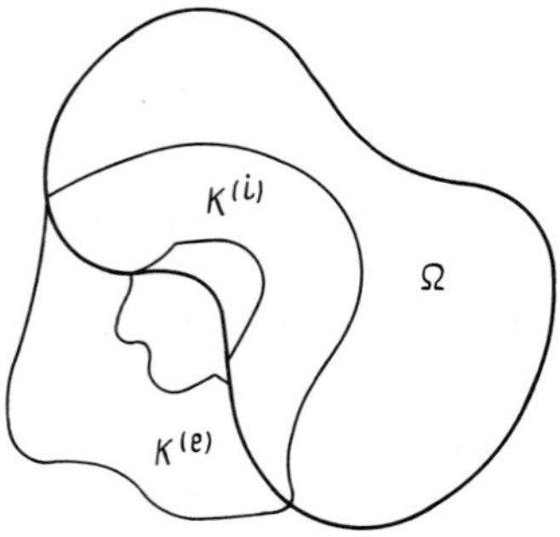

Fig. 13

with the class of quasicircles. The domains of this class are characterized by the following requirement. Any two points $x, y \in \Omega$, $|x-y| < \delta$, can be joined by rectifiable arc $\Gamma \subset \Omega$ satisfying the inequalities:

$$l(\Gamma) \leq |x-y|/\varepsilon$$

and

$$\operatorname{dist}(z, \partial\Omega) \geq \frac{\varepsilon|x-z||y-z|}{|x-y|} \text{ for all } z \in \Gamma.$$

Here $l(\Gamma)$ is the length of Γ and δ and ε are positive constants.

It is not hard to see that domains $\Omega \subset \mathbb{R}^n, n > 2$, the boundaries of which have isolated "cusps", directed inside Ω, satisfy the above requirement. Simple examples show that plane domains with interior peaks on the boundary do not belong to $EW_p^{(l)}$ for $p > 1$. This is also true for domains Ω with exterior cusps for $p \geq 1$, $n \geq 2$. However, for any of these domains there exists a bounded linear extension operator acting from $W_p^{(l)}(\Omega)$ into the weighted *space* $W_{p,\sigma}^{(l)}(\mathbb{R}^n)$ with the norm

$$\left(\sum_{|\alpha| \leq l} \int_{\mathbb{R}^n} |\sigma(x) D^\alpha u(x)|^p dx \right)^{1/p},$$

where σ is a nonnegative function separated away from zero outside an arbitrary neighborhood of the origin 0. We will not enter into the exact formulation but content ourselves to a general description of results of this nature (V.G. Maz'ya, S.V. Poborchiĭ).

The point 0 is by definition *a vertex of a peak directed into the exterior of* Ω if there exists a neighborhood U of this point such that $U \cap \Omega = \{x = (x', x_n) : 0 < x_n < 1, x'/\varphi(x_n) \in \omega\}$. Analogously, 0 is *the vertex of an inner peak* if there exists a neighborhood U of 0 such that $U \backslash \bar{\Omega} = \{x : x_n \in (0,1), x'/\varphi(x_n) \in \omega\}$. Here ω is a domain in $\mathbb{R}^{n-1}$ with compact closure and Lipschitz boundary, while φ is a continuously differentiable increasing function on $[0, 1]$, $\varphi(0) = \varphi'(0) = 0$. Moreover, φ has to be subject to certain "regularity" conditions, which do not exclude power functions. Off the vertex of a peak $\partial\Omega$ is Lipschitz.

In the hypothesis that 0 is the vertex of a peak directed into the exterior of Ω one can as the best weight σ near 0 take one of the following functions

$$(\varphi(|x|)/|x|)^{\min\{l,(n-1)/p\}} \text{ for } lp \neq n-1, p \in [1, \infty],$$

$$(\varphi(|x|)/|x|)^{l} |\log(\varphi(|x|)/|x|)|^{1-1/p} \text{ for } lp = n-1, p \in [1, \infty).$$

This implies, in particular, that $\Omega \in EW_\infty^{(l)}$.

For domains with peaks, not belonging to $EW_p^{(l)}$, one can construct extension operators to $\mathbb{R}^n$ which diminish the number of derivatives or the degree of summability (V.I. Burenkov; V.M. Gol'dshteĭn and V.N. Sitnikov; V.G. Maz'ya and S.V. Poborchiĭ). Consider, for instance, a bounded domain in $\mathbb{R}^n$ with a vertex of an exterior peak at the boundary. Then there exists a continuous linear extension operator $W_p^{(l)}(\Omega) \to W_q^{(l)}(\mathbb{R}^n)$, where $1 \leq q < p < \infty$, $lq \neq n-1$, iff the integral

$$\int_0^1 \left(\frac{t^\beta}{\varphi(t)}\right)^{n/(\beta-1)} t^{-1} dt$$

is convergent. Here β is defined by the equation $1/q - 1/p = l(\beta-1)/(1+\beta(n-1))$ for $lq < n-1$ and by the equation $1/q - 1/p = (n-1)(\beta-1)/np$ for $lq > n-1$. If $lq = n-1$ then the integrand of this integral has to be supplemented by the factor $|\log(\varphi(t)/t)|^{-\alpha}$, where $\alpha = (1-1/q)(1/q-1/p), \beta = (np-q)/q(n-1)$.

Let us turn to the properties of the boundary values of functions with a finite Dirichlet integral defined in a domain with a vertex of a peak at the boundary (V.G. Maz'ya 1983).

Take for example $n = 3$. If 0 is the vertex of a peak directed into the exterior of Ω then one has the relation

$$\inf_{\{u: u|_{\partial\Omega} = f\}} \|u\|^2_{W_2^{(1)}(\Omega)} \sim \int_\Omega \int_\Omega (f(x) - f(\xi))^2 [g(\frac{r}{E})]^2 \frac{ds(x)ds(\xi)}{r^n}$$
$$+ \int_{U \cap \partial\Omega} |f(x)|^2 g\left(\frac{x_n}{\varphi(x_n)}\right) ds(x) + \int_{\partial\Omega \setminus U} |f(x)|^2 ds(x), \qquad (1.70)$$

where $r = |x - \xi|$, $g(t) = (t+1)/\log(t+2)$, $x = (x', x_n)$, $\xi = (\xi', \xi_n)$, $E = \max\{\varphi(x_n), \varphi(\xi_n)\}$ if $x, \xi \in U$ and $E = \varphi(1)$ if $x \notin U$ or $\xi \notin U$.

If 0 is the vertex of a peak directed into the exterior of Ω, $\Omega \subset \mathbb{R}^n$, $n \geq 3$, then

$$\inf_{\{u: u|_{\partial\Omega} = f\}} \|\text{grad } u\|^2_{L_2(\Omega)} \sim \int\int_{\{x, \xi \in \partial\Omega : r < E\}} |f(x) - f(\xi)|^2 \frac{ds(x)ds(\xi)}{r^n}.$$

The traces of functions in $W_p^1(\Omega)$, where $p \in (1, \infty)$ and Ω is a plane domain with turning points at the boundary, were described by G.N. Yakovlev (1965).

Concluding this section, let us write down the space of traces to $\partial\Omega$ of functions in $W_1^{(1)}(\Omega)$ for domains with peaks (V.G. Maz'ya, S.V. Poborchiĭ 1986).

Let Ω be a domain in $\mathbb{R}^n, n \geq 3$, with a vertex of an exterior peak at the boundary. Then

$$\inf_{\{u: u|_{\partial\Omega} = f\}} \|u\|_{W_1^{(1)}(\Omega)} \sim \int_{U \cap \partial\Omega} \varphi'(x_n)|f(x)|ds(x) + \int_{\partial\Omega \setminus U} |f(x)|ds(x)$$

$$+ \iint_{\{x, \xi \in U \cap \partial\Omega : |x_n - \xi_n| < \varphi(x_n)\}} [\varphi(x_n)]^{1-n} |f(x) - f(\xi)| ds(x) ds(\xi). \tag{1.71}$$

We remark that from (1.71) follow the inclusions $L(\partial\Omega) \subset TW_1^{(1)}(\Omega) \subset L(\partial\Omega, \sigma)$, where $TW_1^{(1)}(\Omega)$ is the space of traces with the norm defined by the left hand side of (1.71) and σ a weight function such that $\sigma(x) \leq c\varphi'(x_n)$ for small $x_n > 0$. These inclusions are exact, which shows that $TW_1^{(1)}(\Omega)$ cannot coincide with any space $L(\partial\Omega, \sigma)$.

In the case of domains with an interior peak, the description of the trace space turns out to be less interesting. Namely, if 0 is the vertex of an interior peak on the boundary of the domain $\Omega \subset \mathbb{R}^n$, $n \geq 2$, then

$$\inf_{\{u: u|_{\partial\Omega} = f\}} \|u\|_{W_1^{(1)}(\Omega)} \sim \|f\|_{L(\partial\Omega)}.$$

This relation follows from the preceding criterion (1.66) by Anzellotti and Guiaquinta (cf. Theorem 4.4, §4, Chap. 1) and can also easily be verified directly.

§6. Sobolev Spaces in Domains Which Depend in a Singular Manner on Parameters

The last topic which will be discussed in this chapter concerns norm estimates for imbedding and extension operators in domains depending on parameters. It is easy to imagine various situations when the boundary of the domain loses smoothness where the parameters tend to zero or to infinity. Also, the limit domain can then be bad from the point of view of the classical theory of Sobolev spaces. We will be concerned in the speed with which the imbedding or extension operators degenerate. This problem requires that one determines explicitly the dependence of suitable characteristics of these operators on the parameters. The necessity to analyze this problem arises, in particular, in attempts to justify the formal asymptotics of solutions of boundary value problems in domains with singularly perturbed boundaries.

We begin with an estimate, due to S. V. Poborchiĭ and the author, of the norm of the extension operators $W_p^{(l)}(\mathscr{B}_\rho\backslash\bar{\Omega}_\varepsilon) \to W_p^{(l)}(\mathscr{B}_\rho)$, $W_p^{(l)}(\Omega_\varepsilon) \to W_p^{(l)}(\mathscr{B}_\rho)$, $1 \le p \le \infty$, $l = 1, 2, \ldots$. Here ε is as small positive parameter with $\varepsilon \le \rho \le \infty$. By Ω_ε we denote a small domain $\{x \in \mathbb{R}^n, x/\varepsilon \in \Omega\}$, obtained from the bounded Lipschitz domain Ω by contracting by a factor $1/\varepsilon$ and contained together with its closure in the ball $\mathscr{B}_\rho = \{x \in \mathbb{R}^n : |x| < \rho\}$.

The following theorem concerns extension into the small domain.

Theorem 6.1. *For $n \ge 2$ there exists a continuous linear extension operator $W_p^{(l)}(\mathscr{B}_\rho \setminus \bar{\Omega}_\varepsilon) \to W_p^{(l)}(\mathscr{B}_\rho)$ with norm uniformly bounded in ε, ρ.*

Proof. To fix the ideas, assume that $\bar{\Omega} \subset \mathscr{B}_1$. It suffices to construct a linear extension operator $\mathscr{E}$: $W_p^{(l)}(\mathscr{B}_\varepsilon\backslash\bar{\Omega}_\varepsilon) \to W_p^{(l)}(\mathscr{B}_\varepsilon)$. Introduce the map κ : $\mathbb{R}^n \ni x \mapsto \varepsilon x$. To each function $u \in W_p^{(l)}(\mathscr{B}_\varepsilon\backslash\bar{\Omega}_\varepsilon)$ we associate a corresponding polynomial P in $\mathbb{R}^n$ of degree $l-1$, whose coefficients are linear functionals in u, such that

$$\|u \circ \kappa - P\|_{W_p^{(l)}(\mathscr{B}_1\backslash\bar{\Omega})} \le c\|\nabla_l(u \circ \kappa)\|_{L_p(\mathscr{B}_1\backslash\bar{\Omega})}.$$

Then

$$\|\nabla_s(u - P \circ \kappa^{-1})\|_{L_p(\mathscr{B}_\varepsilon\backslash\Omega_\varepsilon)} \le c\varepsilon^{l-s}\|\nabla_l u\|_{L_p(\mathscr{B}_\varepsilon\backslash\Omega_\varepsilon)},$$

where $0 \le s \le l$. Let E be the extension operator obtained by the Stein construction from the domain $\mathscr{B}_1\backslash\bar{\Omega}$ to $\mathbb{R}^n$ and set $E_\varepsilon u = (E(u \circ \kappa)) \circ \kappa^{-1}$. Put

$$\mathscr{E}u = P \circ \kappa^{-1} + E_\varepsilon(u - P \circ \kappa^{-1}).$$

It is clear that $\mathscr{E}u|_{\mathscr{B}_\varepsilon\backslash\bar{\Omega}_\varepsilon} = u$ for all $u \in W_p^{(l)}(\mathscr{B}_\varepsilon\backslash\bar{\Omega}_\varepsilon)$. Further, using the boundedness of $E : W_p^{(s)}(\mathscr{B}_1\backslash\bar{\Omega}) \to W_p^{(s)}(\mathbb{R}^n)$ for $s \le l$, we obtain:

$$\|\nabla_s E_\varepsilon(u - P \circ \kappa^{-1})\|_{L_p(\mathbb{R}^n)} \le$$

$$\le c\sum_{k=0}^{s} \varepsilon^{k-s}\|\nabla_k(u - P \circ \kappa^{-1})\|_{L_p(\mathscr{B}_\varepsilon\backslash\Omega_\varepsilon)} \le c\varepsilon^{l-s}\|\nabla_l u\|_{L_p(\mathscr{B}_\varepsilon\backslash\Omega_\varepsilon)}.$$

Moreover,

$$\|\nabla_s(P \circ \kappa^{-1})\|_{L_p(\mathscr{B}_\varepsilon)} \le c\|\nabla_s(P \circ \kappa^{-1})\|_{L_p(\mathscr{B}_\varepsilon\backslash\Omega_\varepsilon)} \le$$

$$\le c\|\nabla_s u\|_{L_p(\mathscr{B}_\varepsilon\backslash\Omega_\varepsilon)} + c\varepsilon^{l-s}\|\nabla_l u\|_{L_p(\mathscr{B}_\varepsilon\backslash\Omega_\varepsilon)}.$$

Thus, $\|\mathscr{E}u\|_{W_p^{(l)}(\mathscr{B}_\varepsilon)} \le c\|u\|_{W_p^{(l)}(\mathscr{B}_\varepsilon\backslash\bar{\Omega}_\varepsilon)}$ and we have the desired extension operator $\mathscr{E}$.

The proof is complete.

Let us state a theorem concerning extension to the exterior from a small domain.

Theorem 6.2. (i) *The norm of an arbitrary extension operator* $\mathscr{E} : W_p^{(l)}(\Omega_\varepsilon) \to W_p^{(l)}(\mathscr{B}_\rho)$ *satisfies the inequalities*

$$\|\mathscr{E}\| \geq \begin{cases} c\varepsilon^{-n/p} \min\{\rho^{n/p}, 1\} & \text{if } lp > n, \\ c\varepsilon^{-n/p} \min\{\rho^{n/p}, |\log \varepsilon|^{(1-p)/p}\} & \text{if } lp = n, \\ c\varepsilon^{-n/p} \min\{\rho^{n/p}, \varepsilon^{n/p-l}\} & \text{if } lp < n. \end{cases} \tag{1.72}$$

(ii) *To each pair* ε, ρ *there corresponds a linear extension operator* $\mathscr{E}$ *whose norm satisfies inequalities which are opposite to those formulated in (i).*

An analogous statement holds for the extension operator from a narrow cylinder into a larger one. Set

$$G_\varepsilon = \{x = (y, z) \in \mathbb{R}^{n+s} : y \in \Omega_\varepsilon \subset \mathbb{R}^n, z \in \mathbb{R}^s\},$$

$$D_\rho = \{x = (y, z) \in \mathbb{R}^{n+s} : y \in \mathscr{B}_\rho \subset \mathbb{R}^n, z \in \mathbb{R}^s\}.$$

Theorem 6.3. (i) *The norm of an arbitrary extension operator* $\mathscr{E} : W_p^{(l)}(G_\varepsilon) \to W_p^{(l)}(D_\rho)$ *satisfies the inequalities (1.72).*

(ii) *For each pair* ε, ρ *there exists a linear extension operator* $\mathscr{E} : W_p^{(l)}(G_\varepsilon) \to W_p^{(l)}(D_\rho)$ *whose norm satisfies inequalities opposite to those in (1.72).*

Theorems 6.1–6.3 allow us to get twosided estimates of the norms of extension operators for concrete domains depending on parameters. Let us consider some examples which illustrate the possibilities arising.

Example 6.1. On Fig. 14 there is depicted a boa constrictor (of unit length and thickness ε), that has swallowed an elephant.[1] We may assume that this has been taken place in $\mathbb{R}^n$. One can then show that

$$\inf \|\mathscr{E}\|_{W_p^{(l)}(\Omega_\varepsilon) \to W_p^{(l)}(\mathbb{R}^n)} \sim \begin{cases} \varepsilon^{-\min\{l, (n-1)/p\}} & \text{for } lp \neq n-1, \\ \varepsilon^{-l} |\log \varepsilon|^{(1-p)/p} & \text{for } lp = n-1. \end{cases}$$

The symbol $a \sim b$, in this section, means uniform equivalence in the parameters.

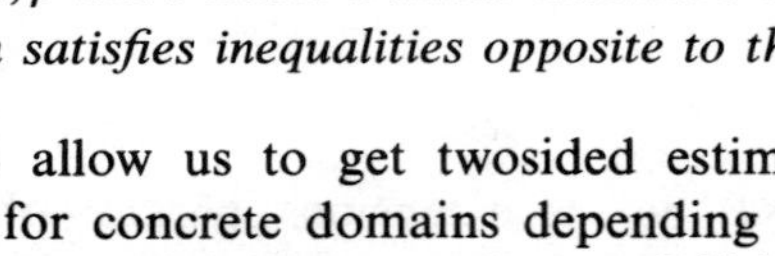

Fig. 14

[1]The illustration is taken from the book "Le petit prince" by A. de Saint-Exupéry (Gallimard, Paris 1957).

Example 6.2. Let Ω_ε be a truncated peak $\{x = (x', x_n) : \varepsilon < x_n < 1, x'/\varphi(x_n) \in \omega\}$. (The notation is the same as in § 5 of Chap. 1). Then

$$\inf \|\mathscr{E}\|_{W_p^{(l)}(\Omega_\varepsilon) \to W_p^{(l)}(\mathbb{R}^n)} \sim \begin{cases} \left(\dfrac{\varepsilon}{\varphi(\varepsilon)}\right)^{\min\{l,(n-1)/p\}} & \text{for } lp \neq n-1, \\ \left(\dfrac{\varepsilon}{\varphi(\varepsilon)}\right)^{l} \left|\log \dfrac{\varphi(\varepsilon)}{\varepsilon}\right|^{(1-p)/p} & \text{for } lp = n-1. \end{cases}$$

In the next example we consider a domain depending on two parameters ε and δ.

Example 6.3. Let $x_1 + ix_2 = r\exp(i\theta)$ and $x' = (r - \delta, x_3, \ldots, x_n)$. We denote by $\mathscr{B}_{\varepsilon,\delta}$ a small and narrow "boublik"[2] $\{x : \theta \in [0, 2\pi), x'/\varepsilon \in \omega\}$, where ω is a domain in $\mathbb{R}^{n-1}$ with compact closure and Lipschitz boundary (cf. Fig. 15).

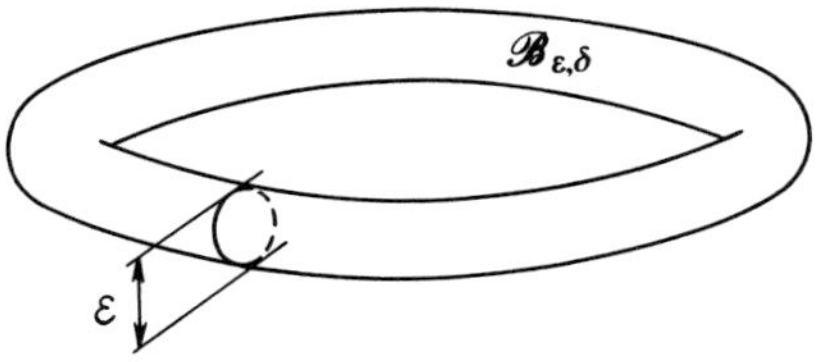

Fig. 15

We assume that $0 < \varepsilon \ll \delta \leq 1$. The appearance of the second parameter increases the number of variants, namely:

$$\inf \|\mathscr{E}\|_{W_p^{(l)}(\mathscr{B}_{\varepsilon,\delta}) \to W_p^{(l)}(\mathbb{R}^n)} \sim$$

$$\sim \begin{cases} \varepsilon^{-l} & \text{for } lp < n-1, \\ \varepsilon^{-l} \left|\log \dfrac{\delta}{\varepsilon}\right|^{(1-p)/p} & \text{for } lp = n-1, \\ \varepsilon^{-(n-1)/p} \delta^{-l+(n-1)/p} & \text{for } n-1 < lp < n, \\ \varepsilon^{-(n-1)/p} \delta^{-1/p} |\log \delta|^{(1-p)/p} & \text{for } lp = n, \\ \varepsilon^{-(n-1)/p} \delta^{-1/p} & \text{for } lp > n\,. \end{cases}$$

If we do not aim at an elementary formulation of the answer, we can write this shorter as follows:

$$\inf \|\mathscr{E}\|_{W_p^{(l)}(\mathscr{B}_{\varepsilon,\delta}) \to W_p^{(l)}(\mathbb{R}^n)} \sim \left(\frac{p\text{-cap}_{B_1}(\mathscr{B}_{\varepsilon,\delta})}{\text{mes}_n(\mathscr{B}_{\varepsilon,\delta})}\right)^{1/p},$$

where $\mathscr{B}_\rho = \{x : |x| < \rho\}$.

[2] *Translator's note.* A kind of roll.

Let us pass to twosided estimates of the norms of imbedding operators. Making concrete the general results formulated in §2 of Chap. 1 we can indeed obtain estimates for the most varied domains.

If Ω is the peak $\{x = (x', x_n) : 0 < x_n < 1, x'/x_n^\kappa \in \omega\}$, where $\kappa > 1$ and ω is a domain in $\mathbb{R}^{n-1}$ with compact closure and Lipschitz boundary, then $W_p^{(l)}(\Omega)$ is not imbedded into $L_q(\Omega)$ for $1/q = 1/p - l/n > 0$ (cf. Example 2.5).

Let us perturb Ω near the point O in two ways. Namely, put $\Omega_\varepsilon^{(1)} = \Omega \backslash \bar{\mathscr{B}}_\varepsilon$ and $\Omega_\varepsilon^{(2)} = \Omega \cup \mathscr{B}_\varepsilon$ (cf. Fig. 16).

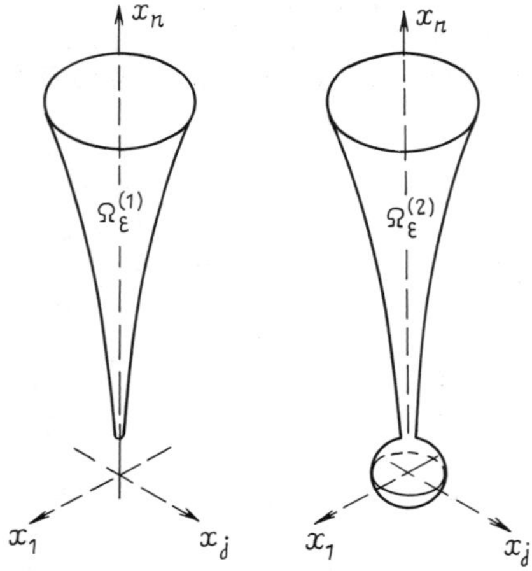

Fig. 16

The boundaries of the domains $\Omega_\varepsilon^{(j)}$, $j = 1, 2$, are Lipschitz and therefore $W_p^{(l)}(\Omega_\varepsilon^{(j)}) \subset L_q(\Omega_\varepsilon^{(j)})$. Upon applying Theorem 2.6 one can show that

$$\sup_u \frac{\|u\|_{L_q(\Omega_\varepsilon^{(j)})}}{\|u\|_{W_p^{(l)}(\Omega_\varepsilon^{(j)})}} \sim \begin{cases} \varepsilon^{-l(n-1)(\kappa-1)/n} & \text{for } j = 1, \\ \varepsilon^{-l(n-1)(\kappa-1)/n} + (\varepsilon^l + \varepsilon^{(n-1)(\kappa-1)/p})^{-1} & \text{for } j = 2. \end{cases}$$

Let us consider one more question for domains depending on parameters. Let Ω_ε be a domain with smooth boundary for each $\varepsilon \in (0, 1)$ and let $T[W_p^{(l)}(\Omega_\varepsilon)]$ be the space of boundary values of functions in the Sobolev space $W_p^{(l)}(\Omega_\varepsilon)$, $1 < p < \infty$, $l = 1, 2, \ldots$, equipped with the norm

$$\|f\|_{T[W_p^{(l)}(\Omega_\varepsilon)]} = \inf_{\{u \in W_p^{(l)}(\Omega_\varepsilon) : u|_{\partial\Omega_\varepsilon} = f\}} \|u\|_{W_p^{(l)}(\Omega_\varepsilon)}. \tag{1.73}$$

As is well-known, $T[W_p^{(l)}(\Omega_\varepsilon)]$ coincides with the space $W_p^{(l-1/p)}(\partial\Omega_\varepsilon)$. However, the ratio between the usual norm in $W_p^{(l-1/p)}(\partial\Omega_\varepsilon)$ and the norm (1.73) is not separated from zero or bounded as $\varepsilon \to 0$. We are interested in the question whether one can introduce in $T[W_p^{(l)}(\Omega_\varepsilon)]$ a norm, described without aid of the extension, which is equivalent to the norm (1.73), uniformly in ε.

The following two theorems due to the author (1983) give an affirmative answer for the space of functions with a finite Dirichlet integral inside or outside the narrow "boublik" $\mathscr{B}_\varepsilon = \mathscr{B}_{\varepsilon,1}$, as defined in Example 6.3.

Theorem 6.4.

$$\|f\|_{T[W_2^{(1)}(\mathscr{B}_\varepsilon)]} \sim \int_{\partial\mathscr{B}_\varepsilon}\int_{\partial\mathscr{B}_\varepsilon} |f(x)-f(y)|^2 \frac{ds(x)ds(y)}{r^n}$$

$$+\varepsilon \int_{\partial\mathscr{B}_\varepsilon} |f(x)|^2 ds(x),\ r = |x-\xi|.$$

Theorem 6.5. *Let g be the same function as in (1.70). Then*

$$\|f\|_{T[W_2^{(1)}(\mathbb{R}^n\setminus\bar{\mathscr{B}}_\varepsilon)]} \sim$$

$$\sim \begin{cases} \int_{\partial\mathscr{B}_\varepsilon}\int_{\partial\mathscr{B}_\varepsilon} |f(x)-f(\xi)|^2 \frac{g(r/\varepsilon)}{r^3} ds(x)ds(\xi) + \frac{1}{\varepsilon|\log\varepsilon|}\int_{\partial\mathscr{B}_\varepsilon} |f(x)|^2 ds(x) & \textit{for } n=3; \\ \int_{\partial\mathscr{B}_\varepsilon}\int_{\partial\mathscr{B}_\varepsilon} |f(x)-f(\xi)|^2 \frac{ds(x)ds(\xi)}{r^n} + \frac{1}{\varepsilon}\int_{\partial\mathscr{B}_\varepsilon} |f(x)|^2 ds(x) & \textit{for } n>3. \end{cases}$$

The circle of ideas described in this section has so far not received much attention, but it undoubtedly deserves a better fate.

Chapter 2

Inequalities for Potentials and Their Applications to the Theory of Spaces of Differentiable Functions

§ 1. Estimates for Riesz and Bessel Potentials

1.1. On Summability in Measure of Functions in the Spaces $h_p^{(l)}, b_p^{(l)}$ etc. Denote by $L_q(\mu)$ the space of those functions in $\mathbb{R}^n$ which are measurable and qth summable with respect to μ. Put

$$\|u\|_{L_q(\mu)} = \left(\int |u|^q d\mu\right)^{1/q}.$$

Define the *Riesz potential of order l* by

$$(I_l f)(x) = \int |x-y|^{l-n} f(y)dy.$$

Introduce further the *Bessel potential* $J_l = G_l * f$, where $G_l(x) = \text{const} \cdot |x|^{(l-n)/2} K_{(n-l)/2}(|x|)$, K_ν being the modified Bessel function of the third kind. Equivalent definition of the Bessel potential: $J_l f = cF^{-1}(1+|\xi|^2)^{-l/2} Ff$, where F is the Fourier transform in $\mathbb{R}^n$. Therefore (up to a constant factor) I_l and J_l are the inverses of the lth powers of the operators $(-\Delta)^{1/2}$ and $(-\Delta+I)^{1/2}$.

We define *the space $h_p^{(l)}$* as the completion of C_0^∞ in the norm $\|(-\Delta)^{l/2}u\|_{L_p}$. Using in this L_p-norm instead the function $(-\Delta+I)^{l/2}u$, we get *the space $H_p^{(l)}$*.

For l integer and $p \in (1,\infty)$ the norms on $h_p^{(l)}$ and $H_p^{(l)}$ are equivalent to the norms $\|\text{grad}_l u\|_{L_p}$ and $\|\text{grad}_l u\|_{L_p} + \|u\|_{L_p}$.

One can show that if $lp < n$, $p \in (1,\infty)$ then u belongs to $h_p^{(l)}$ iff it can be represented as a Riesz potential of a function in L_p. Similarly, $u \in H_p^{(l)}$, $1 < p < \infty$, iff $u = J_l f$, where $f \in L_p$. By the same token, the imbeddings of the spaces $h_p^{(l)}$ and $H_p^{(l)}$ into $L_q(\mu)$ are equivalent to the continuity of the operators I_l and $J_l : L_p \to L_q(\mu)$, respectively.

We formulate and prove a generalization of the imbedding theorems by S.L. Sobolev and V.P. Il'in due to D.R. Adams (1971).

Theorem 1.1. *Let $l > 0, 1 < p < q < \infty, lp < n$. The Riesz potential I_l gives rise to a continuous map $L_p \to L_q(\mu)$ iff the function*

$$\mathbb{R}^n \ni x \mapsto F(x) \stackrel{\text{def}}{=} \sup_{\rho>0} \rho^{-s}\mu[\mathcal{B}_\rho(x)], \quad s = q(n-pl)/p,$$

is bounded.

For the proof of the necessity it suffices to take in the inequality

$$\|I_l f\|_{L_q(\mu)} \le C\|f\|_{L_p} \tag{2.1}$$

the function f equal to the characteristic function of the ball $\mathcal{B}_\rho(x)$.

The sufficiency is obtained from the classical Marcinkiewicz interpolation theorem, along with the inequality

$$t\mu(\mathcal{L}_t)^{1/q} \le c \sup F(x)^{1/q} \|f\|_{L_p}, \tag{2.2}$$

where $\mathcal{L}_t = \{y : (I_l|f|)(y) > t\}, t > 0$. The last fact we establish as follows. Let μ_t be the restriction of the measure μ to the set $\mathcal{L}_t$ and let r be a positive number to be fixed later on. It is clear that

$$t\mu(\mathcal{L}_t) \le \int_{\mathbb{R}^n} |f(x)| \int_{\mathbb{R}^n} |x-y|^{l-n} d\mu_t(y)dx =$$

$$= (n-l) \int_{\mathbb{R}^n} \int_0^\infty |f(x)| \mu_t[\mathscr{B}_\rho(x)] dx \rho^{l-n-1} d\rho.$$

We write the $[0, \infty)$ integral as a sum $A_1 + A_2$, where A_1 is integration over $[0, r]$. As $\mu_t[\mathscr{B}_\rho(x)] \leq (\mu_t[\mathscr{B}_\rho(x)])^{(p-1)/p} F(x)^{1/p} \rho^{s/p}$, we get

$$A_1 \leq \sup F(x)^{1/p} \|f\|_{L_p} \int_0^r \left(\int \mu_t[\mathscr{B}_\rho(x)] dx \right)^{(p-1)/p} \rho^{l-n-1+s/p} d\rho.$$

If we remark that

$$\int \mu_t[\mathscr{B}_\rho(x)] dx = v_n \rho^n \mu(\mathscr{L}_t),$$

we get the estimate

$$A_1 \leq c \sup F(x)^{1/p} \|f\|_{L_p} \mu(\mathscr{L}_t)^{(p-1)/p} r^{l-(n-s)/p}.$$

Analogously:

$$A_2 \leq \|f\|_{L_p} \mu(\mathscr{L}_t)^{1/p} \int_r^\infty \left(\int \mu_t[\mathscr{B}_\rho(x)] dx \right)^{(p-1)/p} \rho^{l-n-1} d\rho =$$

$$= c \|f\|_{L_p} \mu(\mathscr{L}_t) r^{l-n/p}.$$

Consequently

$$t \mu(\mathscr{L}_t)^{1/p} \leq c \|f\|_{L_p} \left(\sup F(x)^{1/p} r^{l-(n-s)/p} + \mu(\mathscr{L}_t)^{1/p} r^{l-n/p} \right).$$

Taking here $r^s = \mu(\mathscr{L}_t) / \sup F(x)$ we get (2.2).

A simple modification of this proof shows that the Bessel potential J_l induces a continuous map $L_p \to L_q(\mu)$, $1 < p < q$, $lp < n$, iff the function

$$\mathbb{R}^n \ni x \mapsto \sup_{1 > \rho > 0} \rho^{-s} \mu[\mathscr{B}_\rho(x)]$$

is bounded.

An analogous result in the case $n = pl$ was established by S.P. Preobrazhenskiĭ and the author (1981). Namely, one has

Theorem 1.2. *Let $l > 0$, $1 < p < q < \infty$, $lp = n$. The Bessel potential J_l induces a continuous map $L_p \to L_q(\mu)$ iff the function*

$$\mathbb{R}^n \ni x \mapsto \sup_{1 > \rho > 0} |\log 2\rho|^{q(p-1)/p} \mu[\mathscr{B}_\rho(x)]$$

is bounded.

For $pl > n$ the necessary and sufficient condition for the continuity of the operator $J_l : L_p \to L_q(\mu)$, where $q \geq p > 1$, takes the simple form:

$$\sup\{\mu(\mathscr{B}_1(x)) : x \in \mathbb{R}^n\} < \infty.$$

The restriction $p < q$ in Theorems 1.1 and 1.2 is essential. The condition for μ in these theorems is not sufficient for $p \geq q$. For such p and q one also knows necessary and sufficient conditions for the validity of inequality (2.1) but they look somewhat more complicated. Before stating such criteria let us introduce the notion of *capacity of a set* $E \subset \mathbb{R}^n$ *generated by the potential* I_l. Namely, let us set

$$c_{p,l}(E) = \inf\{\|f\|_{L_p}^p : f \geq 0, I_l f \geq 1 \text{ on } E\}.$$

The theory and the applications of this and similar set functions have being payed much attention in the past years (cf. Gol'dshteĭn-Reshetnyak [1983], Maz'ya [1985], Adams [1981] etc.). We have the following partial generalization of inequality (1.21) in Chap. 1 (only for functions in the entire space $\mathbb{R}^n$).

Theorem 1.3. *We have the inequality*

$$\int_0^\infty c_{p,l}(\{x : (I_l f)(x) \geq t\}) t^{p-1} dt \leq c\|f\|_{L_p}^p. \tag{2.3}$$

Here $p \in (1, n/l)$, f *being an arbitrary function in* L_p.

The existence of an inequality of this type was discovered by the author in 1972 for $l = 1$ and 2. Adams (1976) extended this result to the case of arbitrary integer l. Subsequently variants of inequality (2.3) were proved with norms in various function spaces by the author and by Dahlberg and Hansson.

One of the proofs of (2.3) for integer l, due to Adams, is based on the following *lemma of Hedberg* (1972).

Lemma. *Let* $0 < \theta < 1, 0 < r < n$, *let* $I_r g$ *be the Riesz potential of order* r *of a nonnegative function* g *and let* $\mathscr{M}$ *be* the Hardy-Littlewood maximal operator,

$$(\mathscr{M}g)(x) = \sup_{\rho > 0} \frac{1}{|\mathscr{B}_\rho|} \int_{\mathscr{B}_\rho(x)} g(y) dy.$$

Then

$$(I_{r\theta} g)(x) \leq c[(I_r g)(x)]^\theta [(\mathscr{M}g)(x)]^{1-\theta}. \tag{2.4}$$

Proof. For each $\delta > 0$

$$\int_{\mathbb{R}^n \setminus \mathscr{B}_\delta(x)} g(y) \frac{dy}{|x-y|^{n-r\theta}} \leq$$

$$\le \delta^{r(\theta-1)} \int_{\mathbb{R}^n \setminus \mathscr{B}_\delta(x)} g(y) \frac{dy}{|x-y|^{n-r}} \le \delta^{r(\theta-1)} (I_r g)(x).$$

The remaining part of $(I_{r\theta})(x)$ is estimated as follows:

$$\int_{\mathscr{B}_\delta(x)} g(y) \frac{dy}{|x-y|^{n-r\theta}} = \sum_{k=0}^{\infty} \int_{\mathscr{B}_{2^{-k}\delta}(x) \setminus \mathscr{B}_{2^{-k-1}\delta}(x)} g(y) \frac{dy}{|x-y|^{n-r\theta}} \le$$

$$\le c \sum_{k=0}^{\infty} (\delta 2^{-k})^{r\theta} (\delta 2^{-k})^{-n} \int_{\mathscr{B}_{2^{-k}\delta}(x)} g(y) dy \le c \delta^{r\theta} (\mathscr{M} g)(x).$$

Consequently

$$(I_{r\theta} g)(x) \le c \left(\delta^{r\theta} (\mathscr{M} g)(x) + \delta^{r(\theta-1)} (I_r g)(x) \right).$$

Putting here $\delta^r = (I_r g)(x)/(\mathscr{M} g)(x)$, we obtain (2.4).

A simple consequence of (2.4), obtained by differentiating, is the inequality

$$|\nabla_l F(I_l g)| \le c Q (\mathscr{M} g + |\nabla_l I_l g|), \tag{2.5}$$

where F is a function in $C^l(0, \infty)$ such that

$$t^{k-1} |F^{(k)}(t)| \le Q, \quad 0 \le k \le l, Q = \text{const}.$$

Now (2.3) is derived as follows.

Set $u = I_l f$ and $v = I_l |f|$. Putting $t_j = 2^j, j = 0, \pm 1, \ldots$ and using the obvious inequality $v(x) \ge |u(x)|$, we find that the left hand side of (2.3) does not exceed

$$c \sum_{j=-\infty}^{+\infty} (t_{j+1} - t_j)^p c_{p,l}(\{x : v(x) \ge t_j\}). \tag{2.6}$$

Pick $\gamma \in C^\infty(\mathbb{R}^1)$, $\gamma(\tau) = 0$ for $\tau < \varepsilon$, $\gamma(\tau) = 1$ for $\tau > 1 - \varepsilon$, where ε is arbitrary. Introduce a function $v \mapsto F \in C^\infty((0,\infty))$ which coincides on the interval $[t_j, t_{j+1}]$ with the function

$$F_j(v) = t_j + (t_{j+1} - t_j) \gamma\left(\frac{v - t_j}{t_{j+1} - t_j}\right).$$

By one of the properties of capacity and inequality (2.5), the quantity (2.6) does not exceed

$$\sum_{j=0}^{+\infty} \|F_j(v)\|^p_{h_p^{(l)}} = \|F(v)\|^p_{h_p^{(l)}} \le c(\|\mathscr{M}|f|\|^p_{L_p} + \|\text{grad}_l I_l |f|\|^p_{L_p}).$$

As the operator $\mathscr{M}$ and the singular integral operator $\nabla_l I_l$ are continuous on L_p, the last sum is less than $c\|f\|^p_{L_p}$, which we set out to prove.

From (2.3) we obtain at once the following result, analogous to Theorem 1.2 in Chap. 1.

Theorem 1.4. *Let $l > 0, 1 < p < \infty, lp < n$. The estimate*

$$\|I_l f\|_{L_p(\mu)} \le C \|f\|_{L_p} \tag{2.7}$$

holds true iff for each Borel set E, $E \subset \mathbb{R}^n$, holds the inequality

$$\mu(E) \le \text{const} \cdot c_{p,l}(E).$$

In 1984 Kerman and Sawyer observed that the necessary and sufficient condition for the validity of (2.7) can be formulated without using arbitrary sets. Namely, one has

Theorem 1.5. *Let $l > 0, 1 < p < \infty, lp < n$. The estimate (2.7) holds true iff for each ball $\mathscr{B}_\rho(y)$ one has the following inequality*

$$\|I_l \mu_{\mathscr{B}_\rho(y)}\|_{L_{p'}} \le \text{const} \cdot [\mu(\mathscr{B}_\rho(y))]^{1/p'}, \tag{2.8}$$

where μ_E is the restriction of the measure μ to E.

A few words about the proof. It is obtained from an analogous result pertaining to the case of an absolutely continuous measure μ: $d\mu(x) = u(x)dx$. By duality (2.7) is equivalent to the following estimate:

$$\int (I_l g(x))^{p'} dx \le c \int (g(x))^{p'} v(x) dx \text{ for all } g \ge 0,$$

where $v(x) = u(x)^{1-p'}$. Introduce the *"fractional" maximal function $\mathscr{M}_l g$*:

$$(\mathscr{M}_l g)(x) = \sup_{\rho > 0} \frac{1}{|\mathscr{B}_\rho|^{1-l/n}} \int_{\mathscr{B}_\rho(x)} g(y) dy.$$

By a result by Muckenhoupt and Wheeden (cf. Adams [1981]), the $L_{p'}$-norms of the functions $I_l g$ and $\mathscr{M}_l g$ are equivalent.

By a theorem of Sawyer [1982] the twosided estimate

$$\int ((\mathscr{M}_l g)(x))^{p'} w(x) dx \le C \int (g(x))^{p'} v(x) dx$$

holds true iff

$$\int_{\mathscr{B}_\rho(y)} ((\mathscr{M}_l(\chi_{\rho,y} v^{1-p}))(x))^{p'} w(x) dx \le C \int_{\mathscr{B}_\rho(y)} (v(x))^{1-p} dx \text{ for all } \mathscr{B}_\rho(y),$$

where $\chi_{\rho,y}$ is the characteristic function of the ball $\mathscr{B}_\rho(y)$. If $v^{1-p} = u$, $w = 1$ the last criterion takes the form

$$\int_{\mathscr{B}_\rho(y)} ((\mathscr{M}_l(\chi_{\rho,y}u))(x))^{p'} dx \leq C \int_{\mathscr{B}_\rho(y)} u(x)dx,$$

which, as is readily seen, is equivalent to (2.8).

The following theorem, pertaining to the case $q < p$, was proved by the author (cf. Maz'ya [1985]).

Theorem 1.6. *Let $l > 0, 0 < q < p < \infty, lp < n$. Let further $\{g_j\}$ be an arbitrary family of open subsets such that $\overline{g_{j+1}} \subset g_j$ and set $\mu_j = \mu(g_j)$ and $\gamma_j = c_{p,l}(g_j)$. Inequality (2.1) holds iff*

$$\sum_{j=-\infty}^{\infty} \left(\frac{(\mu_j - \mu_{j+1})^{1/q}}{\gamma_j^{1/p}} \right)^{pq/(p-q)} \leq \text{const.}$$

From this follows the simpler condition:

$$\int_0^\infty \left(\frac{t}{\kappa(t)} \right)^{q/(p-q)} dt < \infty,$$

where $\kappa(t) = \inf\{c_{p,l}(E) : \mu(E) \geq t\}$.

Exactly similar results have likewise been established for Bessel potentials. Then the condition $lp < n$ can be removed and the role of $c_{p,l}$ is taken by the capacity

$$C_{p,l}(E) = \inf\{\|f\|_{L_p}^p : f \geq 0, J_l f \geq 1 \text{ on } E\}.$$

In the case $lp > n, p > q$ a necessary and sufficient condition for the continuity of the operator $J_l : L_p \to L_q(\mu)$ can be written in a significantly simpler form:

$$\sum_i [\mu(Q_i)]^{p/(p-q)} < \infty,$$

where $\{Q_i\}$ is the sequence of cubes of side length 1 generated by the coordinate lattice in $\mathbb{R}^n$.

As was remarked at the beginning of this section, the results on inequalities of the type (2.1) set forth can be interpreted in terms of the spaces $h_p^{(l)}$ and $H_p^{(l)}$. Recall now that the traces of the spaces $h_p^{(l+1/p)}(\mathbb{R}^{n+1})$ and $H_p^{(l+1/p)}(\mathbb{R}^{n+1})$ on $\mathbb{R}^n$ generate the *Besov spaces* $b_p^{(l)}(\mathbb{R}^n)$ and $B_p^{(l)}(\mathbb{R}^n)$, defined as the completion of $C_0^\infty(\mathbb{R}^n)$ in the norms

$$\|u\|_{b_p^l} = \left(\int \|\Delta_y^2 \text{grad}_m u\|_{L_p}^p |y|^{-n-p\mu} dy \right)^{1/p}$$

$$(l = m + \mu, \mu \in (0, 1], m = 0, 1, \ldots, \Delta_y^2 f(x) = f(x+y) - 2f(x) + f(x-y))$$

and

$$\|u\|_{B_p^{(l)}} = \|u\|_{b_p^{(l)}} + \|u\|_{L_p},$$

respectively. This property, along with what has been said in this section about estimates of the potentials I_l and J_l, leads to continuity criteria for the imbedding operators from $b_p^{(l)}$ and $B_p^{(l)}$ into $L_p(\mu)$. We limit ourselves to stating such results for $b_p^{(l)}$ which follow from Theorem 1.1 of this chapter.

Corollary. *Let $l > 0, 1 < p < q < \infty, lp < n$. The inequality*

$$\|u\|_{L_q(\mu)} \leq C\|u\|_{b_p^{(l)}}, \quad u \in C_0^\infty,$$

holds true iff

$$\mu[\mathscr{B}_\rho(x)] \leq \text{const} \cdot \rho^s, \quad s = q(n - lp)/p. \tag{2.9}$$

The same statement is true for $1 = p \leq q < \infty, l \leq n$. Moreover, condition (2.9) is necessary and sufficient for the validity of the inequality

$$\|u\|_{L_q(\mu)} \leq C\|\text{grad}_l u\|_{L_1}$$

for $1 = p \leq q < \infty, l \leq n$ (cf. Maz'ya [1985]).

1.2. Weighted Estimates for Potentials. In 1974 Muckenhoupt and Wheeden found a necessary and sufficient condition for the validity of the inequality

$$\|(I_l f)v\|_{L_q} \leq C\|fv\|_{L_p}, \tag{2.10}$$

where q is the Sobolev exponent and f an arbitrary function in C_0^∞. Let us state this result.

Theorem 1.7. *Let $0 < l < n, 1 < p < n/l$ and $1/q = 1/p - l/n$. Inequality (2.10) holds true iff*

$$\sup\left(\frac{1}{|\mathscr{B}_\rho|}\int_{\mathscr{B}_\rho(x)} v(x)^q dx\right)^{1/q}\left(\frac{1}{|\mathscr{B}_\rho|}\int_{\mathscr{B}_\rho(x)} v(x)^{-p/(p-1)} dx\right)^{1/p'} < \infty,$$

where the supremum is taken over all balls $\mathscr{B}_\rho(x)$. (In other words, v^q belongs to the Muckenhoupt class A_r with $r = p(n-l)/(n-pl)$ (cf. formula (1.26).*)*

An exhaustive characterization of all pairs of weight functions w, v in the inequality

$$\|(I_l f)w\|_{L_q} \leq C\|fv\|_{L_p} \tag{2.11}$$

is not known. In 1981 Sawyer obtained the following condition for the existence of a function $w \geq 0, w \not\equiv 0$, satisfying inequality (2.11) with a fixed weight v.

Theorem 1.8. *Let $1 < p \leq q < pn/(n - pl), 0 < l < n$ and $v(x) \geq 0$. A function w enjoying the above properties exists iff*

$$\int [v(x)(1 + |x|)^{n-l}]^{-p/(p-1)} dx < \infty.$$

For $q \geq p$ and in the supplementary assumption $v^{-p'} \in A_\infty = \bigcup_{p<\infty} A_p$ one has a complete analogy with the case $v = 1$, as described in Theorems 1.1, 1.4 and 1.5 (Sawyer 1982, Adams 1985). For instance there is the following result due to Adams.

Theorem 1.9. *Assume $1 < p < q < \infty$ and that $v^{-p'} \in A_\infty$ and let μ be a regular Borel measure on $\mathbb{R}^n$. Then the following conditions are equivalent:*

(i) there exists a constant C such that

$$\|I_l f\|_{L_q(\mu)} \leq C\|fv\|_{L_p}$$

for all Lebesgue measurable functions $f \geq 0$;

(ii) there exists a constant C_0 such that

$$\mu(\mathscr{B}_\rho(y)) \leq C_0 H(y;\rho)^{-q/p'}$$

for all $y \in \mathbb{R}^n$ and $\rho > 0$, where

$$H(y;\rho) = \int_\rho^\infty r^{(l-n)p'} \int_{\mathscr{B}_r(y)} (v(x))^{-p'} dx \frac{dr}{r}.$$

We do not consider the analogous results for singular integral operators, the Hardy-Littlewood maximal operator, the fractional maximal operator etc. For the reader interested in this circle of ideas we recommend the recent surveys Dyn'kin-Osilenker [1983] and Sawyer [1982].

§ 2. Multipliers in Spaces of Differentiable Functions

The criterion in § 1 of Chap. 2 for the validity of integral estimates has found applications to the theory of multipliers in Sobolev spaces, spaces of Bessel potentials, Besov spaces etc. The book Maz'ya-Shaposhnikova [1986] is devoted to these questions. Here we restrict ourselves to mentioning just a few results.

By *multipliers* acting from a function space S_1 into another one S_2 we mean functions such that multiplication by them sends S_1 into S_2. Thus, with the two spaces S_1 and S_2 we associate a third one, *the space of multipliers* $M(S_1 \to S_2)$. If $S_1 = S_2 = S$ we will employ the notation $M(S)$.

The norm of the operator of multiplication by the multiplier γ is taken as the norm of γ in $M(S_1 \to S_2)$.

We turn now to equivalent normings of multipliers in pairs of Sobolev spaces.

Let $m > l, mp < n$ and further either $q > p > 1$ or $q \geq p = 1$. Then

$$\|\gamma\|_{M(W_p^{(m)}(\mathbb{R}^n)\to W_q^{(l)}(\mathbb{R}^n))} \sim$$

$$\sim \sup_{x\in\mathbb{R}^n, 0<r<1} (r^{m-n/p}\|\mathrm{grad}_l\gamma\|_{L_q(\mathscr{B}_r(x))} + \|\gamma\|_{L_1(\mathscr{B}_1(x))}). \tag{2.12}$$

For the formulation of the analogous result for $q = p > 1$ we require the capacity $C_{p,l}$, as introduced in § 1.

If $m > l$ and $p > 1$ then

$$\|\gamma\|_{M(W_p^{(m)}(\mathbb{R}^n)\to W_p^{(l)}(\mathbb{R}^n))} \sim$$

$$\sim \sup_{\{e\subset\mathbb{R}^n, \mathrm{diam} e\leq 1\}} \frac{\|\mathrm{grad}_l\gamma\|_{L_p(e)}}{[C_{p,m}(e)]^{1/p}} + \sup_{x\in\mathbb{R}^n} \|\gamma\|_{L_1(\mathscr{B}_1(x))}. \tag{2.13}$$

For $m = l$ the second term to the right in the relations (2.12) and (2.13) has to be replaced by $\|\gamma\|_{L_\infty(\mathbb{R}^n)}$.

If $mp = n$ then the first term to the right in (2.12) takes the form

$$\sup_{x\in\mathbb{R}^n, 0<r<1} |\log 2/r|^{(p-1)/p}\|\mathrm{grad}_l\gamma\|_{L_q(\mathscr{B}_r(x))},$$

and if $mp > n$ the space $M(W_p^{(m)}(\mathbb{R}^n) \to W_q^{(l)}(\mathbb{R}^n))$ coincides with *the space* $W_q^{(l)}(\mathbb{R}^n, \mathrm{unif})$ of functions with the finite norm

$$\sup_{x\in\mathbb{R}^n} \|\gamma\|_{W_q^{(l)}(\mathscr{B}_1(x))}.$$

Multipliers have a number of useful properties. For instance, the traces of multipliers are multipliers of traces. More exactly, each function in $MW_p^{(l-1/p)}(\partial\Omega)$ can be continued to Ω as a function in $MW_p^{(l)}(\Omega)$ (we assume that the boundary of Ω is smooth). The converse statement, that traces of functions in $MW_p^{(l)}(\Omega)$ belong to $MW_p^{(l-1/p)}(\partial\Omega)$, is trivial.

In spaces of multipliers one can search for solutions of elliptic boundary problems. Consider *the Dirichlet problem*

$$-\sum_{i,j=1}^{n} \frac{\partial}{\partial x_i}\left(a_{ij}(x)\frac{\partial u}{\partial x_j}\right) = 0, \quad u|_{\partial\Omega} = \varphi,$$

where Ω is a domain in $\mathbb{R}^n$ with compact closure and Lipschitz boundary and the coefficients a_{ij} are measurable, bounded and define a uniformly positive definite $n \times n$ matrix. This problem has a unique solution in $MW_2^{(1)}(\Omega)$ for each $\varphi \in MW_2^{(1/2)}(\partial\Omega)$.

Because of their algebraic properties multipliers are useful objects if we want to extend to nonsmooth functions the main facts of differential calculus.

For instance, the *implicit function theorem* can be formulated in this way in terms of multipliers.

Theorem 2.1. *Consider* $G = \{(x,y) : x \in \mathbb{R}^{n-1}, y > \varphi(x)\}$, *where* φ *is a function on* $\mathbb{R}^{n-1}$ *satisfying a uniform Lipschitz condition. Assume that* φ *satisfies the equation* $u(x,\varphi(x)) = 0$ *and that one has*

$$\frac{\partial u(x,\varphi(x)+0)}{\partial y} \geq \text{const} > 0 \quad \text{for a. a.} \quad x \in \mathbb{R}^{n-1}.$$

Then it follows from $\operatorname{grad} u \in MW_p^{(l-1)}(G)$, *where* l *is an integer,* $l \geq 2$, *that* $\operatorname{grad}\varphi \in MW_p^{(l-1-1/p)}(\mathbb{R}^{n-1})$.

With the aid of multipliers one can describe some classes of differentiable nonsmooth transformations which preserve Sobolev spaces. Let U be an open subset of $\mathbb{R}^n$ and set $W_p^{(l)}(U,\text{loc}) = \{u : u\eta \in W_p^{(l)}(\mathbb{R}^n) \text{ for all } \eta \in C_0^\infty(U)\}$. We say that a map $\kappa : U \to V \subset \mathbb{R}^n$ is a (p,l)-diffeomorphism, if κ is a bi-Lipschitz map and if the elements of the Jacobi matrix belong to $MW_p^{(l-1)}(U,\text{loc})$.

The transformations defined in this way map $W_p^{(l)}(U,\text{loc})$ onto $W_p^{(l)}(U,\text{loc})$. If $\kappa, \kappa_1, \kappa_2$ are (p,l)-diffeomorphisms, so are κ^{-1} and $\kappa_1 \circ \kappa_2$.

Chapter 3

Imbedding Theorems for Spaces of Functions Satisfying Homogeneous Boundary Conditions

§ 1. Inequalities for Functions with Noncomplete Homogeneous Cauchy Data

We saw already in Sect. 2.3 of Chap. 1 that supplementary information about the behavior of the functions at the boundary of the domain allows an extension of the class of domains for which the imbedding theorems hold true.

Let us consider the following question. What assumptions about Ω are needed in order to assure that inequalities of Sobolev type hold for functions in

$W_p^{(l)}(\Omega)$ vanishing on $\partial\Omega$ along with all derivatives of order up to $k-1$ for some $k < l$? More exactly, it is question of functions in the space $W_p^{(l)}(\Omega) \cap \mathring{W}_p^{(k)}(\Omega)$, where $\mathring{W}_p^{(k)}(\Omega)$ is the completion of $C_0^\infty(\Omega)$ in the norm of the space $W_p^{(k)}(\Omega)$.

As every function in $\mathring{W}_p^{(l)}(\Omega)$ can be extended to $\mathbb{R}^n$ with preservation of the classs $W_p^{(l)}$, no restrictions on the boundary are needed if $k = l$. It turns out that this is the case also k is not less than $l/2$. The condition $l \leq 2k$ cannot be weakened: on Fig. 17 is depicted a domain Ω such that the space $W_p^{(l)}(\Omega) \cap \mathring{W}_p^{(k)}(\Omega)$, $l > 2k$, is not imbedded in $L_\infty(\Omega)$ for $pl > n > 2pk$ and not imbedded in $L_{pn/(n-pl)}(\Omega)$ for $n > pl$.

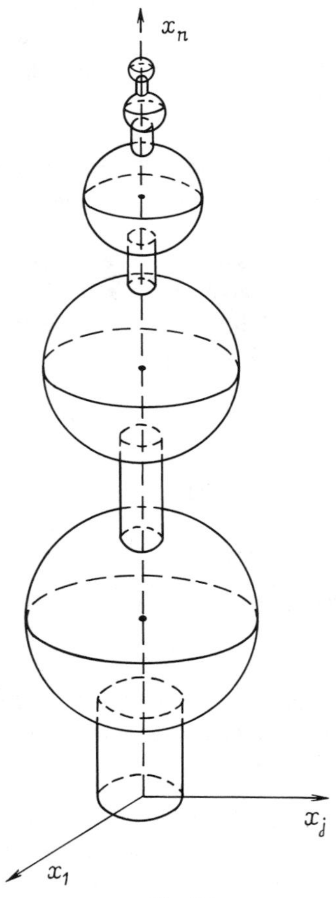

Fig. 17

That Sobolev's inequality is preserved if $l \leq 2k$ follows from an integral representation of differentiable functions in an arbitrary bounded domain; the following theorem is about this.

Theorem 1.1. *Let $u \in W_1^{(l)}(\Omega) \cap \overset{\circ}{W}{}_1^{(k)}(\Omega)$ where l is an integer with $1 \leq l \leq 2k$. Then for all $x \in \Omega$ holds the identity*

$$D^{\gamma}u(x) = \sum_{\{\beta:|\beta|=l\}} \int_{\Omega} K_{\beta\gamma}(x,y) D^{\beta}u(y) \frac{dy}{|x-y|^{n-1}}. \tag{3.1}$$

Here γ is any multi-index of order $l-1$ and $K_{\beta\gamma}$ a measurable function on $\Omega \times \Omega$ such that $|K_{\beta\gamma}(x,y)| \leq c(n,l)$.

For the proof of (3.1) one uses the representation of the solution of the first boundary problem for the equation $y^{(2[l/2])} = f$ on the segment situated within Ω of any line passing through the point x. It is precisely this circumstance that leads to the condition $l \leq 2k$. For comparison let us recall that the usual integral representation of Sobolev is based on the Newton-Leibniz formula on an arbitrary ray issuing from x.

Next we pass to the applications of Theorem 1.1.

Theorem 1.2. *Let m, l and k be integers, $0 \leq m < l \leq 2k$, let $p > 1$ and let $u \in W_p^{(l)}(\Omega) \cap \overset{\circ}{W}{}_1^{(k)}(\Omega)$ with $n < p(l-m)$. Then $u \in C^m(\Omega)$,* $\operatorname{grad}_m u \in L_\infty(\Omega)$ *and we have the estimate*

$$\|\operatorname{grad}_m u\|_{L_\infty(\Omega)} \leq c(\operatorname{diam}\Omega)^{l-m-n/p} \|\operatorname{grad}_l u\|_{L_p(\Omega)}.$$

The imbedding operator $W_p^{(l)}(\Omega) \cap \overset{\circ}{W}{}_1^{(k)}(\Omega) \to W_\infty^{(m)}(\Omega)$ is compact.

Let μ be a measure on Ω such that for each ball $\mathscr{B}_\rho(x)$

$$\mu(\mathscr{B}_\rho(x) \cap \Omega) \leq Q\rho^s, \quad Q = \text{const}, \quad 0 < s < n. \tag{3.2}$$

We denote by ω an open subset, $\bar{\omega} \subset \Omega$. We assume that $p(l-m) > n-s$ and write $q = ps(n - p(l-m))^{-1}$. Put further $L_q(\Omega, \mu, \text{loc}) = \prod_\omega L_q(\omega, \mu)$.

It follows from Theorem 1.1 in Chap. 1 that there exists one and only one linear map $\gamma : W_p^{(l-m)}(\Omega) \to L_q(\Omega, \mu, \text{loc})$ such that (i) if v is a function in $W_p^{(l-m)}(\Omega)$ which is smooth on $\bar{\omega}$, then $\gamma v = v$ on $\bar{\omega}$; (ii) the operator $\gamma : W_p^{(l-m)}(\Omega) \to L_q(\omega, \mu)$ is continous, this for all choices of ω.

The following two theorems follow immediately from the integral representation (3.1) and Theorem 1.1 in Chap. 2.

Theorem 1.3. *Let m, l and k be integers, $0 \leq m \leq l < 2k$, $p > 1$ and μ a measure on Ω satisfying condition (3.2). Then holds for each $u \in W_p^{(l)}(\Omega) \cap \overset{\circ}{W}{}_1^{(k)}(\Omega)$ the estimate*

$$\|\gamma(\operatorname{grad}_m u)\|_{L_q(\Omega,\mu)} \leq cQ^{1/q} \|\operatorname{grad}_l u\|_{L_p(\Omega)},$$

where $q = ps(n - p(l-m))^{-1}$.

Theorem 1.4. *Let m, l, k, p, q be as in Theorem 1.3 and μ a measure on Ω subject to the condition*

$$\lim_{\rho\to 0} \sup_{x\in\mathbb{R}^n} \rho^{-s}\mu(\mathscr{B}_\rho(x)\cap\Omega) = 0. \tag{3.3}$$

Then the operator $\gamma \circ \operatorname{grad}_m : W_p^{(l)}(\Omega) \cap \overset{\circ}{W}{}_1^{(k)}(\Omega) \to L_q(\Omega, \mu)$ is compact.

If (3.2) and (3.3) are replaced by the conditions

$$\mu(\mathscr{B}_\rho(x)\cap\Omega) \le Q|\log 2\rho|^{q(1-p)/p} \text{ for all } \rho \in (0,1),$$

$$\lim_{\rho\to 0} \sup_{x\in\mathbb{R}^n} |\log\rho|^{q(p-1)/p}\mu(\mathscr{B}_\rho(x)\cap\Omega) = 0,$$

where $q > p > 1$, then the statements of Theorems 1.3 and 1.4 remain in force also for $n = p(l-m)$.

§ 2. "Massive" Sets of Zeros of Functions in Sobolev Spaces

It is not necessary to point out that the inequality

$$\|u\|_{L_p(\mathscr{B}_1)} \le K \sum_{k=1}^{l} \|\operatorname{grad}_k u\|_{L_p(\mathscr{B}_1)}, \tag{3.4}$$

$1 < p < \infty$, does not hold true for all smooth functions in the ball $\mathscr{B}_1$. Besides this it is well-known and easy to prove that if we impose on the function u the requirement $u = 0$ on $\partial\mathscr{B}_1$, we obtain (3.4) with a constant K depending on n, p and l. There arises the question how large a compact set F, $F \subset \bar{\mathscr{B}}_1$, must be in order that the inequality in question remains in force for all u vanishing on F. An exhaustive answer, a necessary and sufficient condition for the validity of (3.4), is the positivity of the capacity $C_{p,l}(F)$. Moreover, in terms of $C_{p,l}(F)$ it is likewise possible to give an estimate for the best constant K in (3.4). The history of such estimates is set forth in the comments to Chap. 10 in the book Maz'ya [1985].

Theorem 2.1 (cf. Hedberg [1981], Maz'ya [1985]). *Let F be a closed subset of $\mathscr{B}_1$ and u any function in $C^\infty(\bar{\mathscr{B}}_1)$, equal to zero on F together with all its derivatives up to order $l-1$. Then*

$$\|u\|_{L_p(\mathscr{B}_1)} \le c\left(\frac{1}{[C_{p,l}(F)]^{1/p}}\|\operatorname{grad}_l u\|_{L_p(\mathscr{B}_1)} + \|\operatorname{grad} u\|_{L_p(\mathscr{B}_1)}\right),$$

where c is a constant depending on n, p and l.

Let us give the proof of this theorem for $l = 1$, in which case it can be simplified considerably.

First of all let us remark that

$$C_{p,1}(F) \leq c\|v\|^p_{W_p^{(1)}(\mathscr{B}_1)}, \tag{3.5}$$

where v is an arbitrary function in $C^\infty(\bar{\mathscr{B}}_1)$ equal to one on F. This is easily seen if we extend v to $\mathbb{R}^n$ in such a way that the inequality

$$\|v\|_{W_p^{(1)}(\mathbb{R}^n)} \leq c\|v\|_{W_p^{(1)}(\mathscr{B}_1)},$$

is fulfilled and then apply the inequality

$$C_{p,1}(F) \leq c\|v\|^p_{W_p^{(1)}(\mathbb{R}^n)},$$

which follows from the definition of $C_{p,1}$.

Denote by $\bar{u}$ the average of u over $\mathscr{B}_1$. Without loss of generality we may assume that $\bar{u} \geq 0$. Set further $N = \|u\|_{L_p(\mathscr{B}_1)}$. By (3.5)

$$C_{p,1}(F) \leq c\|1 - N^{-1}u\|^p_{W_p^{(1)}(\mathscr{B}_1)} = cN^{-p}\|\operatorname{grad} u\|^p_{L_p(\mathscr{B}_1)} + c\|1 - N^{-1}u\|^p_{L_p(\mathscr{B}_1)},$$

that is,

$$N^p C_{p,1}(F) \leq c(\|\operatorname{grad} u\|^p_{L_p(\mathscr{B}_1)} + \|N - u\|^p_{L_p(\mathscr{B}_1)}). \tag{3.6}$$

As $\bar{u} \geq 0$ then

$$|N - \bar{u}| = \|u\|_{L_p(\mathscr{B}_1)} - \|\bar{u}\|_{L_p(\mathscr{B}_1)} \leq \|u - \bar{u}\|_{L_p(\mathscr{B}_1)}.$$

Hence

$$\|N - u\|_{L_p(\mathscr{B}_1)} \leq \|N - \bar{u}\|_{L_p(\mathscr{B}_1)} + \|u - \bar{u}\|_{L_p(\mathscr{B}_1)} \leq 2\|u - \bar{u}\|_{L_p(\mathscr{B}_1)}. \tag{3.7}$$

Upon applying (3.6) and (3.7), along with the Poincaré inequality

$$\|u - \bar{u}\|_{L_p(\mathscr{B}_1)} \leq c\|\operatorname{grad} u\|_{L_p(\mathscr{B}_1)},$$

we obtain the desired inequality.

If the capacity $C_{p,l}(F)$ is small, one has also a converse to Theorem 2.1 of this chapter.

Theorem 2.2. *Assume that $C_{p,l}(F) \leq \gamma$, where γ is a sufficiently small constant, $\gamma = \gamma(n, p, l)$. If inequality (3.4) holds true for all functions u in $C^\infty(\bar{\mathscr{B}}_1)$ vanishing on F, then the exact constant K in (3.4) admits the estimate*

$$K^{-1} \leq c[C_{p,l}(F)]^{1/p}, \tag{3.8}$$

where $c = c(n, p, l)$.

Without the requirement that $C_{p,l}(F)$ is small, the last inequality is not true. In the case $C_{p,l}(F) \geq \gamma$ one has to formulate twosided estimates for K in terms of the so-called (p, l)-inner diameter $D_{p,l}(\mathscr{B}_1 \backslash F)$, i.e. the inner diameter of the difference $\mathscr{B}_1 \backslash F$ up to sets of small capacity. Indeed, one has the relation

$$K \sim [D_{p,l}(\mathscr{B}_1 \backslash F)]^l.$$

Theorem 2.1 plays a major role in Hedberg's solution [1981] to the celebrated spectral synthesis problem for Sobolev spaces. It is about the following result. Let $u \in W_p^{(l)}(\mathbb{R}^n)$ and let F be a closed subset of $\mathbb{R}^n$. One can define in a natural way the traces on F of u and its derivatives $D^\alpha u$, $|\alpha| \leq l - 1$. If all these traces vanish, then $u \in \overset{\circ}{W}{}_p^{(l)}(\mathbb{R}^n \backslash F)$, that is, there exists a sequence $\{\varphi_j\}_{j \geq 1}$ of functions in $C_0^\infty(\mathbb{R}^n \backslash F)$ converging to u in $W_p^{(l)}(\mathbb{R}^n)$. If $p = 2$ this result is equivalent to the following uniqueness theorem for Dirichlet's problem. Let Ω be a bounded domain in $\mathbb{R}^n$ and u a function in $W_2^{(l)}(\mathbb{R}^n)$ satisfying the equation $\Delta^l u = 0$ on Ω and the boundary condition $D^\alpha u|_{\partial\Omega} = 0$, $0 \leq |\alpha| \leq l - 1$. Then $u = 0$.

Concluding this section let us turn to the characterization of sets of uniqueness for the space $H_p^{(l)}$. By this we intend a set $E \subset \mathbb{R}^n$ enjoying the following property. From the condition $u \in H_p^{(l)}, u(x) = 0$ for all $x \in \mathbb{R}^n \backslash E$, outside a suitable subset of zero $C_{p,l}$-capacity, follows that $u = 0$.

The description of set of uniqueness in $H_p^{(l)}$ in Theorem 2.3 below was given by Hedberg (1972) and Polking (1972). The first result of this kind, for the space $H_2^{(1/2)}$ on the circle, was obtained in 1950 by Ahlfors and Beurling.

Theorem 2.3. *Let E be a Borel subset of $\mathbb{R}^n$. The following conditions are equivalent:*

(i) *E is a uniqueness set for $H_p^{(l)}$;*

(ii) $C_{p,l}(G \backslash E) = C_{p,l}(G)$ *for each open set G;*

(ii) *for allmost all x*

$$\overline{\lim_{\rho \to 0}}\, \varrho^{-n} C_{p,l}(\mathscr{B}_\rho(x) \backslash E) > 0.$$

If $lp > n$ then E is a uniqueness set for $H_p^{(l)}$ iff it does not have inner points.

§ 3. Imbedding of the Space $\overset{\circ}{L}{}_p^{(l)}(\Omega)$ into Other Function Spaces

Denote by $\overset{\circ}{L}{}_p^{(l)}(\Omega)$ the completion of $C_0^\infty(\Omega)$ in the norm $\|\operatorname{grad}_l u\|_{L_p(\mathbb{R}^n)}$. If $n > pl$ for $p > 1$ and $n \geq l$ for $p = 1$, then for each function $u \in C_0^\infty(\Omega)$ holds *Sobolev's inequality*

$$\|u\|_{L_q(\Omega)} \leq c \|\operatorname{grad}_l u\|_{L_p(\mathbb{R}^n)}, \tag{3.9}$$

where $q = pn/(n - pl)$. In view of (3.9) the map $C_0^\infty(\Omega) \ni u \mapsto u \in L_q(\Omega)$ is continuous. As $C_0^\infty(\Omega)$ is dense in $\overset{\circ}{L}{}_p^{(l)}(\Omega)$, this map can be extended to $\overset{\circ}{L}{}_p^{(l)}(\Omega)$ with preservation of continuity; such an extension is unique. It is not hard to show that the map constructed is one to one. Hence each element in $\overset{\circ}{L}{}_p^{(l)}(\Omega)$ for $n > pl, p > 1$ and for $n \geq l, p = 1$ can be identified with a function in $L_q(\Omega)$ and the identity map $\overset{\circ}{L}{}_p^{(l)}(\Omega) \ni u \mapsto u \in L_q(\Omega)$ is a topological imbedding (i.e. injective, linear and continuous).

If $n \leq lp, p > 1$ or $n < l, p = 1$ and if inequality (3.9) holds for some $q > 0$, with a constant c, not depending on $u \in C_0^\infty(\Omega)$, then we may draw the same conclusion. However, for this range of n, p, l, inequality (3.9) is not always fulfilled and moreover $\overset{\circ}{L}{}_p^{(l)}(\Omega)$ cannot always be imbedded topologically into the space of distributions. Let, for example, $n < pl$, $\eta \in C_0^\infty(\mathbb{R}^n)$, $\eta(0) = 1$ and $\eta_m(x) = m^\varepsilon \eta(x/m)$, $\varepsilon > 0$. As $\|\operatorname{grad}_l \eta_m\|_{L_p(\mathbb{R}^n)} = cm^{\varepsilon-(lp-n)/p}$, then $\lim_{m\to\infty} \eta_m = 0$ in $\overset{\circ}{L}{}_p^{(l)}(\mathbb{R}^n)$ for sufficiently small ε. On the other hand, for each $\varphi \in C_0^\infty(\mathbb{R}^n)$ we have $\lim \int \eta_m \varphi(x)dx = \infty$ and, consequently, $\overset{\circ}{L}{}_p^{(l)}(\mathbb{R}^n)$ with $lp > n$ is not imbedded in $[C_0^\infty(\mathbb{R}^n)]^*$.

Let us state a theorem giving necessary and sufficient conditions for the imbedding of $\overset{\circ}{L}{}_p^{(l)}(\Omega)$ into $[C_0^\infty(\Omega)]^*$ or, what is the same thing, the imbedding of $\overset{\circ}{L}{}_p^{(l)}(\Omega)$ into $L_p(\Omega, \text{loc})$. (Concerning the history of this and other theorems of the present section, see Maz'ya [1985].)

Theorem 3.1. *The space $\overset{\circ}{L}{}_p^{(l)}(\Omega)(1 \leq p < \infty)$ is topologically imbedded in $[C_0^\infty(\Omega)]^*$ iff one of the following conditions is fulfilled:*

1) $n > pl$ for $p > 1$, $n \geq l$ for $p = 1$;

2) $\mathbb{R}^n \backslash \Omega$ has positive (inner) capacity $C_{p,n/p}$ if $n = pl, p > 1$;

3) $\mathbb{R}^n \backslash \Omega$ is nonempty, provided $n < pl$ and n/p is noninteger;

4) $\mathbb{R}^n \backslash \Omega$ has positive (inner) capacity $C_{p,n/p}$ and does not lie in an $(n-1)$-dimensional hyperplane, provided $n < pl$, n/p integer.

The following theorem yields another application of the notion of a set of zero capacity.

Theorem 3.2. *A necessary and sufficient condition for the space $C_0^\infty(\Omega)$ to be dense in $C_0^\infty(\mathbb{R}^n)$ in the metric $\overset{\circ}{L}{}_p^{(l)}(\mathbb{R}^n)$ is that the complement of Ω has zero $C_{p,l}$-capacity.*

Let us also remark that a closed set F has zero capacity $C_{p,l}$ iff the space of distributions $W_{p/(p-1)}^{(-l)}(\mathbb{R}^n)$, dual to $W_p^{(l)}(\mathbb{R}^n)$, does not contain elements with support in F other than the zero element.

In comparison with Theorem 3.1 more stringent conditions on Ω are necessary and sufficient for the validity, for all $u \in C_0^\infty(\Omega)$, of inequality (3.9) with $q \neq pn/(n-pl)$, which is equivalent to the imbedding of $\overset{\circ}{L}{}_p^{(l)}(\Omega)$ in $L_q(\Omega)$. Indeed, one has the following

Theorem 3.3. *Let $p > 1$ and assume that q satisfies one of the conditions*
(i) $q \in [p, np(n-pl)^{-1}]$ *if* $n > pl$;
(ii) $q \in [p, \infty)$ *if* $n = pl$;
(iii) $q \in [p, \infty]$ *if* $pl > n$.

The space $\overset{\circ}{L}{}_p^{(l)}(\Omega)$ is imbedded in $L_q(\Omega)$ iff $D_{p,l}(\Omega) < \infty$, where $D_{p,l}(\Omega)$ is the (p,l)-inner diameter, which if $n < pl, p > 1$ coincides with the usual inner diameter.

The exact constant C in (3.9) satisfies the relation

$$C \sim [D_{p,l}(\Omega)]^{l-n(p^{-1}-q^{-1})}.$$

Let us also give a necessary and sufficient condition for the imbedding $\overset{\circ}{L}{}_p^{(l)}(\Omega) \subset L_q(\Omega)$, formulated in other terms and a consequence of Theorems 2.1 and 2.2.

Theorem 3.4. *The space $\overset{\circ}{L}{}_p^{(l)}(\Omega)$ is imbedded in $L_q(\Omega)$ for $n \geq pl$, $p \leq q < pn/(n-pl)$ iff for some $r > 0$ holds*

$$\inf_{x \in \mathbb{R}^n} C_{p,l}(\mathscr{B}_r(x) \backslash \Omega) > 0.$$

In the case $n < pl$ this criterion means that the set Ω does not contain arbitrary large balls. The same thing is true also for $n = pl$ provided the complement of Ω is connected.

In the hypothesis $p > q$ one has likewise obtained necessary and sufficient conditions for the imbedding $\overset{\circ}{L}{}_p^{(l)}(\Omega) \subset L_q(\Omega)$. These have a more complicated formulation and mean that, up to a "small error", Ω is a sum of balls, having finite intersection, whose radii $\{r_i\}_{i \geq 1}$ satisfy the condition

$$\sum_{i=1}^{\infty} r_i^{n+lpq/(p-q)} < \infty.$$

For instance, for the domain $\Omega = \{x : 0 < x_n < \infty, (x_1^2 + \ldots + x_{n-1}^2)^{1/2} < \varphi(x_n)\}$, where φ is a bounded nonincreasing function (Fig. 6), the imbedding of $\overset{\circ}{L}{}_p^{(l)}(\Omega)$ into $L_q(\Omega)$, $q < p$, holds iff

$$\int_0^{\infty} [\varphi(t)]^{\alpha} dt < \infty, \quad \alpha = n - 1 + plq/(p-q).$$

The imbedding operator of $\overset{\circ}{L}{}_p^{(l)}(\Omega)$ into $L_q(\Omega)$ for $q < p$ is compact iff it is continuous. If $n > pl, p \leq q < pn(n-pl)$, the necessary and sufficient condition for the compactness of the imbedding is that for each $r > 0$

$$\lim_{\rho\to\infty} \inf_{x\in\mathbb{R}^n\setminus\mathscr{B}_\rho(0)} c_{p,l}(\mathscr{B}_r(x)\setminus\Omega) > cr^{n-pl},$$

where $c = \text{const} > 0$ and $c_{p,l}$ is the Riesz capacity, as defined in Sect. 1.1 of § 1 of Chap. 2. If $n < pl$ and, if the complement of Ω is connected, likewise for $n = pl$, for the compactness of the imbedding under consideration it is necessary and sufficient that the set Ω does not contain an infinite sequence of nonintersecting balls of the same size. The proof of these compactness criteria is based on Theorems 2.1 and 2.2.

§ 4. On the Space $\overset{\circ}{H}{}_p^{(l)}(\Omega, \nu)$

A large portion of the results mentioned in § 3 can be carried over to the case of $\overset{\circ}{H}{}_p^{(l)}(\Omega, \nu)$, that is the completion of $C_0^\infty(\Omega)$ in the norm

$$\left(\int_{\mathbb{R}^n} |(-\Delta)^{l/2}u|^p dx + \int_\Omega |u|^p d\nu\right)^{1/p},$$

where $p > 1$, l is a positive integer and ν a nonnegative measure on Ω (V.G. Maz'ya (1973), V.G. Maz'ya and M. Otelbaev (1977)). Here the role of the (p, l)-inner diameter is taken by the number

$$D = D_{p,l}(\Omega, \nu) = \sup\{r : r^{n-pl} \geq \inf_{e\in N(\mathscr{B}_r)} \nu(\mathscr{B}_r\setminus e)\},$$

where $N(\mathscr{B}_r)$ is the family of all (p, l)-nonessential subsets of the ball $\mathscr{B}_r$, i.e. subsets satisfying the inequality $C_{p,l}(E) \leq \gamma C_{p,l}(\mathscr{B}_r)$ with a sufficiently small constant $\gamma = \gamma(n, p, l)$, and $\{\mathscr{B}_r\}$ is the set of all balls of radius r whose intersection with $\mathbb{R}^n\setminus\Omega$ is (p, l)-nonessential. If $pl > n$ the only nonessential set is the empty set and so

$$D = \sup_{\{\mathscr{B}_r : \mathscr{B}_r\subset\Omega\}} \{r : r^{n-pl} \geq \nu(\mathscr{B}_r)\}.$$

The exact constant C in the inequality

$$\|u\|_{H_q^{(m)}(\mathbb{R}^n)} \leq C\|u\|_{\overset{\circ}{H}{}_p^{(l)}(\Omega,\nu)},$$

where $0 \leq m \leq l, p \leq q < \infty, l - m > n(1/p - 1/q)$, satisfies the relation

$$C \sim D^{l-n(1/p-1/r)} \max\{D^{-m}, 1\}.$$

Consequently, the necessary and sufficient condition for the boundedness of the imbedding operator from $\overset{\circ}{H}{}_p^{(l)}(\Omega, v)$ into $H_q^{(m)}(\mathbb{R}^n)$ for given p, q, l, m has the form $D < \infty$. This criterion is equivalent to the following: for r sufficiently large holds for all balls $\mathscr{B}_r$

$$\inf_{e \in N(\mathscr{B}_r)} v(\mathscr{B}_r \backslash e) > 0.$$

Analogously, the same imbedding operator is compact iff

$$\lim \inf_{e \in N(\mathscr{B}_r)} v(\mathscr{B}_r \backslash e) = \infty$$

when the balls $\mathscr{B}_r$ go out to infinity.

For $p = q = 2, l = 1, m = 0$ the last condition reduces to a known criterion for the discretness of the spectrum of the Schrödinger operator $-\Delta + p(x), \int_e p(x)dx = v(e)$, due to A.M. Molchanov (1952).

Let E be the identity map on $C_0^\infty(\Omega)$ considered as an operator from $\overset{\circ}{H}{}_p^{(l)}(\Omega, v)$ into $H_q^{(m)}(\mathbb{R}^n)$. We associate with E its essential norm, i.e. the quantity

$$\inf_{\{T\}} \|E - T\|,$$

where $\{T\}$ is the set of all compact linear operators $\overset{\circ}{H}{}_p^{(l)}(\Omega, v) \to H_p^{(m)}(\mathbb{R}^n)$.

Set further

$$\overset{\infty}{D} = \overset{\infty}{D}_{p,l}(\Omega, v) = \lim_{R \to \infty} D_{p,l}(\Omega \backslash \mathscr{B}_R(0), v).$$

Theorem 4.1. *If* $0 \le m < l, 1 < p \le q < \infty, l - m > n(1/p - 1/q)$, *then*

$$\inf_{\{T\}} \|E - T\| \sim \overset{\infty}{D}{}^{l-n(1/p-1/r)} \max\{\overset{\infty}{D}{}^{-m}, 1\}.$$

From this we see at once that the imbedding from $\overset{\circ}{H}{}_p^{(l)}(\Omega, v)$ into $H_q^{(m)}(\mathbb{R}^n)$ is compact iff $\overset{\infty}{D} = 0$.

The function

$$\mathbb{R}^n \ni x \mapsto D_{p,l}(x) = \sup\{r : r^{n-pl} \ge \inf_{e \in N(\mathscr{B}_r(x))} v(\mathscr{B}_r(x) \backslash e)\}$$

turns out to be useful if we want to obtain estimates for diameters (in the sense of Kolmogorov) of the space $\overset{\circ}{H}{}_p^{(l)}(\Omega, v)$ in $L_p(\Omega)$. Let us give the corresponding definition and state the result.

Let $\mathbf{B}_1$ and $\mathbf{B}_2$ be Banach spaces, assuming that $\mathbf{B}_1$ is imbedded in $\mathbf{B}_2$. Let B_1 be the unit ball in $\mathbf{B}_1$. The quantity

$$d_k(\mathbf{B}_1, \mathbf{B}_2) = \inf_{\{\mathbf{L}_k\}} \sup_{f \in B_1} \inf_{g \in \mathbf{L}_k} \|f - g\|_{\mathbf{B}_2},$$

where $k = 1, 2, \ldots$, $\{\mathbf{L}_k\}$ denoting the family of all linear subspaces of $\mathbf{B}_2$ of dimensions $\leq k$, is called the *kth diameter in the sense of Kolmogorov* of $\mathbf{B}_1$ in $\mathbf{B}_2$.

We say that the imbedding operator from $\mathbf{B}_1$ into $\mathbf{B}_2$ belongs to the class l_θ if

$$\sum_{k\geq 1} [d_k(\mathbf{B}_1, \mathbf{B}_2)]^\theta < \infty.$$

Theorem 4.2 (M. Otelbaev 1979). *Let $N(\lambda)$ be the number of diameters of $\overset{\circ}{H}{}_p^{(l)}(\Omega, v)$ in $L_p(\Omega)$, $p > 1$, greater than λ, $\lambda > 0$. There exists a constant $c \in (1, \infty)$, not depending on Ω, v and λ, such that*

$$c^{-1}N(c\lambda) \leq \lambda^{-n/l}\mathrm{mes}_n\{x : D_{p,l}(x) \geq \lambda^{1/l}\} \leq cN(c^{-1}\lambda).$$

This theorem shows that the imbedding operator from $\overset{\circ}{H}{}_p^{(l)}(\Omega, v)$ into $L_p(\Omega)$ is in l_θ if $\theta l > n$ and

$$\int_\Omega [D_{p,l}(x)]^{l\theta - n} dx < \infty.$$

As an immediate application of the theorems of this section one obtains twosided estimates for the lower bound of the spectrum of the operator $(-\Delta)^{l/2} + p(x)$ and, further, criteria for the nuclearity of the resolvent of this operator.

References*

The present Part can only in a weak sense serve as an orientation on the history of the results mentioned. The reader who is interested in this aspect of our subject and, likewise, in more detailed proofs may consult the books and papers quoted below.

The classical book Pólya-Szegö [1950] considers various isoperimetric inequalities, rigorous or heuristic methods for their proof, and applications to mathematical physics. The ideas and concepts discussed there subsequently turned out to be useful also in the theory of Sobolev spaces.

In the monograph Burago-Zalgaller [1980] inequalities between the basic geometric characteristics of domains on surfaces or of submanifolds of Euclidean spaces and Riemannian manifolds are studied. Among other things, this monograph considers the classical isoperimetric problem, its generalizations as well as applications to integral inequalities for differentiable functions.

Also the interesting survey by Osserman [1978] has basically a similar objective. There considerable attention is payed to applications of isoperimetric inequalities to mathematical physics, in particular, to the theory of imbedding and to estimates of eigenvalues of boundary problems.

---

*For the convenience of the reader, references to reviews in Zentralblatt für Mathematik (Zbl.), compiled using the MATH database, have, as far as possible, been included in this bibliography.

The quite readable paper by Aubin [1976] contains a study of the exact constants in inequalities of Sobolev type on Riemannian manifolds.

The monograph Gol'dshteĭn-Reshetnyak [1983] is devoted to the study of connections between Sobolev spaces, capacities and special classes of mappings: quasiconformal mappings and quasi-isometric mappings. In particular, it contains a systematic treatment of the extension of a function from a domain to $\mathbb{R}^n$ under preservation of differentiability properties.

The works Dynkin-Osilenker [1983] and Sawyer [1982], each having its merit, are surveys of the contemporary state what regards problems of weighted estimates for various spaces. In the former, in particular, much attention is given to the Muckenhoupt condition (1.25) and its consequences, while the latter contains proofs of some definitive results on inequalities with two different measures.

The book Maz'ya-Shaposhnikova [1986] treats results in the theory of multipliers in spaces of differentiable functions and its applications to functional analysis and partial differential equations.

The lectures by Adams [1981] are devoted to recent achievements in nonlinear potential theory and related topics in the theory of Sobolev spaces and generalizations thereof.

Hedberg's paper [1981], which was awarded a prize by the Royal Swedish Academy of Sciences, is devoted to the solution of the spectral synthesis problem in the space $W_p^{(l)}$, as discussed in §2 of Chap. 2. In Hedberg [1981] there is, in particular, a proof of Theorem 2.1 of Chap. 2 (cf. also Maz'ya [1985]). This paper gives a presentation of the depth of some questions in the theory of Sobolev spaces having in itself a simple formulation.

In the monograph Maz'ya [1985] various questions in the theory of Sobolev spaces are considered. The first chapter there can be read as a modern introduction to the theory. In the following eleven chapters it is set forth in which minimal hypotheses on the domain the imbedding theorems remain in force, but also such questions as how they change if the properties of the boundary deteriorate and how the space of admissible domains depends on supplementary requirements concerning the behavior of the function near the bounadry etc. A characteristic feature is the wide use of the isoperimetric inequalities, which connect various measures and capacities.

Adams, D. R.

[1981] *Lectures on L^p-potential theory.* Dep. Math. Univ Umeå Publ., 1981, no. 2.

Aubin, T.

[1976] Problèmes isopérimétriques et espaces de Sobolev. J. Differ. Geom. *11*, 573-598. Zbl. 371.46011

Burago, Yu. D., Zalgaller, V. A.

[1980] *Geometric inequalities.* Nauka, Leningrad. Zbl. 436.52009. English translation: Grundlehren 285. Springer, Berlin - Heidelberg - New York 1988

Dyn'kin, E. M., Osilenker, B. P.

[1983] Weighted estimates for singular integrals and their applications, in: *Itogi Nauki Tekhn.*, Mat. Analiz., *21*, 42-129. Zbl. 568.42009. English translation: J. Sov. Math. *30*, 2094-2154 (1985)

Gol'dshteĭn, V. M., Reshetnyak, Yu. G.

[1983] *Introduction to the theory of functions with generalized derivatives and their applications.* Nauka, Moscow [Russian]. Zbl. 591.46021

Hedberg, L. I.

[1981] Spectral synthesis in Sobolev spaces and uniqueness of solutions of the Dirichlet problem. Acta Math. *147*, 237-264. Zbl. 504.35018

Maz'ya, V. G.

[1985] *Sobolev spaces.* LGU, Leningrad. English translation: Springer, Berlin - Heidelberg - New York - Tokyo 1985

Maz'ya, V. G., Shaposhnikova, T. O.

[1985] *Multipliers in spaces of differentiable functions.* LGU: Leningrad. English translation: Monographs and Studies in Mathematics 23. Pitman, Boston 1985. Zbl. 645.46031

Osserman, R.
[1978] The isoperimetric inequality. Bull. Am. Math. Soc. *84*, 1182-1238. Zbl. 411.52006
Pólya, G. Szegö, G.
[1951] *Isoperimetric inequalities in mathematical physics.* Princeton University Press, Princeton. Zbl. 44, 383
Sawyer, E. T.
[1982] Two weight norm inequalities for certain maximal and integral operators, in: Lecture Notes Math. *908*, pp. 102-127. Springer, Berlin - Heidelberg - New York. Zbl. 508.42024

Author Index

Adams, D. R. 152, 190, 192, 194, 197, 210
Adams, R. A. 128
Agalarov, S. I. 93
Ahlfors, L. 181, 204
Akilov, G. P. 128, 133
Amanov, T. I. 52, 84, 128
Anzellotti, G. 179, 184
Arcela, C. 16
Aronszajn, N. 53, 60
Ascoli, G. 16
Aubin, T. 176, 210

Babich, V. M. 53, 76, 128
Bakhvalov, N. S. 84, 128
Banach, S. 13, 15-16, 21, 24, 28, 30, 42, 103-105, 107, 112-113, 116, 118-120, 122, 124, 160, 208
Benedek, A. 119, 128
Bénilan, Ph. 108, 128
Behrens, H. 105, 128, 130
Bergh, J. 105, 128
Berkolaĭko, M. Zh. 120-121, 124-125, 128
Bernshteĭn, S. N. 47, 52
Besicovitch, A. S. 148
Besov, O. V. 4, 38-39, 41, 52-57, 61, 64-65, 68-70, 74-76, 78-79, 82-86, 94, 98-99, 103, 105, 108, 120, 123-124, 128-129, 195, 197
Bessel, F. W. 60, 101-103, 120, 190-191, 195, 197
Beurling, A. 75, 129, 204
Birnbaum, Z. 110
Bliss, G. A. 146
Borel, E. 12, 149, 154, 194, 197, 204
Brezis, H. 108, 128
Browder, F. E. 115
Brudnyĭ, Yu. A. 78, 120, 123, 129
Brunn 143-144
Bugrov, Ya. S. 52, 76-77, 92, 119, 121, 124, 129
Bukhvalov, A. V. 41, 119-120, 123-124, 130
Bunyakovskiĭ, V. Ya. 5, 19
Burago, Yu. D. 144, 174, 179-180, 209-210
Burenkov, V. I. 4, 56, 69-70, 76-83, 92, 124, 128, 130, 183
Buser, P. 173
Butzer, P. L. 105, 128, 130

Caccipolli, R. 177
Calderón, A. P. 39, 60, 76, 84, 108-109, 120, 122-124, 130
Campanato, S. 31, 84
Cauchy, A. 5, 10, 17-19, 24, 156, 199
Chebyshev, P. L. 23
Cheeger, J. 173
Choquet, G. 149
Ciesielski, Z. 41
Courant, R. 157-158
Crandall, M. G. 108, 128

Dahlberg, B. 192
Darboux, G. 103
DeGiorgi, E. 177
Demidenko, G. V. 54, 128, 140
Deny, J. 80, 91, 130
Dirichlet, P. G. L. 5-6, 75, 92, 97, 149, 157, 164, 173, 183, 198, 204
Douglas, A. 75, 131
Dubinskiĭ, Yu. A. 42, 131
Dunford, N. 128, 131
Dyn'kin, E. M. 197, 210
Dzhabrailov, A. D. 39, 84-85, 93, 131
Dzhafarov, A. S. 65, 93, 131

Euler, L. 88, 103, 127

Fabes, E. B. 154
Faĭn, B. L. 78-79, 130-131
Federer, H. 144, 177
Fefferman, Ch. 68, 131
Fischer, R. 6
Fillipov, V. M. 31
Fleming W. H. 144, 178
Fokht, A. S. 93, 131
Fomin, S. V. 128, 134
Fourier, J. 45, 60-62, 64, 68, 96, 98, 101-103, 190

Friedrichs, K. 157
Fubini, G. 119
Fucik, S. 31, 41, 128, 135

Gagliardo, E. 53, 81-82, 85, 105, 143
Gallouet, Th. 108, 131
Gauss, C. F. 157, 174, 177-179
Giaquinta, M. 179, 184
Gil'derman, Yu. I. 131
Gil-Medrano, O. 176
Globenko, I. G. 41, 73, 131
Gol'dman, K. K. 56, 65-67, 69, 130-131
Gol'dshteĭn, V. M. 39, 77, 128, 132, 181, 183, 192, 210
Golovkin, K. K. 56, 74, 120, 124, 132
Gossez, J. P. 115, 132
Green, G. 149
Grisvard, P. 56, 105

Haar, A. 29
Hamel, G. 21
Hansson, K. 192
Hardy, G. H. 38, 51, 67-68, 123, 150, 155, 192, 197
Hausdorff, F. 8, 104-105, 171, 177
Hedberg, L. I. 192, 202, 204, 210
Hermite, C. 23
Hestenes, M. R. 76
Hilbert, D. 6, 17-19, 21-23, 25, 28, 68, 157-158
Hoffman, D. 175
Hölder, O. L. 14-15, 19, 26, 30, 41, 46-49, 67, 84, 93, 113, 146, 171

Il'in, V. P. 38-39, 41, 54, 56-57, 70, 73-76, 78-79, 82-86, 93, 120, 128-129, 132, 190

Jackson, D. 52
Jacobi, C. G. J. 199
John, C. 31, 41, 128, 135
John, F. 31, 84
Jones, P. W. 77-78, 84, 132, 181

Kalinichenko, D. F. 98
Kalyabin, G. A. 65-68, 84, 96, 98, 132
Kazaryan, G. G. 83, 133
Kantorovich, L. A. 128, 133
Kenig, C. E. 154
Kerman, R. 194
Kipriyanov, I. A. 101, 103, 133
Klimov, V. S. 120, 133
Kocharli, A. F. 86, 93-94, 133
Kolmogorov, A. N. 16-17, 114, 119, 128, 134, 208-209
Kokilashvili, V. M. 99, 118, 128, 133
Kondrashov, V. I. 39, 41, 90, 134
Konyushkov, A. A. 54, 134
Korotkov, V. B. 41, 134
Krasnosel'skiĭ, M. A. 128, 134
Kreĭn, S. G. 105, 128, 134
Kudryavchev, L. D. 52, 77, 80, 86, 89, 91-95, 98-100, 126, 128, 134, 143
Kufner, A. 31, 41, 82, 91, 99, 128, 135
Kuznetsov, Yu. V. 79, 135

Lacroix, M.-Th. 118, 135
Laguerre, E. 23
Laplace, P. S. 173
Latfullin, T. G. 181
Lebesgue, H. 6-7, 10, 12, 15, 19, 22, 28, 30-33, 36, 99, 102, 105, 108, 110-111, 113-114, 119-120, 166, 197
Legendre, A. M. 22
Levi, B. 31
Leibniz, G. W. 32, 45, 201
Lions, J.-L. 41, 61, 80, 91, 105, 130, 135, 137
Liouville, J. 45, 58-59, 61, 64, 68, 84, 120, 125
Lipschitz, R. 35, 38, 41, 73, 76, 82, 92, 110, 116-117, 122-123, 144, 166, 177, 181, 183, 185, 187-88, 199
Littlewood, J. E. 38, 51, 64-65, 67-68, 123, 192, 197
Lizorkin, P. I. 4, 35, 41, 46, 54, 58, 60-62, 64-65, 67-69, 86, 91-93, 99, 103, 105, 108, 125, 135-137
Lobachevskiĭ, N. I. 174
Löfström, J. 105, 128
Lorentz, G. G. 66, 99, 106-108, 119, 121
Luxemburg, W. A. 113, 128, 136

Macdonald, H. M. 60, 120
Magenes, E. 135-136
Marchaud, A. 62
Maz'ya, V. G. 4, 41, 59, 73-74, 128, 136, 152, 179-180, 182-184, 192, 195-197, 202, 205, 207, 210
Marcinkiewicz, J. 106-108, 119, 136, 190
McKean, H. P. 174
Meyers, N. G. 80, 136
Michael, J. H. 174
Mikhaĭlov, V. P. 84, 136
Minkowski, H. 14-15, 80
Mityagin, B. S. 108
Molchanov, A. M. 208
Morel, J.-M. 108, 131
Morrey, Ch. B. 30-31, 84, 123, 136

Muckenhoupt, B. 154, 194, 196, 210

Nagel, J. 59, 136
Nagumo, M. 119
Natanson, I. P. 128, 136
Nečas, J. 39, 81, 91, 114, 136
Neumann, C. 169, 173
von Neumann, J. 6
Newton, I. 32, 45, 201
Nikodym, O. 157, 168
Nirenberg, L. 84-85
Nikol'skaya, N. S. 64, 137
Nikol'skiĭ, S. M. 35, 38-39, 41, 46-47, 49-50, 52-59, 61, 63-65, 68-70, 74, 76, 78-79, 81-86, 90-93, 98, 103, 105, 108, 120-122, 126, 128-129, 136-137, 143
Nikol'skiĭ, Yu. S. 89, 93-94, 98, 137

Orlicz, W. 66, 110-116, 118-119
Osilenker, B. P. 197, 210
Osserman, R. 209, 211
Ostrogradskiĭ, M. V. 157, 177-178
Otelbaev, M. 99, 136, 207, 209
Ovchinnikov, V. I. 128, 137

Paley, R. E. A. 64-65
Panzone, R. 119, 128
Parseval, M.-A. 102
Peetre, J. 56, 65, 105, 108, 137
Perepelkin, V. G. 54, 93-94, 128, 137, 140
Petunin, Yu. I. 105, 128, 134
Pigolkina, T. S. 93-94, 138
Poborchiĭ, S. V. 182-184
Poincaré, H. 157-158, 160, 174, 203
Poisson, S. D. 103
Pokhoshaev, S. I. 113, 138
Polking, J. G. 204
Pólya, G. 100, 138, 150, 160, 209, 211
Popova, E. A. 81
Portnov, V. R. 83, 99, 138
Potapov, E. 52, 84, 119, 138
Poulsen, E. 91, 99, 138
Preobrazhenskiĭ, S. P. 191

Radyuno, Ya. V. 44, 138
Rellich, F. 157
Reshetnyak, Yu. G. 39, 70, 75, 77, 128, 132, 138, 192, 210
Richel, R. W. 178
Riemann, B. 5-6, 173-175, 209-210
Riesz, F. 6, 16, 114
Riesz, M. 61-62, 99, 104, 190, 192, 207
Rutitskiĭ, Ya. B. 128, 134

Samko, S. G. 61-62, 128, 138
Sawyer, E. T. 194, 197, 210-211
Schauder, J. 21, 29
Schmidt, E. 6
Schrödinger, E. 151, 208
Schwartz, J. 128, 130
Schwartz, L. 61
Schwarz, H. A. 5, 19
Sedaev, A. A. 108
Sedov, V. N. 92, 99, 138
Semenov, E. M. 105, 108, 118, 128, 134
Serapioni, R. P. 154
Serrin, J. 80, 136
Shalashov, V. K. 123, 129
Shaposhnikova, T. O. 125, 138, 197, 210
Shubochkina, T. A. 132
Shvartsman, P. A. 78-79
Simon, L. M. 174
Sitnikov, V. N. 183
Slobodetskiĭ, L. N. 53-54, 76, 81, 128, 138
Smith, K. T. 39, 60
Sobolev, S. L. 29, 31, 33, 37-42, 46, 53-54, 58-61, 64-65, 69-70, 73, 76, 79, 81-83, 86, 93-94, 99, 110, 113-118, 120-122, 124-125, 138, 143, 145-146, 156-157, 166, 170-172, 176, 184, 188, 190, 197, 199, 201-202, 204, 209-210
Solntsev, Yu. K. 74
Solonnikov, V. A. 39, 54, 57, 74-76, 85, 132, 139
Sparr, G. 108
Spruk, J. 175
Stein, E. M. 68, 76, 93, 105, 128, 131, 139, 185
Steklov, V. A. 16, 28-29, 86
Szegö, G. 150, 160, 209, 211

Tandit, B. 81, 96, 139
Terekhin, A. P. 85, 139
Thorin, G. O. 104, 139
Triebel, H. 64-65, 67-69, 91, 93-94, 99, 103, 105, 108-109, 125, 128, 139
Troisi, M. 93, 140
Tulaĭkov, A. N. 16

Ul'yanov, P. L. 54, 56, 65-66, 140
Uspenskiĭ, S. V. 52, 54, 75-77, 85, 90-94, 96, 99, 128, 140

Vasharin, A. A. 91-93, 140
Vishik, M. I. 115
Vodop'yanov, S. K. 77, 181

Weierstrass, K. 19-20

Weiss, G. 105, 128, 139
Weyl, H. 45
Wheeden, R. 194, 196

Yamabe, H. 175-176
Yakovlev, G. N. 76, 78, 140, 184
Yudovich, V. I. 74
Young, W. H. 105, 110-114, 116-118

Zalgaller, V. L. 144, 174, 209-210
Zygmund, A. 47-49, 52, 84

Subject Index

Analogue of Hölder's inequality 113
–, of Marcheau derivatives 62
–, of Riesz compactness criterion 86
–, of the theorem on the limit of uniformly converging sequences of functions with uniformly convergent sequence of derivatives 32
Approximation, best, local 123
Average, Steklov 16, 29
Axioms of distance 9

Base for topology 8
–, –, local 9
Basis, Hamel 21
–, orthogonal 21
–, Schauder 21
–, unconditional 21
Boundary of a domain, Lipschitz 38
–, reduced 177

Capacity, electrostatic 149
–, generated by a potential 192
–, harmonic 149
–, inner 149
–, outer 149
p-capacity 149
(p, Φ)-capacity 153
Class, anisotropic $H_p^r(G)$ 50
–, –, $W_p^l(G)$ 50
–, Orlicz 110
–, $EW_p^{(l)}$ 181
–, $H_{p,\alpha}$ 165
–, $H_p^{(r)}(G)$ 47
–, $I_{p,\alpha}$ 164
–, J_α 159
–, $\overset{o}{J}_\alpha$ 161
Classes, isotropic 49
H-classes 47
–, anisotropic 50
Closure 11
Completion, of a space 11, 18
Condition, Ahlfors 181
–, cone 156
–, flexible cone 75
–, –, for a λ-horn 75
–, infinite cone 78
–, integral, Hölder 48
–, –, Zygmund 48
–, Muckenhoupt's A_p 154
–, Pólya 100
–, λ-horn 70
–, (1,2)-horn 76
Δ_2-condition 112
ε-δ-condition 77
Convergence, in mean 15
–, in measure 13
–, in norm 13
–, in square norm 15
–, in Φ-mean 113
–, pointwise 9
–, uniform 15
–, weak-* 27
Conductor 163
p-conductance 163
Constant, imbedding 21
Continuity, (global) average 28
–, equicontinuous in the L_p-metric 1
–, average, with respect to Φ-means 114
Corridor 75
Covering 12
–, finite 12
Criterion, compactness
–, –, for a space 11
–, –, of Kolmogorov 17

Degeneracy, strong 92
–, weak 89
Derivative, fractional of order $\alpha > 0$, in the sense of Liouville 45
–, generalized 31, 34
–, generalized Sobolev 32

–, –, of order r 32
–, –, $\partial f/\partial x_i$ 33
–, in the sense of Weyl of order β 45
Diameter of a space 11
k-th diameter in the sense of Kolmogorov 209
Differentiation, fractional 45
Dimension of a Hilbert space 21
–, of a vector (linear) space 7
Distance 9
Domain, starshaped with respect to a ball 39

Equation, linear elliptic 127
–, of Euler-Poisson-Darboux type 103

Family, of functions, equicontinuous 16
–, –, uniformly bounded 16
Formula, Gauss-Ostrogradskiĭ 177
–, Newton-Leibniz 32
Function, averaging, Steklov 16
–, conjugate 112
–, convex 110
–, distribution 106
–, entire, of exponential type 46
–, essentially bounded 29
–, finite (of compact support) 29
–, generalized (distribution) 33
–, locally integrable (summable) 7
–, marginal 58
–, maximal 68
–, –, fractional 194
–, –, Hardy-Littlewood 67
–, stabilizing, of order m to a function 99
–, weight 22, 87
–, Young 11
–, P_Ω 177
–, $P_{C\Omega}$ 178
–, $\tau_\Omega(E)$ 180
Functional 23
–, linear, bounded 24
–, –, homogeneous 24
–, –, unbounded 24
Functions, conjugate, in the sense of Young 112
–, equimeasurable 106
–, equivalent 14
–, –, in the sense of Young 113
–, weight 22

Imbedding of seminormed spaces 20
Inequalities, isoperimetric, for p-capacity 150
–, for entire functions 47
Inequality, Bernshteĭn's 47
–, Cauchy-Bunyakovskiĭ(-Schwarz) 17
–, Cauchy's 17
–, Cheeger's 173
–, Choquet's 149
–, of Friedrichs 157
–, Hölder's 15
–, isoperimetric 143, 173
–, Mc Kean's 174
–, Minkowski's
–, –, generalized 15
–, –, integral, usual 14
–, Poincaré's 157
–, –, generalized 160
–, Sobolev's 145, 204
–, Steklov's 87
Integral, Dirichlet 5
–, energy 88
–, fractional, Weyl, of order β 45
–, $I^a_\alpha f(x)$, of order α 45
Integration, fractional 45
–, –, Weyl 45
Intersection, of a Banach pair 104
Isomorphism, of normed linear spaces 18

Kernel, averaging, of functions 28
–, Bessel-Macdonald 60

Lemma, Borel-Lebesgue 12
–, Hedberg's 192
–, Rellich's 157
Limit, of a sequence 10
–, of Banach spaces
–, –, inductive 43
–, –, projective 42

Mean, Sobolev 29
–, with a kernel 28
Measure, complete 7
–, vector 176
Metric, 9
–, integral 9
Method, interpolation, of linear operators 103
–, dimension 58
–, of integral representations 38
–, of smoothing, of functions 39
K-method 105
Modulus of continuity 35
–, local averaged 68
–, of order k, in the direction of a vector 35
Multiplier 197

Neighborhood, of a point 8
ε-neighborhood 10
ε-net 11
Norm 12
–, of an operator 27
–, Luxemburg 113
–, mixed 30
–, monotone 118
–, uniform 14
p-norm 14
Normal, in the sense of Federer 177
Norms, equivalent 13

Open 8
Operator 27
–, bounded 27
–, –, from a Banach pair into a Banach pair 104
–, continuous 27
–, linear 27
–, maximal, Hardy-Littlewood 192
–, N_q 169
Order of growth, polynomial 115
Orthogonal 21

Pair, Banach 104
–, of conjugate functions 117
Parallelogram rule 18
Pasting 78
Perimeter, of a measurable set 155
Point, interior (inner) 10
–, limit 11
Polynomial, Chebyshev 23
–, Hermite 23
–, Laguerre 23
–, Legendre 22
Potential, Bessel 190
–, Riesz 190
–, –, generalized 61
Principle, Dirichlet 5
–, –, with prescribed level surfaces 150
Problem, Dirichlet 198
–, Neumann 169
–, of pasting functions 78
–, Yamabe 175
Product, quasi-inner 17
–, semi-inner 17
–, inner (scalar) 17
Property, monotonicity, of an Orlicz space 112
–, interpolation 104

Quasidisk 181

Rearrangement, of a function 106
Representation, for the variation of a function of n variables 144
–, integral, of Sobolev 69

Seminorm 12
–, induced by a given inner product 17
Set, admissible 143
–, closed 11
–, dense 11
–, –, in a seminormed space 20
–, open 8
–, precompact 12
–, totally bounded 11
–, uniformly continuous, in the L_p-metric 28
Sequence, convergent 10
–, fundamental 10
–, strongly convergent 26
–, weakly convergent 26
–, weak-* convergent 27
Space, anisotropic $B^r_{p,q}$ 54
–, Banach 13
–, –, ideal (Banach lattice) 118
–, –, ideal, complete 118
–, –, ideal, symmetric (rearrangement invariant) 118
–, –, intermediate 104
–, Bessel potential 60
–, bounded 11
–, compact 11
–, complete 10
–, dual (conjugate) 24
–, finite dimensional 7
–, function 4
–, –, weighted 87
–, Hausdorff 8
–, Hilbert 18
–, infinite dimensional 18
–, interpolation 104
–, linear (vector), complex 6
–, –, real 6
–, metric 9
–, Morrey $L^M_{p,\lambda}$ 30
–, normed 12
–, of multipliers 198
–, Orlicz $L_\Phi(X)$ 112
–, pre-compact 12
–, pre-Hilbert 18
–, quasinormed 17
–, reflexive 25
–, of Riesz potentials 62
–, seminormed 12
–, separable 11

Space,
–, Slobodetskiĭ $W_p^{(r)}(G)$ 53
–, Sobolev $W_p^{(l)}(G)$ 37
–, Sobolev-Liouville, isotropic 59
–, Sobolev-Orlicz $W^{(l)}L_\Phi$ 115
–, symmetric 118
–, topological 8
–, with mixed norm 119
–, $B(E)$ 7
–, $BC(E)$ 8
–, $B_p^{(l)}(\mathbb{R}^n)$ 55, 195
–, $b_p^{(l)}(\mathbb{R}^n)$ 195
–, $B_{p,q}^{(r)}(\mathbb{R}^n)$ 54
–, $B_{p,q}^{(Q,s)}$ 65
–, $B_{p,q}^{\omega(\cdot)}$ 65
–, $BM(E,\mu)$ 7
–, $BV(\Omega)$ 176
–, $\mathbb{C}(E)$ 7
–, $C(E)$ 7
–, $C^\infty(E)$ 29
–, $C^{(l)}(G)$ 39
–, $\overset{o}{C}{}^\infty(G)$ 29
–, $\check{C}^\infty(G)$ 81
–, $C\mathscr{D}^{(l)}(E)$ 8
–, $CL_p(E)$ 19
–, $\mathscr{D}^{(l)}(E)$ 8
–, $H_p^{(l)}$ 190
–, $h_p^{(l)}$ 190
–, $\overset{o}{H}{}_p^{(l)}(\Omega,\nu)$ 207
–, $L_p(E,\mu)$ 7
–, $L_{2,\nu}(\overset{+}{\mathbb{R}}{}^{n+1})$ 102
–, $L_p^{(l)}(\Omega)$ 156
–, $\overset{o}{L}{}_p^{(l)}(\Omega)$ 204
–, $L_{p,\lambda}^M$ 30
–, $L_{p,q}(X)$ 106
–, $L_{p,q}^{(N,S)}$ 67
–, $L_{p,\varphi_j}^{(r_j)}$ 95
–, $L_{p,\varphi_j,\varphi_k}^{(r_j,r_k)}$ 95
–, $L_{p,\varphi,\psi}^{(r,l)}$ 96
–, l_q 63
–, $L_\infty(E,\mu)$ 7
–, $\mathscr{L}(X,Y)$ 28
–, $M(E,\mu)$ 7
–, $MW_p^{(l)}(\Omega)$ 198
–, $\mathbb{R}(E)$ 7
–, $TW_p^{(l)}(\Omega)$ 184
–, $T[W_p^{(l)}(\Omega_\varepsilon)]$ 188
–, $\overset{o}{W}{}_p^{(l)}(G)$ 82
–, $w_p^{(l)}(G)$ 40
–, $\tilde{W}_p^{(l)}(G)$ 81
–, $\overset{o}{\tilde{W}}{}_p^{(l)}(G)$ 82
–, $W_p^{(l)}(\Omega)$ 37, 156
–, $\overset{o}{W}{}_p^{(k)}(\Omega)$ 200
–, $\tilde{W}_p^{(l)}(\Omega)$ 156
–, $W^{(k)}E_\Phi(G)$ 116
–, $W^{(k)}L_\Phi(G)$ 116
–, $W_{p,\sigma}^{(l)}(\mathbb{R}^n)$ 182
–, $w_{p,\alpha}^{(r)}(G)$ 88
–, $W_{p,\alpha}^{(r)}(G)$ 89
–, $W_{p,\alpha}^{(r)}(\mathbb{R}^n|\infty)$ 93
–, $W_{2,\nu}^{(s)}(\overset{+}{\mathbb{R}}{}^{n+1})$ 102
–, $W_p^{(l)}(U,\mathrm{loc})$ 199
–, $W_q^{(l)}(\mathbb{R}^n,\mathrm{unif})$ 198
–, $W^\infty\{a_\alpha,p_\alpha\}(G)$ 42
–, $\overset{o}{W}{}^\infty\{a_\alpha,p_\alpha\}(G)$ 42
Spaces, Besov 54, 195
–, Besov-Nikol'skiĭ 54
–, Besov, generalized $B^{(r)}(X,E)$ 123
–, Calderón 122
–, isometric 10
–, Kipriyanov, weighted 101
–, Lebesgue 7
–, Lipschitz 122
–, Lizorkin-Triebel, generalized 67
–, –, $L_{p,q}^{(r)}$ 64
–, Lorentz, $L_{p,q}$ 106
–, –, $\Lambda_{\psi,q}$ 107
–, Marcinkiewicz, M_p 106
–, –, $M_\psi(X)$ 107
–, nontrivial 42
–, Sobolev, of infinite order 42
–, Sobolev-Liouville $L_p^r(\mathbb{R}^n)$ 58
–, Sobolev-Slobodetskiĭ 53
–, weighted, anisotropic 94
–, weighted, function 87
–, S_p^rH 83
–, S_p^rB 84
–, S_p^rL 84
–, S_p^rW 84
–, $W_{p,q,\kappa}^r(G)$ 84
Subspace 7
Sum, Dirichlet 97
–, of Banach pairs 104
System, complete 20
–, Haar, orthogonal 29
–, linearly independent 21
–, trigonometric 20
–, of indices, admissible 100
–, –, complete 100

Theorem, Arcela-Ascoli 16
–, of Besicovitch 148
–, Burago-Maz'ya 180
–, convexity, of Riesz-Thorin 104
–, DeGiorgi-Federer 177
–, direct, of the type of Bernshteĭn, Jackson and Zygmund 52
–, Fleming-Richel 178
–, Hausdorff-Young 105
–, implicit function 199
–, interpolation, for Lorentz spaces 107
–, inverse, of the type of Bernshteĭn, Jackson and Zygmund 52
–, Littlewood-Paley 64
Theorems, of Besov for the classes $B^r_{p,q}(G)$ 55
–, imbedding 15, 59, 64
–, –, converse (inverse) 49
–, –, for isotropic H-classes 50
–, –, for Sobolev-Liouville spaces 59-60
–, –, of Nikol'skiĭ, for anisotropic H-classes 50-51
–, –, of Sobolev 39
–, –, of Triebel 64
Topology of a space 8
–, weak 27
–, weak-* 27
Trace of a function 36
–, in the sense of almost everywhere convergence 36
–, in the sense of approximation with continuous functions 36
–, in the sense of L_p convergence 36
–, rough 178
–, of f ; $W^\infty\{a_\alpha, p_\alpha\}(G)$ 43
Transform, Fourier-Bessel 102

Version, multidimensional, of the isoperimetric property of the circle 143
Vertex of a peak, directed into the exterior of a domain 182
Vertex, of an inner point 182

Weight 22, 87